GENE SHARING AND EVOLUTION

Frontispiece: The idea of gene sharing—that the meaning (function) of a structure is context-dependent—is not restricted to biology. This expressive bird-shaman Inuit carving by Tommy Sevoga (Baker Lake, Nunavut) illustrates the artistic impact of placing familiar body parts in a foreign arrangement. The fusion of a human face, animal ears, and a bird body has created a shaman possessing greater powers than the living forms from which the parts were derived. (Sculpture from the author's collection; photograph courtesy of John Burdick, Burdick Gallery, Washington, DC, and reproduced with the permission of Burdick Gallery and Tommy Sevoga.)

JORAM PIATIGORSKY

GENE SHARING

AND

EVOLUTION

The Diversity of Protein Functions

HARVARD UNIVERSITY PRESS

Cambridge, Massachusetts

London, England

2007

To Lona, my wife and best friend

Library of Congress Cataloging-in-Publication Data

Piatigorsky, Joram.
Gene sharing and evolution : the diversity
of protein functions / Joram Piatigorsky.
p. cm.
Includes bibliographical references and index.
ISBN-13: 978-0-674-02341-3 (cloth : alk. paper)
ISBN-10: 0-674-02341-2 (cloth : alk. paper)
1. Genetic regulation.   2. Proteins—Evolution.   3. Eye—Molecular aspects.
4. Crystalline lens.   I. Title.

QH450.P53   2007
572'.6—dc22   2006045769

Contents

Illustrations

Preface

This book explores gene sharing—a term originally coined to describe the observation that structural proteins (called crystallins) responsible for the transparent and refractive properties of the cellular lens in the eye also perform nonrefractive stress-related or metabolic functions in other tissues. In short, a lens crystallin has distinct molecular functions that share the identical gene-coding sequence. The term "gene sharing" emphasizes that a protein's function may be directly affected by the differential expression of its gene.

Although crystallins are clearly not alone among proteins that serve more than one biological role, I was uncertain as to what extent multiple molecular functions are common among proteins. Determining whether a protein engages in gene sharing is complicated by the need to distinguish between two alternatives: Does the protein perform distinct molecular (cf. biochemical/ biophysical) functions, or does it exploit precisely the same properties (a specific enzymatic reaction, for example) with different phenotypic consequences? This can be a tricky question to answer, especially because mechanisms of protein action are often not known in detail. As I read the extensive scientific literature on proteins, I became increasingly convinced of the widespread use of the same protein for different molecular functions. Protein specialization and molecular diversity seem to go hand-in-hand as a general rule and be a fundamental characteristic of evolution. My original idea was to provide an extensive list of multifunctional proteins. I soon realized the futility of that approach: I would have had to include most, if not all, proteins!

A theme of this book, then, is that distinct specialized functions required for life are carried out by the same proteins cast as different characters, much as in a Peter Sellers movie. Specialization of an individual protein neither fixes its functional boundaries in a molecular sense nor prevents it from changing functions under different circumstances. In addition, the fact that the molecular function of a protein may be dictated by the differential expression of its gene affects the very definition of its gene. How should we identify a gene—by its structure or by its function? Do the boundaries of a gene include its regulatory elements? What is the influence of gene expression on natural selection of protein functions, and how is variation in gene expression selected in evolution? These are neither new nor resolved questions in biology. I wrote this book to examine these questions from the vantage point of the principle of gene sharing. I believe that the extensiveness of gene sharing and protein multifunctionality argues against rigid compartmentalization in biology and introduces a fluid concept of genes and proteins.

The multiple functions of an organism's proteins, neatly encoded within the genome, ultimately lead to the visible phenotypes of its cells, tissues, organs, body, and behavior. This book concerns the vast biological space between genotype and phenotype. I have considered implications of gene sharing, but there is a lot of ground to cover between molecular and systems biology. Therefore, forgive me if I have missed an important issue or dwelled on a lesser one. Some topics may be a bit detailed, but I felt that it was important to document carefully because the thrust of the book is to explore. I recognize that I have left gaps in detail and scope, but I hope that I have shed some light on the complex interrelationships between gene regulation, protein functions, and evolution. There are many topics covered here that I had to learn as I wrote, and I take responsibility for the errors that are no doubt present. I hope that my efforts will be useful to students and seasoned scientists alike who are interested in the interplay between pragmatism and serendipity within living systems and evolutionary processes.

This book is a considerable expansion of three Russell Marker Lectures on Evolutionary Biology that I was privileged to deliver at Pennsylvania State University in April 2002. I thank Masatoshi Nei for his role in my being selected to give the lectures and for encouraging me to expand upon them in the form of a book. I acknowledge Alex Keynan from the Hebrew University in Jerusalem; he first suggested that I write a book on gene sharing and then insisted that I do not procrastinate! I am grateful to Alan Walker at Pennsylvania State University for introducing me to my editor, Michael Fisher, at Harvard University Press, whose prompt responses and helpful encouragement and advice throughout my writing are greatly appreciated. I also acknowledge the helpful, expert editorial assistance of Elizabeth Collins, Kate Brick, and Anne McGuire.

Many colleagues and members of my laboratory have shaped my knowledge and have impacted the ideas expressed here. There are too many to list individually, but I am grateful to each. I would like to single out a few for many years of friendship, stimulating scientific discussions, patient listening, helpful critical comments on the manuscript or on ideas therein, and supplying figures. Among these are the late Frederick Bettelheim, Richard Burian, A. H. Jay Burr, Gareth Butland, Chi-Hing "Chris" Cheng, Alex Cvekl, Janine Davis, Russell D. Fernald, Joseph Horwitz, Eugene Koonin, Zbynek Kozmik, Zdenek Kostrouch, Barbara Norman, Joseph Edward Rall, the late R. Gerald Robison, Jr., Alan Schechter, Jacob Sivak, Shivalingappa Swamynathan, Dan S. Tawfik, Stanislav Tomarev, Vasilis Vasiliou, Eric Wawrousek, and Peggy Zelenka. The anonymous reviewers, including Austin L. Hughes, who revealed his identity, provided valuable suggestions that I gratefully followed. I am indebted to Alan Hoofring for most of the illustrations in this book.

Acknowledgments are also in order to my employers at the National Institutes of Health, where I have conducted research since 1967. In particular, I acknowledge Alfred J. Coulombre, Charles Lowe, and Philip Leder at the National Institute of Child Health and Human Development; more recently I am grateful to Jin H. Kinoshita, Carl Kupfer, Sheldon Miller, and Paul Sieving at the National Eye Institute. They all gave me the wonderful gift of freedom to explore.

One can never repeat too often that trivial truth, that small causes can produce great results.

—José Saramago, *The Double* (2004)

The few really big steps in evolution clearly required the acquisition of new information. But specialization and diversification took place by using differently the same structural information.

—François Jacob, *The Possible and the Actual* (1982)

1

What Is "Gene Sharing?"

We live in a world of specialization, and intentional subdivision of labor is embedded into our beliefs. Within a typical day we might visit the dentist, buy milk at the grocery store, and prepare tomorrow's lecture on electromagnetism for physics class. Moreover, each of these categories has an ordered group of subspecializations with parts neatly arranged and identified by their function. The dentist drills with power tools crafted for that purpose, the milk is a tiny part of an organized industry to feed us, and electromagnetism is incomprehensible without years of schooling. We accept specialized functions as a reality of design and efficiency. Yet, we recognize that we must do multiple specialized tasks under different circumstances. Indeed, we could not survive without diverse skills. This book argues that the same may be true for genes and proteins.

Although fiction, it would be easy for us to understand the following. An alien from space lands in a university campus, walks down the hall of a building, and sees a spring door held open by a sturdy little chair with rubber caps on its legs as the students stream into the classroom. "Ah," the alien says to himself, "what a clever device to keep the entrance to that hole in the wall open," as he skips along his way. If he had peered into the classroom, the alien might have been confused when he saw students sitting on devices similar to the door stop. His confusion would have compounded if he had seen someone standing on the seat of a chair changing a light bulb or had his way barred from entering the restroom by two of the same devices supporting a wooden plank.

Although the alien might confuse the functions of a chair, we would not recognize the alien if he (or she) happened to look superficially like us earthlings. After a glance we would assume, without further thought, that he evolved from our ancestors and that he is one of us. It would be assumed,

1

again without consideration, that he has a function connected with the large and complex university. In short, it would be taken for granted that he intends to be where he is and that he is walking to a specific place for a reason. This book applies these considerations to the cellular "society" of genes and proteins. Genes and proteins are finely adapted for their specialized tasks, yet most have a surprising number of diverse roles, both within a species and in different species.

NEW FUNCTIONS FOR OLD PROTEINS AND THE QUESTION
OF GENE DUPLICATION

Functional diversity is counter to the idea of specialization, and the notion that specialization narrows expertise and limits the acquisition of a multitude of skills is intuitively easy to accept. Such reasoning highlights the importance of gene duplications for functional diversification of proteins: Two (or more) genes encoding similar polypeptides would free one of the siblings to develop a new function while the other carried on the original function.[1-4] This attractive idea has been much debated[5,6,1117] (see Chapter 9).

Various lines of evidence, however, have indicated that making two copies of a gene does not "free" either one.[6] For example, analysis of the rates of nucleotide changes in duplicate genes in the allotetraploid frog *Xenopus laevis,* and between orthologous genes in humans and rodents, has shown that sibling genes are as constrained in evolution as single-copy genes and may diversify independently.[7,1118,1119] This raises concern as to the necessity of gene duplication for the innovation of a new protein function. The concept of gene sharing explored in this book states that a polypeptide may develop more than one molecular function without gene duplication.[8-10,1117]

The idea that single-copy genes encode polypeptides with two or more molecular functions is not new and implies that gene duplication is not required for functional innovation of proteins (see Chapter 10).[1117] At the level of protein:protein interaction, Goodman and colleagues[11] have speculated that hemoglobin homotetramers (comprising four identical polypeptides) existed before the ancestral gene duplicated to give rise to α- and β-globin genes. The multitude of protein:protein interactions known to exist today creates a network that is critical for a modern understanding of cellular metabolism and that provides an entry point for a systems-level approach to regulatory processes and many functions played by individual proteins.[12] Jensen[13] was a pioneer in considering multifunctionality of individual proteins by emphasizing substrate ambiguity of enzymes, and Jensen and Byng[14] proposed that gene duplication was important for narrowing substrate specificity of the resulting sibling polypeptides rather than innovating new biochemical reactions. Multifunctional proteins were considered beneficial because they bestowed a selective advantage to organisms, allowing them to adapt under changing environmental conditions and facilitating survival and evolution. Modern studies relating multiple conformations to multiple functions of individual proteins support these early views of Jensen[15,16] (see Chapter 10). Finally, Orgel[17] was another investigator who generalized the idea some time ago that protein multifunctionality precedes rather than results from gene duplication. The concept of gene sharing discussed in this book is about the extent, biological significance, and implications of multifunctional proteins.

ORIGIN OF THE TERM "GENE SHARING"

Studies in the 1980s discovered that duck ε-crystallin was similar to lactate dehydrogenase B_4,[18] that chicken δ-crystallin was similar to argininosuccinate lyase, and that turtle τ-crystallin was similar to α-enolase.[19] These early findings suggested that a protein with enzyme potential could act as a structural refractive protein in the lens if its concentration was high enough. However, these observations did not establish whether the identical gene was expressed highly in the lens when its protein functioned as a crystallin (and possibly an enzyme as well) and at a lower level in other tissues when the same protein functioned strictly as an enzyme. Showing gene *identity* for the enzyme and crystallin, not just *similarity* of homologous (or sibling) genes, was critical, because neofunctionalization (evolution of a new function) can arise in a duplicated gene (Figure 1.1) (see Chapter 9).

Gene identity was established for chicken δ2-crystallin and argininosuccinate lyase[20] and for duck ε-crystallin and lactate dehydrogenase B_4.[21] The term "gene sharing" was coined to describe the fact that the major water-soluble structural proteins of the transparent eye lens responsible for refraction—the crystallins—are also expressed at lower levels in other tissues of the same animal, where they have nonrefractive, enzymatic roles (see Chapter 4).[20] The critical fact was that the gene expression pattern of these so-called en-

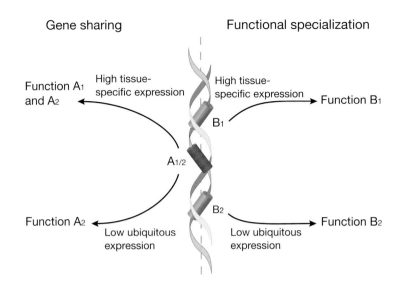

Figure 1.1. Gene sharing versus gene duplication followed by functional specialization. The left side of the figure illustrates one gene (A) encoding a protein with the ability to perform two functions (A_1 and A_2). The polypeptide performs functions A_1 and A_2 when gene A is expressed highly in a specific tissue, but only function A_1 when expressed at a low level. The right side of the figure illustrates functional specialization of sibling genes (B_1 and B_2) after gene duplication. In this case function B_1 and B_2 are performed by distinct polypeptides encoded in duplicated genes.

zyme-crystallins dictated the functions of their encoded proteins. It is now known that numerous, perhaps all, families of proteins that function as lens crystallins fit neatly into this story of evolution by a gene-sharing process.[22–24]

GENE SHARING: GENERAL DEFINITION AND IMPLICATIONS

In general, the term "gene sharing" means that one gene produces a polypeptide that has more than one molecular function: Two or more entirely different functions of a polypeptide share the identical gene. This book is about the generality of the gene-sharing principle, which ultimately raises the question of "what is a gene?" This turns out to be no easy matter: Is a "gene" structure, function, regulation, or an open combination of these parameters? Chapter 3 is devoted to the elusive concept of a "gene," with attention given to its application to gene sharing.

The gene-sharing concept postulates that protein function is determined not only by primary amino acid sequence, which remains the same in the multiple functions that are performed by the protein, but also by the micro-environment within the cell and by the expression of its gene. Awareness of gene sharing cautions against assuming that a protein will be used in the same way wherever or whenever it is present, or that it has always done what it is doing at any given moment. The functions of genes and proteins are context dependent.

The existence of gene sharing makes it erroneous to assume that the "primary" function of a protein is either the first one that was discovered or the one that we recognize most often. That view minimizes the importance of an infrequently performed function, or what might be considered an esoteric function in little-studied creatures. It also obviates radical changes in the function of a protein that might have occurred not directly dependent upon changes in protein sequence during evolution.

Gene sharing encompasses both an existing situation and a dynamic process that contributes to the innovation of protein functions. The existing situation is when a gene encodes a polypeptide that performs more than one molecular function, such as enzymatic and refractive in the case of crystallins. As a process, gene sharing "tests" for new protein functions. A gene may be engaged in gene sharing with its encoded polypeptide performing the same multifunctional tasks for eons of time. Alternatively, gene sharing may be a transient event during evolution: The set of specific biochemical or biophysical tasks deployed by a polypeptide may be modified during evolution by the gene-sharing process. Thus, in addition to being a statement of multifunctionality, the concept of gene sharing refers to an evolutionary process that can come and go, promoting functional shifts over time.

Gene sharing reflects the complexity and pragmatism of biology. To appreciate the significance of gene sharing is to recognize the versatility of genes and proteins and to marvel at the resourceful mechanisms that arise as the same polypeptides are used and reused during evolution.

PROTEIN LOCATION AND GENE REGULATION

Changes in protein function are typically known to occur by mutagenesis resulting in modifications in amino acid sequence. Such mutations in the coding sequence of genes are unequivocally a major driving force of evolutionary changes in protein structure and function. Gene sharing, by contrast, concerns functional shifts and additions in proteins associated with changes in intracellular environment. This does not preclude, of course, the occurrence of mutations in the coding sequence that contribute to the gene-sharing process during evolution.

It is common knowledge that a polypeptide can be expressed in different tissues and at different levels by the regulated expression of its gene. More recently it has become established that the same polypeptide can occupy different cellular compartments by various mechanisms that do not involve changes in its amino acid sequence. Mechanisms of intracellular protein compartmentation include dynamic competition between molecular events such

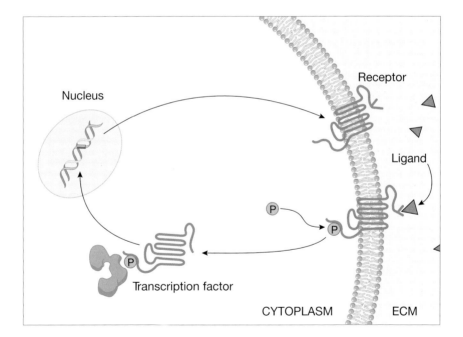

Figure 1.2. Two functions for one polypeptide. This diagram shows a polypeptide acting as a receptor for a ligand when the polypeptide is inserted into the cell membrane. Ligand binding causes the receptor to be phosphorylated and internalized into the cytoplasm, where it binds a transcription factor, moves to the nucleus, and activates gene expression.

as protein folding, posttranslational modification, and protein:protein inter-action.[25] Changes in protein location can affect function because a protein interacts with other molecules in its microcosm. Changing the immediate intracellular environment of a polypeptide or its concentration within that environment affects its probability of interacting with other polypeptides, the nature of the specific polypeptides (or other molecules) with which it partners, and even the posttranslational modifications to which it may be subjected. Thus, a polypeptide may express different properties in the cytoplasm, in the nucleus of a cell, or in different tissues; these different micro-environments may promote entirely different molecular functions displayed by the polypeptide (Figure 1.2).

The gene-sharing concept requires that a polypeptide engaged in a specific molecular function associated with a particular intracellular location or tissue also engage in another function(s). This other function may be associated with a different cellular microenvironment, tissue, or concentration. These alternate functions do not need to be carried out simultaneously. Any cellular change in intracellular or tissue prevalence resulting in an alternate molecular function for a polypeptide is part of the gene-sharing process.

Thus, gene sharing may result from the complexity of gene expression and biochemical processes within the cellular environment, which contains a dynamic flux of many proteins. The multiple functions of proteins generated by a gene-sharing process may cross-talk at some level or be completely unrelated. Gene sharing is a pragmatic example of using a polypeptide in whatever way is possible and beneficial because it is in a complex environment; it is a fundamental cause of expanding and modifying protein functions without necessarily tampering only with protein-coding sequences during evolution.

WHY THE TERM "GENE SHARING"?

While the intent of a concept can be stated in various ways, the precise words that are used have influence. This was eloquently expressed by Simpson[26] (p. 160) many years ago regarding use of the word "opportunism" in evolution. He wrote, "Opportunism is, to be sure, a somewhat dangerous word to use . . . It may carry a suggestion of conscious action or of prescience in exploitation of the potentialities of a situation . . . words too often carry undertones appropriate to the human scene and misleading in discussion of the grander scene in which men have so late, so brief, and yet so important a part." Consequently, the words used to describe functional shifts or multiple functions for a protein are important.

Because the term "gene sharing" refers to situations in which a polypeptide performs more than one biochemical or biophysical function, one may ask why not use the term "*protein* sharing"? This is an important question that concerns many issues considered in this book, such as the definition of a gene, the complexity of its regulation, and the dependence of protein function on gene expression and cellular context. Apart from historical considerations,[20,23] I have chosen the term "gene sharing" rather than "protein sharing" because the gene may be considered the least common denominator for the many functions of its polypeptide (Figure 1.3). A polypeptide may undergo structural changes as it performs different functions: a polypeptide with a given amino acid sequence adopting different conformations when performing different molecular functions is not strictly "identical" in the two biochemical roles. Similarly, covalent posttranslational changes such as phosphorylations affect both function and structure, yet the gene encoding the modified and unmodified polypeptide is the same. Even the concepts of polypeptide and protein are subject to ambiguities: A polypeptide with a single amino acid sequence performing one function as a homodimer and another function as a heterodimer forms two different proteins. Would this be a case of protein sharing? In short, a given polypeptide is not the same entity when engaged in two different molecular functions. This is not true for the DNA nucleotide sequences encoding the polypeptide: The gene remains invariant despite the function of the polypeptide (although see Chapter 3 for difficulties in defining a "gene").

Also important, the term "gene sharing" acknowledges gene expression as a factor for differential protein function. The term "protein sharing" does not incorporate this aspect of multifunctionality. Indeed, differential gene expression was at the heart of the finding, from which the term "gene sharing" originated, that an active enzyme serves as a refractive protein only when expressed highly in the eye lens. Differential gene expression is thus a primary cause for gene sharing to occur. This means that proteins can be selected for specific and multiple functions as a consequence of mutations affecting gene expression as well as protein structure.

"Pleiotropy" is another term that conveys the idea that a protein may have multiple functions. However, pleiotropy refers more to having single genes control multiple traits[27] (pp. 117–118). It means "many ways" and has been used to indicate that mutations affect multiple traits. Thus, pleiotropy is more closely connected to phenotype than protein function *per se* (for example, see Hekerman et al.,[28] Kullo et al.,[29] and Sung et al.[30]) The focus of gene sharing is protein function, not cell or tissue phenotype.

Before employing the term "gene sharing," we called the use of enzymes as structural crystallins "recruitment,"[19] a much-used conceptual term in evolu-

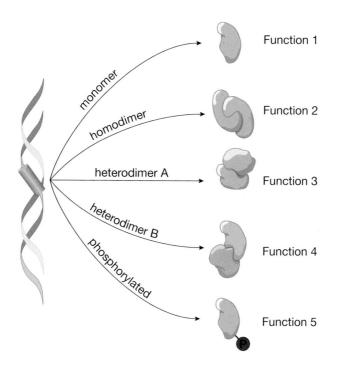

Figure 1.3. Multiple functions for one polypeptide via gene sharing. The diagram shows five ways in which a hypothetical polypeptide can theoretically perform different molecular functions (engage in gene sharing) by changes in protein:protein interaction or by posttranslational modification. In all cases the polypeptide has an identical amino acid sequence. The same polypeptide may have subtle differences in tertiary structure (functions 2, 3, and 4), depending on its interactions with other polypeptides. Posttranslational phosphorylation (function 5) also changes the chemical nature of the polypeptide. Thus, the only invariant structural feature of the polypeptide is the gene sequences encoding it.

tion and functional genomics.[27] "Hijacking"[31] and "co-option"[32] also litter the scientific literature to describe processes that have led to proteins acquiring new functions. One difficulty with these terms is that recruitment, hijacking, and co-option are ambiguous as to whether the protein in question is still performing its original function or any other function; is a recruited, hijacked, or co-opted protein serving more than a single function? This is an important point: Gene sharing *requires* that more than one function is operative.

"Moonlighting" is an imaginative term that incorporates the notion that the protein carries out more than one function rather than having been recruited, hijacked, or co-opted from one function to another.[33] However, moonlighting connotes a sense of illegitimacy (as does hijacking), as if the protein is doing something it should not be doing, or at least was not meant to do, or is doing it only as a side job.

An unsatisfying aspect of all these terms (recruitment, hijacking, co-option, and moonlighting) for protein multifunctionality is that they presuppose knowledge of original use. The term "gene sharing" claims no priority of one function over another, nor presumes which function came first. It remains conjecture, of course, whether an ancestral function was lost during evolution. Ancestral functions may or may not remain today, either partially or completely. From what function, then, was a protein recruited, hijacked, or co-opted? While it is possible to make an educated guess on the basis of phylogenetics and consideration of the specialized tasks being performed by an extant protein, we still must infer the evolutionary history of any gene's function. Thus, while perhaps not perfect, the term "gene sharing" is limited to the observation—two or more functions are shared by a single polypeptide encoded by one nucleotide sequence—without any assumptions as to evolutionary history or alternate protein structures.

MECHANISMS FOR DIVERSIFYING GENE FUNCTIONS

Gene sharing is only one mechanism for diversifying gene function. The complex organization of a gene leads to the generation of many variant polypeptides (Figure 1.4). In general, a generic gene has an enhancer and a promoter in front of (5′ to) the nucleotide sequences (exons and introns) that are transcribed into the primary RNA transcript [pre-messenger RNA (pre-mRNA)]. The enhancer and promoter direct the level and tissue specificity of gene expression (RNA synthesis) but are not transcribed into RNA themselves. There can be (and often are) more than one enhancer and/or promoter for a gene, and these can be located close to, within, or far away from the gene; the promoter is by definition contiguous with the first nucleotide that is transcribed into RNA when that promoter initiates gene expression. The resulting primary transcript comprises RNA sequences of the exons (protein-coding sequences) and introns (noncoding sequences). The first exon also contains some noncoding (untranslated, or not translated into protein) sequences 5' to the codon that initiates translation, or protein synthesis (the start codon); untranslated sequences are also present behind (3' to) the codon that terminates translation (the stop codon). The introns separate the exons and are removed from the RNA primary transcript by RNA processing, a complex biochemical event that also splices together the exon sequences into the final mRNA. The mRNA is translated into a polypeptide that is initiated at the start codon (AUG; adenine, uridine, guanine, specifying the amino acid methionine); mRNA translation is terminated at the stop codon (TGA; thymidine, guanidine, adenine, or TAA or TAG) codon. The untranslated sequences 3' to the stop codon contain a signal (called the polyadenylation signal) to add a stretch of adenines at the end of the mRNA.

The complexity of a gene and its expression allows the production of different polypeptides that can engage in different molecular functions (see Figure 1.4). One of the most common mechanisms for generating polypeptide diversity by expression of a single gene is alternative RNA splicing of its primary transcript. Alternative splicing generates two or more distinct mRNAs by eliminating certain sequence stretches (generally representing one or more of the exons) when processing the primary transcript into the mRNA. The different mRNA sequences are translated by a complex biochemical process into proteins with different amino acid sequences. The resulting proteins can have different expression patterns and variant biological roles.

Another mechanism generating protein diversity involves initiating transcription (RNA synthesis) at alternate sites from a stretch of DNA. This is called alternative promoter utilization because the promoter of a gene lies in front of and attached to the sequence where transcription is initiated. In Figure 1.4 this is represented by having either promoter 1 initiate transcription at the beginning of exon 1 or promoter 2 (located within intron 1) initiate transcription at exon 2. Like alternative RNA splicing, alternative promoter

utilization can result in polypeptides with different amino acid sequences
and functions. The controlling feature of this process involves specialized
enhancer and promoter sequences associated with the gene.

Although infrequent, RNA editing is another mechanism that affects gene
function. Conversion of one nucleotide into another in the protein-coding
sequence of the mRNA itself by an editing process results in a different amino
acid sequence after translation of the edited mRNA. Alternative RNA splicing,
differential promoter utilization, and mRNA editing are some of the examples of mechanisms that amplify the function of a gene. Usually they generate
diverse polypeptides with different amino acid sequences and thus, in themselves, are not examples of gene sharing. However, this need not be the case.

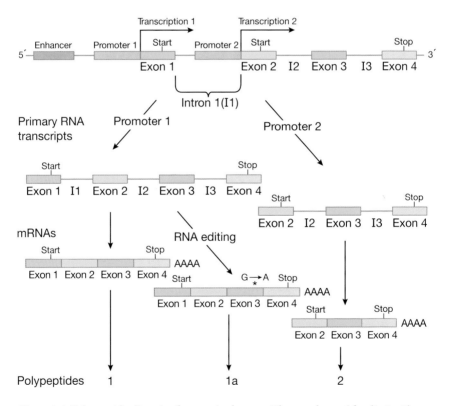

Figure 1.4. Polypeptide diversity from a single gene. Three polypeptides (1, 1a, 2)
with different amino acid sequences are produced from one gene by alternative RNA
splicing and RNA editing. See text for explanation.

The targets of these processes can be confined to noncoding sequences and result in diverse mRNAs that encode the same polypeptide; under such circumstances they may be involved in the gene-sharing process by affecting the amount or location of the synthesized polypeptide (see Chapter 3 for further discussion).

Figure 1.5 diagrams hypothetical examples for generating polypeptide diversity by alternative RNA splicing and mRNA editing. In some cases this results in gene sharing and in other cases it does not. The central point is that gene sharing involves two or more distinct molecular functions for one polypeptide.

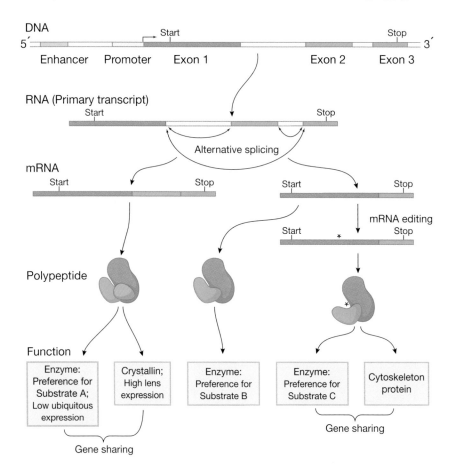

Figure 1.5. Multifunctional polypeptides from a single gene. Among the diverse polypeptides generated by various pathways used for gene expression, some can have multiple functions via a gene-sharing strategy.

POSTTRANSLATIONAL MODIFICATIONS

Posttranslational modifications that do not affect amino acid sequence but modify protein function are included within the gene-sharing concept. These occur abundantly[34,34a]; and, unlike the typical situations in alternative RNA splicing, differential promoter utilization, and mRNA editing, these modifications often do not affect the primary amino acid sequence of a protein. With noted exceptions (for example, protein cleavages or fusion of independent polypeptides by protein splicing),[35] an unmodified and modified protein shares the identical DNA coding sequence. There is also a gray area involving posttranslational modifications that make it impractical to eliminate the possibility of protein modifications being a component of gene sharing: They may remain undetected or occur transitionally as part of a biochemical pathway. Thus, posttranslational modifications that do not change the primary structure of a protein are included in this book as a mechanism that contributes to the functional amplification of proteins during gene sharing.

CONDITIONS FOR INITIATING GENE SHARING

Theoretically many factors could initiate gene sharing. These include changes in the external environment affecting the composition (for example, types of proteins or ions) or physical state (for example, temperature or viscosity) of cells. Concentration fluctuations in metabolites and/or ions may also influence interactions and performances of proteins. Variations, including mutations leading to changes in gene regulation and/or differences in the intracellular location of proteins, can act as major determinants for innovating protein functions (see Figures 1.1 and 1.2). Mutations affecting protein structure can also result in the acquisition of a new protein function. If the original protein function remains, or if two new functions arise, such mutations in the coding sequence would be contributors to the gene-sharing process.

Tinkering with gene regulation occurs at many levels. Changes in promoter regulatory elements, enhancers (DNA sequences that increase gene activity or direct it to specific tissues), silencers (DNA sequences that decrease gene activity), transcription factors (proteins that bind DNA and regulate gene activity) and cofactors (proteins that regulate the activity of the transcription factors), gene rearrangements, and possibly even stochastic variations in gene expression are mechanisms that can drive gene sharing by affecting protein concentration and/or tissue location. Modifications in gene expression may also occur at the posttranscriptional level, such as pre-mRNA processing, mRNA transport or location within the cytoplasm, mRNA stability, or translational efficiency. All could initiate new possibilities for the function of the same protein by changing its concentration and/or location within the cell.

Changes in gene expression may present gene-sharing opportunities for a polypeptide either actively or passively (Figure 1.6). For example, if expression of polypeptide A is enhanced, its elevated concentration could actively initiate a new function in that tissue. This is an example of an active en-

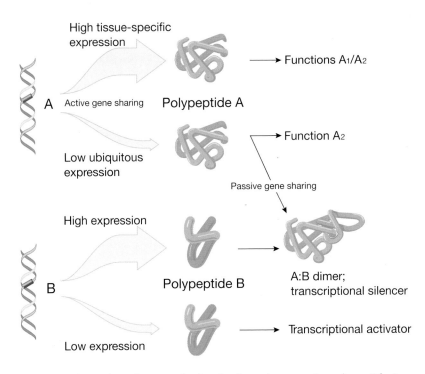

Figure 1.6. Active and passive gene sharing. In the active scenario, polypeptide A engages in two functions by virtue of differential expression of its gene. In the passive scenario, polypeptide A is drawn into a new function by virtue of a change in expression of gene *B*.

gagement in gene sharing because there is a direct relationship between expression of the gene and function of the encoded polypeptide. A passive mechanism would ensue if polypeptide A acquired a new function by an independent change in the expression of polypeptide B. The elevated concentration of polypeptide B may promote its interaction with and add a function to polypeptide A. In this case polypeptide A is induced to engage in gene sharing passively because its expression or tissue location did not change. Polypeptide A, an innocent bystander, finds itself newly or overexposed to polypeptide B, to which it binds serendipitously. Consequently, polypeptide A (and perhaps polypeptide B as well) develops a new function that may persist over time if it is advantageous or at least not harmful. Such a mechanism could involve more than one gene, compounding the complexity for polypeptides to develop new roles without changes in their amino acid sequence.

If the new function for polypeptide A (and/or polypeptide B) becomes fixed and the old function is retained, the dual functions may or may not have a physiological connection. The important consideration is that polypeptide A, an already employed polypeptide, acquires a new, additional biochemical or biophysical use. In the initial evolutionary stages of gene sharing, both functions will be used to some degree. If the altered expression pattern associated with a new polypeptide function obliterates the old expression pattern, leaving the gene with a single function, it is no longer engaged in gene sharing. In this case gene sharing would have played a transient role in evolution for that polypeptide. It is a daunting evolutionary problem to determine which genes and polypeptides have gone through a gene-sharing process to reach their present state. It would not be surprising if most, perhaps even all, polypeptides have done so.

CONTRASTING PHENOTYPE WITH PROTEIN FUNCTION

Gene sharing applies to polypeptides that display at least two molecular functions. Despite its enormous importance, expansion of the biological role of a polypeptide that affects cell or tissue phenotype by widening the use of the same intrinsic function does not fall into the domain of gene sharing. The relationship between polypeptide function and phenotype, such as might result from using the same transcription factor or enzyme reaction in different cells or under different conditions, plunges into the complex area of evolution and development (evo-devo) and has been discussed extensively.[27,35a]

The different biological consequences of the activity of the same enzyme serve as examples of the same molecular function of a protein leading to multiple phenotypes. The use of argininosuccinate lyase activity for urea synthesis in the mammalian liver but only for arginine synthesis in birds (because birds are uricotelic and do not make urea) is not gene sharing because there has been no change in the use of the inherent enzymatic property. On the other hand, use of argininosuccinate lyase as a lens crystallin in birds and reptiles[36] is gene sharing because its refractive role in the lens is an entirely separate molecular function from its metabolic role. Other enzymes, such as extensive members of the aldehyde dehydrogenase superfamily,[37] have broad substrate specificity and may assume different biological roles by changing their expression pattern; but as long as they maintain generally similar enzymatic functions, they will not be considered as cases for gene sharing. Similarly, if the same product of an enzyme is used in different biological roles because that enzyme is expressed in different places at different times, it will not be considered as gene sharing: The many phenotypes (vascular tone, inflammation, neurotransmission, apoptosis, respiration) associated with nitric oxide produced by nitric oxide synthase[38–43] are an example (Figure 1.7). Thus, gene sharing requires a different molecular use of the same polypeptide sequence, not a different phenotype resulting from a similar molecular use of that polypeptide.

There are subtle cases in which it remains ambiguous whether two functions of a polypeptide should be considered as gene sharing. For example, Pax transcription factors may fall into such a gray area. The same Pax protein may bind different DNA regulatory sequences and activate or repress genes; it may also initiate different cascades of developmental expression. These different biological roles employ alternate peptide sequences within the polypeptide (a paired domain or a homeodomain for DNA binding, and different regions of the transactivation domain for activation or inhibition). Thus, there could be justification in considering different biological roles of such a transcription factor as gene sharing by the criterion of requiring an alternate molecular use of different properties or modules (physical stretches of amino acid sequence that form defined structures) of the polypeptide, blurring the boundaries of gene sharing, even though the protein remains as a transcription factor regulating gene expression in its various roles. Thus, mechanism is intimately connected to differences in molecular function.

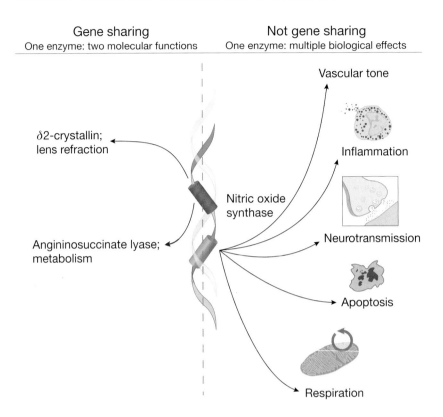

Figure 1.7. Multiple molecular functions versus multiple phenotypes. The use of the same polypeptide for more than one molecular function, such as argininosuccinate lyase for a lens crystallin or an enzyme (left side), is gene sharing. Use of the same function of a polypeptide for the generation of different phenotypes, such as the enzymatic activity of nitric oxide synthase (right side), is not gene sharing.

TAKE-HOME MESSAGE

Gene sharing is a process that expands the number of molecular functions that are performed by a polypeptide. The fact that individual polypeptides with a given amino acid sequence engage in entirely different biochemical and/or biophysical functions leads to an intuitively unexpected paradox: polypeptides are simultaneously specialized for individual functions and functionally diverse.

2

Multiple Functions and Functional Shifts: Echos from the Past

Chapter 1 introduced the concept of gene sharing in broad strokes, emphasizing that a gene may encode a protein with more than one molecular function and that protein function(s) may change as a consequence of alterations in the cellular microenvironment. In short, the gene-sharing idea acknowledges that the many properties of proteins are exploited in various ways in a cellular context, and that proteins may shift molecular functions during evolution in ways that are not entirely dependent on changes in their amino acid sequence.

Relating a single structure to multiple functions and recognizing functional shifts are hardly new ideas in biology. It has been appreciated since Darwin himself that organs performing one function may evolve into structures performing a different function. Clearly the evolutionary processes involved in functional shifts at the organ level and at the molecular level, such as with gene sharing, differ. At best morphological evolution will provide occasional parallels with molecular evolution. Nonetheless, previous as well as more recent recognitions that one structure may perform diverse functions and/or evolve into new functions indicate a long-standing appreciation of functional plasticity in evolutionary thought. In view of the interest and benefit of hearing echoes of past thought in modern speech, the present chapter reviews briefly some of the major ideas relating structure to function and functional shifts in biology, even if they do not apply directly and/or mechanistically to the molecular events of gene sharing.

PREADAPTATION, PROSPECTIVE ADAPTATION, AND HOPEFUL MONSTERS

Darwin was concerned with functional shifts in structures that were originally used for different roles earlier in their evolutionary history. Consider, for example, Darwin's thoughts on the frequent absence of intermediate states in evolution. He writes[44] (p. 204), "we should be very cautious in concluding that none could have existed, for the homologies of many organs and their intermediate states show that wonderful metamorphoses in function are a least possible." Darwin goes on to state that "a swim bladder has apparently been converted into an air-breathing lung" to provide a case for "the same organ having performed simultaneously very different functions, and then having specialized for one function." Darwin's use of the word "simultaneously" is interesting because it encompasses the idea of multifunctional uses of one structure, not just one structure being converted from one function to another over time.

Still in the nineteenth century (1875), Dohrn proposed a "principle of functional change" (see Bock[45]). The idea that a structure with one function could be co-opted to perform another led to the concept of "preadaptation," which became entwined with mutagenesis. In 1933 Goldschmidt (see Goldschmidt,[46] p. 105) went to an extreme when he considered freakish animals as "hopeful monsters"; in his opinion such mutations allowed rapid occupation of a new niche. The development of two eyes on the same side of the head in a flounder, resulting in life at the bottom of the sea, is an example. More subtle mutations could also be perpetuated until changing conditions or proper opportunities suddenly made them useful: In short, the occupation of a new niche could occur by preadaptation horizontally (in space) or temporally (in time). As often occurs when tracing history, the "hopeful monster" idea too came earlier. In 1895, Bonavia speculated that "monstrosities" might occupy new niches in a single evolutionary step, thereby anticipating the concept of preadaptation (see Goldschmidt[46] [p. 391]).

Preadaptation was defined by Simpson[26] (p. 236) as "the random origin, by mutation, of characteristics nonadaptive or inadaptive for the ancestral way of life, but adaptive for some other way of life which happens to be available." He recognized the importance of opportunism in evolution, as opposed to direction, and held to the idea that what could happen often did happen, but this was not necessarily the best or only way things could go (see Simpson[26] ([Chapter 12]). Because of random mutation, Simpson believed that changes in function survived due to selection, making preadaptation just a form of adaptation. He concluded that preadaptation was too broad a concept and thus meaningless. Consequently, he proposed the term "prospective adaptation," hoping that this would place emphasis on the fact that structures can undergo changes in adaptation.[47] Despite its limitations, preadaptation continued to be used by some evolutionists and was defined by Bock[45] (p. 201) as follows: "A structure is said to be preadapted for a new function if its present form which enables it to discharge its original function also enables it to assume the new function whenever need for this function arises." In other words, any property of a structure that allowed it to become employed for another role was a preadaptive property. Thus, preadaptation was invoked by Bock to explain potential utilities inherent in what we consider original uses,[45,48] such as the later expansion of feathers for flight although they were employed earlier as thermoregulators on the forearms of small running dinosaurs (see Gould,[49] p. 1231). The recent concept of "constructive neutral evolution"[50] resonates with the earlier concept of preadaptation. Constructive neutral evolution involves changes that can be acted upon by a complex series of steps to give rise to evolutionary novelty when the opportunity arises.

QUIRKY FUNCTIONAL SHIFTS AND EXAPTATION

Gould[49] considers "quirky functional shifts" between biological structures and distinguishes homology from pragmatic dual utilization. One of his many analogies resonates with the alien walking across campus and observing multiple uses for a chair (see Chapter 1). Gould points out that the thinness of a dime has nothing to do with its commercial use. Yet, because of its particular dimensions, most of us have used a dime at one time or another as a screwdriver. He relates this to the situation in post–Soviet Union Russia, where the five-kopec coin became worthless as currency, but gained value due to its dimensions as a coin that could be specifically used to operate public telephones.

Gould and Vrba[51] invented the term "exaptation" for features that were co-opted for a new use following an origin for a different use. As with Simpson's prospective adaptation, exapted proteins are subject to natural selection as are all biological structures (see Gould[49] for examples and extensive discussion of this point). Gould[49] (p. 1232) distinguished the term exadaptation from the shorter exaptation to emphasize that adaptations are features that were selected for their original use, while aptations are features that have other side effects as a consequence of their existence that were not originally under selective pressure. Gould[49] (p. 86) considered crystallins as "outstanding examples of exaptation . . . in so many vertebrates and from so many independent and different original functions."

Before Gould, Williams distinguished effects from functions in his treatise on natural selection and biological adaptations.[52] Although Williams provides examples of complex functions, phenotypes such as vision or locomotion, the thought process is applicable to genes and proteins. He believed that natural selection worked at the lowest possible level beneficial to the individual organism, rather than the more complex levels of groups and species. Williams was careful to distinguish an effect derived by happenstance from a property that was selected specifically for its function. He believed that it was insufficient for an effect to be beneficial for it to be considered as evidence for adaptation. In his book Williams[52] (p. 11) stated that the concept of adaptation "should not be invoked when less onerous principles, such as those of physics and chemistry or that of unspecific cause and effect, are sufficient for a complete explanation." He thought (p. 12) that "it would be absurd to recognize an adaptation to achieve the mechanically inevitable." Williams famously considers, tongue in cheek, a flying fish returning to water after gliding a nonadaptive consequence of gravity. Conceptually, however, these ideas resonate with the later formulation of exaptation, and Gould[49] (p. 1230) states that "Williams invoked the term 'effect' to designate the operation of a useful character not built by selection for its current role."

SPANDRELS AND GENE SHARING

Gould and Lewontin[53] (p. 147) challenged a strictly adaptationist view of evolution on the basis of its "failure to distinguish current utility from reasons for origin" as well as a number of other limitations connected with the rigidity required to find an adaptive reason for a structure even if none was apparent. They claimed that the usual explanation in such a case would be that there is an adaptive reason for the selection of a protein but we do not know it, without consideration of an alternative explanation. Gould and Lewontin used the colorful metaphor of spandrels in St. Mark's Cathedral in Venice to argue that many biological features derive from physical constraints that originated as a necessary consequence of other, unrelated features. Spandrels (or "spaces left over") are the triangular areas that always appear between the arches that support a central architectural dome. Gould and Lewontin pointed out that these triangular spaces in St. Mark's Cathedral were utilized at a later date for artistic expression because they were there, not because their use as such provided adaptive value connected with the arches. They equate the illogic of saying that the spandrels were placed between the supporting arches for design with the obvious absurdity of Voltaire's Pangloss in *Candide* insisting that "everything is made for the best purpose," such as "our noses being made to carry spectacles" and so on. More recently, Gould[49] associated the spandrel analogy of creative use of a nonadaptive inherent structure with the phenotypic expression of neutral mutations at the DNA level. He called this a cross-level rather than an at-level effect because the consequences of utilizing a spandrel may be much broader than the particular structure involved, similar to the far-reaching and often unpredictable consequences of a mutation.

Although there have been objections to the spandrel metaphor for evolution,[54] the idea that all structures have features that can be and are used in various pragmatic ways over time is important and encompasses gene sharing. The spandrel metaphor referred to obligatory byproducts of structures being exploited innovatively. However, the concept of gene sharing is more inclusive than alternative use of byproducts, as suggested for spandrels. The very idea of byproducts of proteins is a difficult one. Which part of a protein is essential for maximal fitness and which is equally dispensable, if any? For example, experiments on the A chain of insulin in vertebrates illustrate the difficulty in understanding the functional utility of all parts of a protein. This polypeptide has an invariant leucine amino acid residue at position 16, a key internal side chain that does not make contact with the insulin receptor.[55] Mutagenesis of this residue leads to misfolding of the protein but retention of significant receptor-binding activity. The investigators in this study do not claim that this residue is an intrinsic requirement for an insulin-like fold, yet argue that it is under stringent selection due to interlocking evolutionary constraints for maximal fitness. They consider this as an example of natural selection of protein sequences operating at multiple levels during evolution. We are left with the question, What is a byproduct of protein structure?

The gene-sharing concept predicts that innovative, pragmatic adaptations of proteins can arise from any property of the molecule, whether that property is a critical part of its known function (such as an active site for enzyme activity) or whether it is what appears to be a "space left over." In other words, gene sharing addresses any "space" that can be occupied; it does not depend on utilization of a space that does not appear to be part of a known function of the protein. It is implicit in the gene-sharing concept that nature does not limit options; it tinkers with and exploits all possibilities, whether cooperative, neutral, or conflicting. Resolution of emerging difficulties due to conflicting functions is another matter for natural selection to cope with.

GENE REGULATION AND TINKERING

The idea of evolutionary changes in gene regulation being a driving force for biological innovation is not new and is at the center of the burgeoning field of evo-devo (evolution of development).[27,56] Wilson[57,58] drew attention to the adaptive changes of bacterial gene expression in stressful environments by changes in enzyme concentrations. This was extended to animal evolution by correlating large differences in the concentrations of the same enzyme in different species with adaptive evolutionary changes. Wilson and his colleagues suggested that changes in protein structure followed an evolutionary clock with mutations in slowly evolving forms (such as frogs and birds) occurring at the same rate as those in the rapidly diverging mammals. From this it was concluded that the major biological differences between humans and chimpanzees were due to regulatory mutations, further implying that changes in gene regulation play a dominant role in speciation.[59] The hypothesis that evolutionary changes in gene expression were responsible for the loss of ability of interspecific hybridization was used to support the central role of gene regulation for evolutionary novelty.[60] While perhaps mechanistically different, the gene-sharing concept also places great importance on changes in gene regulation for evolutionary novelty of protein function.

In his thoughtful article, Jacob[61] (p. 1165) argued that tinkering with gene regulation plays a key role during evolution. He stated: "What makes one vertebrate different from another is a change in the time of expression and in the relative amounts of gene products rather than the small differences observed in the structure of these products. It is a matter of *regulation* [my italics] rather than structure." Jacob amplified this theme in his subsequent book[62] (p. 41) by writing the following: "Small changes in modifying in time and space the distribution of the same structures are sufficient to affect deeply the form, function, and behavior of the final product: the adult animal. It is always a matter of using the same elements, of adjusting them, of altering here or there, of arranging various combinations to produce new objects of increasing complexity. It is always a matter of tinkering." The gene-sharing concept falls within this description nicely.

Both Wilson and Jacob were pioneers in postulating that changes in gene regulation were at the heart of evolution. Britten and Davidson proposed a theory of gene regulation by which batteries of producer genes associated with sensor and receptor genes were subject to co-regulation by activator RNAs.[63–65] Their novel scheme was guided by the idea[66] (p. 129) "that many of the important events that have occurred in the evolution of genomes are those which have induced change and growth in the regulatory system" and that "a new structure or organ might have arisen in evolution" by "the rearrangement of existing gene regulatory relationships so that certain cells of an antecedent structure gain additional new characteristics."

Clearly, the winds of evolution by gene regulation that were gaining strength in the 1970s had reached gale force. Wilson related enzyme concentration to adaptation and evolutionary advances and correlated speciation with modification of gene regulation, Jacob envisioned biological novelty arising by tinkering with existing structures and with gene regulation, and Britten and Davidson devised an imaginative theory relating developmental and evolutionary changes with modifications in gene expression rather than invention of new structures. Today it is well recognized that changes in gene regulation are at the heart of morphological and other evolutionary changes.[27,35a,56,65] The explosion of new species evident in the Cambrian strata 530 million years ago appears to owe a lot to regulatory modifications and using existing genes in new roles rather than the creation of new genes.[27] This is deduced from the complexity of transcription factors and structural genes used throughout evolution that were already present in the Porifera (sponges) and Cnidaria (jellyfish, corals, anemones).[67,68] Note, however, that Nei[69] (p. 417) cautions reasonably that transcription factors are encoded in structural genes, which complicates the distinction between mutation of structural genes and changes in gene regulation. Moreover, the invention of a few, critical structural genes could in theory have a major impact on evolution. Imagine the power of a few paperclips on organizing scattered sheets of a draft of an evolving novel! This leaves understanding the interrelationships between body plans and genetic architecture, or how "toolboxes" comprising cascades of regulatory factors are themselves deployed to direct different structures, as an important challenge for the future.[70]

TAKE-HOME MESSAGE

The concept of gene sharing involving functional shifts and dual tasks for
proteins has roots in evolutionary thought, albeit not necessarily in molecular
terms, dating back to Charles Darwin in the mid-nineteenth century. The
gene-sharing concept resonates with more recent ideas of nonadaptive, pragmatic use of proteins and tinkering with gene regulation as causes for evolutionary novelty.

3

The Elusive Concept of a "Gene"

Genes have been defined by their absence when an organism manifests a recessive trait, by their damaging potential when mutated, and by their contributions to the extraordinary abilities of Olympic athletes and Nobel Laureates. Genes have been considered root causes for our talents and blamed for our limitations. They have been given independence by their ability to mutate randomly, leaving us at their mercy, and were even charged with a selfish life of their own. They have instilled fear as targets of eugenics and have raised hope by promises of gene therapy. Genes may have been redefined over time but they have not lost their importance to our view of life. All this, and more, and yet there is still no agreement on a complete and satisfying definition for a gene that encompasses both structure and function. This creates a conundrum for a specific definition of the "gene" in gene sharing.

The gene-sharing concept is about the existence of different molecular functions for a single polypeptide encoded in one gene. The function of the encoded polypeptide depends both on the expression levels of its gene and its intracellular and tissue locations. Gene expression in turn is regulated by DNA sequence motifs within the genome, cognate transcription factors encoded in different genes, and innumerable variables and biochemical events between the extracellular environment and the intracellular chromosomes. Protein intracellular location is determined by a dynamic flux of many molecules and the physiological state of the cell. Thus, defining the boundaries of a "gene" in the gene-sharing concept is not straightforward because it ties the concept of a gene to function and there are many agencies for the function of the encoded protein.

From an historical perspective, the concept of a "gene" has changed enormously. (I am indebted to Dr. Richard Burian, Virginia Polytechnic Institute and State University, Blacksburg, VA for his personal communications to me on this subject.) An early Mendelian view of a gene was an ontologically unspecified state, condition, or process that has the effect of specifying a particular cellular property or of directing development in a particular way. Subsequently, the Mendel-Morgan school of genetics formulated the chromosomal theory. This transformed the unknown entity, state, or process considered a "gene" to an undefined "factor" located within or near an identifiable part of a chromosome.[71] Much later, after discovery of DNA as the hereditary material and the chemical substance of the gene, the revolution in molecular biology led to the modern view of genes as annotated nucleotide sequences encoding specific proteins. The gradual switch in the concept of a gene— from an ill-defined program for development and phenotype to a one-dimensional array of nucleotides—is associated with a change in the gene concept from context-dependent to context-independent.

Detailed compartmentalization of chromosomal DNA into functional regions—structural genes, regulatory genes, expressed noncoding and coding DNA sequences, and the like—have resulted in an increasingly complex and quantitative description of the genome. This potential richness of information about the raw material of genes and gene expression has given far more insight into how DNA packages information, underlies development, and reflects phylogenetic kinship among species than could have been imagined not very long ago. However, despite our wealth of knowledge, or perhaps because of it, the quest for a single definition of a "gene" that covers both structure and function remains elusive. Many excellent reviews have been written on the concept of the gene at various stages of its evolution (see, for example, references[72–85]). The present chapter discusses briefly the evolving concepts of the "gene" and ends with an open, abstract version for the idea of gene sharing.

THE CLASSICAL GENE CONCEPT

Early on, the need for a conceptual view of genes was implicit in considerations of ontogeny (embryonic development). As discussed by Moss,[82] a duality between preformationism (the progressive unveiling of form during ontogeny) and epigenesis (the progressive elaboration of form during ontogeny) has been an ongoing theme in the study of biology ever since Aristotle. The broadening of epigenesis beyond the individual organism to include the phylogeny of adaptive changes throughout evolution was a later development. The seventeenth and eighteenth centuries introduced preformationism in biology with deistic connotation. Moss (p. 8) writes that scientists of the time thought that "the embryos of all the organisms which would and could ever be had come into existence with the creation of the world and its first creatures, as so many Russian dolls, full formed miniatures nested and encased one inside the other. Subsequent generations were deemed to 'evolve' from the old on the basis of the purely mechanical unfolding and elaboration, the *inflating* really, of parts already in place."

In the eighteenth century, Immanuel Kant (see Moss,[82] p. 12) championed the idea of *Keime und Anlagen,* which considered epigenesis to be a mechanism giving "rise to stabilized new form . . . preadapted to the environmental pressures faced by the preceding generations." The preadaptive notion put epigenesis in the category of playing out a preformed potential, another form of preformationism. Indeed, the word "evolution" means to unfold or disclose and was considered throughout the eighteenth century as a preformationist notion (see Moss[82]). While these ideas predate knowledge of genes, they indicate the early and deeply rooted conceptual foothold of predetermination and reflect a conflation of epigenesis and preformationism in explaining form and function in biology. Note also, as discussed in Chapter 2, preformation and preadaptation are parallel concepts that have influenced the history of biology.

The concept of Darwinian evolution and new methods of experimentation both mark major innovations in biology in the nineteenth century. As well known by readers of this book, Darwin established the evolutionary concept of gradual modifications of traits resulting from the forces of natural selection acting on numerous, random heritable variations. However, Darwin did not know about chromosomes and genes to explain the inheritance of traits. He put forth a pangenesis theory in 1868, the "provisional hypothesis," in which he visualized that variations were transmitted via invisible "gemmules" that were scattered throughout the body and aggregated in the germ cells where they would be inherited. Darwin proposed this theory primarily to explain how acquired characters are transmitted from parent to offspring.[72] Darwin's pangenesis theory followed an earlier speculation of the presence of "physiological units" by Herbert Spencer in 1863. Like gemmules, physiological units were scattered throughout the body; but, unlike gemmules, they were all identical. The physiological units concept was based on the ability of parts to regenerate a whole organism in some cases.

In 1883 August Weismann challenged the idea of transmitting acquired characteristics (for example, hypertrophy of body parts) and believed in complete separation of the body (soma) and the germ plasm (eggs and sperm) (see Morgan[72]). This was a very important advance for the concept of the "gene." The germ plasm remained segregated and was transmitted from generation to generation. The individual had no influence on the germ plasm, which contained hereditary elements grouped into "idants" composed of self-replicating "ids" (which were further divided into "determinants" and "biophores"). These specified the organism and were discriminated by being different representatives of ancestral individuals. Development proceeded by parceling the ids to different parts of the embryo. Various combinations of these mysterious determinants were present in different eggs as a consequence of fertilization, resulting in the variations observed among individuals. To keep the number of ids at a reasonable level, Weismann postulated that they reduced by half during the formation of eggs and sperm (see Moss[82]).

Following Weismann nearing the end of the nineteenth century, the Dutch botanist Hugo de Vries, influenced by Darwin's pangenesis theory, hypothesized in 1889 the existence of intracellular "pangens" (see De Vries[86]). These were distinct units that were contained within germ cells and were responsible for inheritance and development of traits. A few years later, De Vries formulated a mutation theory stating that species arose by sudden changes due to addition or loss of pangens. The gene concept remained as the bestower of traits. However, the conceptual isolation and continuity of the germ plasm determinants were major contributions by Weismann and De Vries to the concept of the gene, advancing theories of "hard heredity" incrementally to the center of the debate (Burian, personal communication).

The "rediscovery" (by Hugo de Vries, Carl Correns, and Erich von Tschermak in 1900) of the work by the Austrian monk Gregor Mendel on inheritance and segregation of determinants in peas was a well-known pivotal event for genetics (see Stern and Sherwood[87]). The experimental basis for Mendel's conclusions contrasted sharply with the theorizing that came before his rediscovery (see Carlson,[74] Chapter 3). Closely following Darwin's *Origin of the Species* (1859) and still well before knowledge of heredity units as physi-

cal structures, Mendel wrote of "elements" and "unit characters." He considered groups of traits (pod color, seed color, etc.) that were transmitted together in hybridization experiments as pleiotropic effects of a single unit-
trait (see Falk[88]). Although not mechanistically the same, Mendel's multiple
phenotypic traits from undefined single physical units foreshadow the idea of
multiple roles for individual structures, as genes or protein have in the concept of gene sharing.

In the early twentieth century following the rediscovery of Mendel, Wil-
liam Bateson, an Englishman, conceived of "unit-characters" called "allelomorphs" that may be present in antagonistic forms.[89] The root meaning of
allelomorph is "other form." Thus, allelomorphs extended the concept of the
ever-elusive gene to having multiple forms. It was no longer the presence or
absence of a gene (or rather the other terms in use at the time) that was critical for development of specific characters, but the form of the gene. The concept of absence of a trait challenged investigators of the time and led to fascinating scientific debates beyond the scope here (see Carlson,[74] Chapter 8).
Was absence truly an absence, or was it the presence of a repressor or over-
dominance of another trait? Moreover, neither Bateson nor his peers could
distinguish between hypothetical factors and complex traits. Nonetheless, allelomorphs remained as alternative states of a unit-character rather than a
factor or gene until their transformation into "alleles," the term used today for
alternative states of a physical gene. Bateson's term "allelomorphs" relates to
the idea of gene sharing by conceptually transforming a single gene into an
entity ultimately capable of fulfilling multiple roles and is of sufficient importance and interest to quote Bateson's description here[89] (p. 31):

> By crossing two forms exhibiting antagonistic characters, cross-hybrids were
> produced. The generative cells of these cross-breds were shown to be of two
> kinds, each being pure in respect of *one* of the parental characters. This purity
> of germ cells and their inability to transmit both of the antagonistic charac-
> ters is the central fact proved by Mendel's work. We thus reach the conception
> of unit-characters existing in antagonistic pairs. Such characters we propose
> to call *allelomorphs,* and the zygote formed by the union of a pair of opposite
> allelomorphic gametes we shall call a *heterozygote.* Similarly, the zygote
> formed by the union of gametes having similar allelomorphs, may be spoken
> of as *homozygote.* Upon a wide survey, we now recognize that this first princi-

ple has an extensive application in nature. We cannot as yet determine the limits of its applicability, and it is possible that many characters may really be allelomorphic, which we now suppose to be "transmissible" in any degree or intensity. On the other hand, it is equally possible that characters found to be allelomorphic in some cases may prove to be non-allelomorphic in others.

Bateson coined the word "genetics" in 1906 in order to initiate a new field of physiology; Wilhelm Johannsen came up with the term "gene" three years later (see Keller[79]). He was looking for a neutral word that did not carry a connotation of preformationism. Keller (p. 2) quotes Johannsen: "The word 'gene' is completely free from any hypotheses; it expresses only the evident fact that, in any case, many characteristics of the organism are specified in the gametes by means of special conditions, foundations, and determiners which are present in unique, separate, and thereby independent ways—in short, precisely what we wish to call genes." Two years later, Johannsen stated, "The 'gene' is nothing but a very applicable little word" that "may be useful as an expression for the 'unit factors,' 'elements' or 'allelomorphs' in the gametes, demonstrated by modern Mendelian researches . . . As to the nature of the 'genes,' it is as yet of no value to propose any hypothesis; but that the notion of the 'gene' covers a reality is evident in Mendelism." Carlson[74] (p. 22) evaluates Johannsen's contribution as follows: "Johannsen's gene was undefined. If to some geneticists he gave a concept that appeared to lack a material reality, for others he freed the concept from any one theory of action, specificity, or composition. It gave the gene concept an opportunity to evolve and to take on, or discard, definition."

The early phases of the "gene" concept attempted to explain the development and transmission of biological traits. Traits (phenotypes) were indisputably real and measurable; genes (genotypes) were presumed and inaccessible so there was no knowledge of their intrinsic properties. Taken together, genes were hypothetical "somethings" (factors? processes?) that elaborated defined characteristics and were responsible for transmission of traits from parent to offspring. Despite the enormous contributions of Bateson and Johannsen to the development of the concept of a "gene," neither gave the gene a structural foothold (see Burian[78]). Both opposed the chromosome theory described in the next section. Bateson interestingly considered the possibility that genes are akin to stable harmonic resonances, which would be consistent with traits being saltational and able to increase by a process analogous to the addition of nodes on a standing wave.

Two additional points concerning this classical stage for elaborating the concept of a gene are noteworthy with respect to the concept of gene sharing. First, the idea of the gene was grounded in the explanation of traits that depended upon unknown *functions,* not structures. Second, Johannsen recognized the importance of language in influencing the field and settled on the term "gene" because it was devoid of bias. As discussed in Chapter 2, the term "gene sharing" is relatively free of hypothesis or bias; it simply implies *multiple functions,* unlike the terms "recruitment," "co-option," or "moonlighting," which suggest primary roles from which genes are captured and placed in other roles from time to time.

How did these origins of the concept of a gene play out in the twentieth century?

THE MENDEL-MORGAN CHROMOSOMAL THEORY OF THE GENE

The gene concept was advanced to a structural basis in the twentieth century in connection with the effort to characterize and define a "gene" as a cytological-chromosomal entity with alternative states (see Burian[90]). Although his work preceded the discovery of the precise physical nature of genes, Thomas Hunt Morgan was invaluable for modernizing the gene concept. Even though Morgan and his colleagues still used the term "factor" rather than "gene," their textbook of 1915 is considered the standard marker for the establishment of the so-called Mendel-Morgan chromosomal theory of the gene.[71] This theory depended on numerical data derived from mating experiments and had nothing to do with assigning explanatory properties to invisible units, as had been done previously. Morgan's clarity is a pleasure to read; I quote from his Sillman Lectures[72] (p. 25):

> We are now in a position to formulate the theory of the gene. The theory states that the characters of the individual are referable to paired elements (genes) in the germinal material that are held together in a definite number of linkage groups; it states that the members of each pair of genes separate when the germ-cells mature in accordance with Mendel's first law, and in consequence each germ-cell comes to contain one set only; it states that the members belonging to different linkage groups assort independently in accordance

with Mendel's second law; it states that an orderly interchange—crossing-over—also takes place, at times, between the elements in corresponding linkage groups; and it states that the frequency of crossing-over furnishes evidence of the linear order of the elements in each linkage group and of the relative position of the elements with respect to each other.

Morgan predicted[72] (p. 317) that individual genes have widespread effects as he wrote that "each gene may have a specific effect on a particular organ, but this gene is by no means the sole representative of that organ, and it has also equally specific effects on other organs and, in extreme cases, perhaps on all the organs or characters of the body." He anticipated the modern ideas of tinkering and rearranging existing genes as the source of biological complexity as he assessed the data of his era by saying[72] (pp. 319–320), "the evidence that we have furnishes no grounds whatsoever for the view that new genes independently arise . . . If the same number of genes is present in a white blood corpuscle as in all the other cells of the body that constitutes a mammal, and if the former makes only an amoeba-like cell and the rest collectively a man, it scarcely seems necessary to postulate fewer genes for an amoeba or more for a man."

These incisive words preceded our current surprises that vastly different species have roughly comparable numbers of protein-coding genes and the molecular discoveries of differential gene expression, gene interactions, and multiple functions for individual genes and their products. Though it was before his time, Morgan attempted to develop a physical/chemical notion of the ephemeral gene. He speculated[72] (p. 321) that "the order of magnitude of the gene is near that of the larger-sized organic molecules . . . perhaps . . . the gene is not too large for it to be considered as a chemical molecule, but further than this we are not justified in going." He went on to admit that it is "difficult to resist the fascinating assumption that the gene is constant because it represents an organic chemical entity . . . it seems, at least, a good working hypothesis." Indeed it was! But note: The gene was still defined by being both a proposed physical structure and the responsible agent, or process, for developing biological traits.

LATER DEVELOPMENTS: ONE GENE/ONE ENZYME/ONE POLYPEPTIDE
During the Morgan years "genes" became mapped as linear units along chromosomes, described as units of mutations by x-rays and positively identi-

fied as DNA (see Portin[81]). The growing optimism of the reductionists that a unitary concept of the "gene" was close at hand did not develop without challenges (see Dietrich in Beurton et al.[77]). For one, the German biologist Richard Goldschmidt denounced the idea of a corpuscular gene as dead; he was largely influenced by the observations of the importance of position effects (DNA rearrangements of any sort) on phenotype, a number of which had first been found by the Morgan group (see Chapter 13 in Carlson[74]). Goldschmidt envisaged a gene as a continuum rather than particulate, and this impacted the development of the gene concept. Carlson (p. 130) states that "Goldschmidt's outlook . . . had one important influence on the development of the gene concept. His 'continuum' model permitted a climate of acceptance, rather than incredulity, when attempts were made to demonstrate that the gene itself might be a linear continuum."

Genetics underwent a conceptual upheaval after World War II. The genetic material was unequivocally identified as DNA and the "gene" became a discrete unit of double-stranded, helical DNA.[91] Despite the increasingly physical definition of a gene, it also remained as a unit of function as determined by the complementation test. The complementation test depended upon the phenotype of heterozygotes containing separate recessive mutations of a single allele. If a phenotypic mutation remained mutant in a heterozygous individual for distinct recessive mutations, the allele under consideration was considered a single gene. If the two mutations functionally complemented each other in the heterozygote to produce a normal (wild type) individual, the mutations were considered to be situated in separate genes. Ultimately, studies on the fungus *Neurospora* led to this functional definition of a gene being transformed into the one gene/one enzyme hypothesis of Beadle and Tatum[92] and Srb and Horowitz.[93]

Further studies resulted in the subdivision of the discrete gene. Seymour Benzer's investigations on the rII region of bacteriophage T4 DNA were especially important in this connection (see Holmes in Portin[77]). He fractionated a functional gene by virtue of hundreds of mutations into "cistrons" (named after the "cis-trans" test for function of heterozygotes) (see Portin[81]). Cistrons were further resolved by genetic recombination into "recons" (smallest units of recombination) and "mutons" (smallest unit of mutation). Due to these and other studies, the one gene/one enzyme hypothesis gave way to the one gene/one polypeptide hypothesis because an enzyme can be composed of several different kinds of polypeptides: One "gene" ("cistron") was responsible

for the synthesis of one mRNA molecule,[94,95] which in turn was translated to
yield one polypeptide. This ended what Portin called the neoclassical view of
the gene.[81] The years preceding the boom in molecular genetics thus provided
hope that the function of a gene would be reconciled into a single entity.

THE MOLECULAR ERA OF THE GENE: SO MUCH DATA, SO MANY POSSIBILITIES

The molecular era changed everything. Recombinant DNA, DNA cloning,
and all the rest comprising modern molecular genetics made it impossible to
analyze the voluminous sequence data, taken (perhaps naively) as a blueprint
for life, without computers, adding a new language and set of metaphors to
the concept of the gene.[80] It became clear that it would not be possible to
make sense of the genome, development, or heredity by dealing with genes
one at a time; this realization forced investigators to consider how genes interacted even before they had settled on exactly what a single "gene" is. We
now speak of interacting conserved networks of genes and proteins that
are responsible for functions and cellular behavior in prokaryotic[96–98] and
eukaryotic[35a,99–106] cells.

The sea change brought about by molecular biology allowed for the first
time an intrinsic characterization and model of the gene (see Chapter 1; Figures 1.4 and 1.5). The new molecular landscape for studying biology yielded
enormous surprises and unexpected complexities for identifying gene structure and function.[77,81,82,85,107–109,1120–1122] The discoveries of gene families (repeated versions of the same gene), split genes (exons and long introns), pseudogenes (protein-coding genes that contain inactivating mutations), mobile
genes (often called transposons) that hop around the genome, rearranged
genes (for example, immunoglobulins), nested genes (genes within genes),
overlapping genes created by initiating or terminating transcription at multiple sites along the DNA, and genes generating multiple proteins by alternative
RNA splicing (piecing together different exons), mRNA editing, and alternative initiation and termination of mRNA translation created so many new
concepts that the very notion of one gene yielding one functional polypeptide
became untenable as a generality. DNA motifs regulating gene expression
were found within, next to, and very distant from the gene(s) they regulate,
raising the thorny issue as to whether these regulatory sequences, which often
control the expression of more than one gene, should be included as part of

the "gene." The complexity of genomic organization sparked much debate about the nature or even usefulness of a gene concept in the classical or neo-classical sense.

QUANTIFYING GENES BEFORE THE MOLECULAR ERA

The molecular descriptions of the genome raised hopes that it would be possible to establish the numbers of genes for each species. Quantifying the genome in one sense or another has been a tantalizing issue for investigators. An early goal was to correlate genetic information with evolutionary novelty and phenotypic complexity. Williams[52] assumed that the amount of DNA per species was selected to optimize information content. He thought (pp. 41–42) that more DNA would increase "noise and consequent reduction in the adaptive precision of the phenotype. Quantity of information and precision of information are somewhat opposed requirements of the genetic message. The amount of DNA in an organism would presumably reflect the optimum compromise between these opposing values."

Subsequent studies by others have confirmed that there is no simple relationship between species and DNA quantity.[110] The poor correlation between DNA quantity and biological complexity is known as the C-value paradox. C-value means the characteristic amount of DNA within a species. Some unicellular amoebae, fish, amphibians, or plants have considerably more DNA than any mammal, including humans (see Hughes[111]). There have been efforts to relate genome size to variations in the amount of noncoding DNA, average intron sizes, or relationships between cell size, generation, or developmental time, population genetics, or even adaptive lifestyles (e.g., flying). Although intriguing correlations have been suggested, it has not been possible to link the amount of DNA and the apparent biological complexity of a species.

The poor correlation between amount of DNA and phenotypic complexity serves as a reminder that we have much to learn about DNA function despite our present wealth of molecular knowledge, and reinforces the concept that adaptation and evolutionary progress cannot be charted as a straight line leading to humans.[52,112] Williams pointed out the difficulty of evaluating biological complexity.[52] He noted that more ancient species may have developed as many or more adaptive specializations during their lengthy stay on this planet than the apparently more sophisticated, relatively recent newcomers such as humans. Moreover, different lifestyles of individual organisms that must require different gene sets to deal with an independent larval as well as an adult existence, for example, may confound how many problems need to be confronted by genes by any one species. Williams writes thoughtfully in his classic treatise (p. 45) that "until sound arguments are formulated we must be wary of passing judgments on the relative complexities of organisms of very different life cycles."

QUANTIFYING GENES IN THE MOLECULAR ERA: FEWER THAN EXPECTED

The belief that genes could be precisely identified by sequence led to a flurry of activity in the molecular era to convert the ambiguous amount of information in DNA content to unambiguous numbers of genes. However, it became apparent that protein-coding sequences comprise only a fraction of the information content within a genome, with a rule of thumb being that they account only for about 1.5 percent of the total DNA. Another problem is the difficulty of recognizing individual genes: Most are composed of separated modular sequences (exons) and deciphering the complexities of overlapping genes, nested genes, and the like is challenging.[113–115] Nonetheless, much attention was and continues to be given to establishing the number of genes in different organisms.

The gene-counting data obtained indicated that the absolute number of protein-coding genes per genome was lower than anticipated[116] and did not support the notion that more complex species have more genes.[115] Estimates of the number of protein-coding genes for different species include approximately 4–6,000 in the bacteria Escherichia coli K-12[117] or Pseudomonas aeruginosa,[118] 10,000 in the filamentous fungus Neurospora crassa,[119] 6,000 in the yeast S. cerevisiae,[120] 20,000 in the nematode worm C. elegans[121] and 13,600 in the fly D. melanogaster.[122,123] These similar numbers of protein-coding genes in species of greatly different levels of organizational complexity and lifestyles extend to mice[124] and humans.[125–127] The most recent estimates for human protein-coding genes at the time of writing is 20–25,000[115,128,129]— at best twice that of flies and a far cry from the earlier anticipated minimum of 100,000!

The lack of correlation between numbers of protein-coding genes and phenotypic complexity has been reconciled by combinatorial differences in expressing the genome that have resulted in large species differences in protein diversity and function. These allow generally similar amounts of coding information to provide ample resources to account for the phenotypic diversity of species.[127,130] Multiple functions for identical polypeptides by a gene-sharing process also would be expected to contribute to phenotypic complexity without an increase in the number of distinct genes.

The long-time ambiguity of defining a gene by its structure and/or function remains as a nontrivial issue in the molecular era and could alter gene counts considerably. I discuss this issue later in the chapter in terms of the different concepts of a modern gene. According to Epp[131] (p. 537), "the single best molecular definition of the term 'gene' is . . . the nucleotide sequence that stores the information specifies the order of the monomers in a final functional polypeptide or RNA molecule, or set of closely related isoforms." Does it matter how many functions are encoded in a single gene? Are *many* functions the same as *one* function? In attempting to define genes in the genomics era, Snyder and Gerstein[114] (p. 258) considered a gene "a complete chromosomal segment responsible for making a functional product." The authors included both coding and regulatory regions, which further complicates deciding which sequences throughout the genome need to be considered as part of the functional entity or "gene." What if different sets of regulatory sequences are employed when the encoded protein is differentially expressed and carrying out entirely different functions? Do the gene boundaries change in such an instance? Inclusion of regulatory regions as part of the gene could also merge one gene with others because the same DNA sequence motifs may regulate more than one gene. At the end of their article (p. 260), Snyder and Gerstein resorted to function for annotating a gene: "One solution for annotating genes in sequenced genomes may be to return to the original definition of a gene—a sequence encoding a functional product—and use functional genomics to identify them."

In another broad guideline offering choices in case of uncertainty for annotating the human genome, Wain and colleagues[132] (p. 464) defined a gene as "a DNA segment that contributes to phenotype/function. In the absence of demonstrated function a gene may be characterized by sequence, transcription or homology." It would seem according to this definition that a DNA locus that affects phenotype/function should be counted as a gene whether or not it encodes a polypeptide or acts as template for a specific RNA. Interestingly, this definition was used to assess protein-coding gene numbers.[115]

Clearly, many questions remain concerning what defines a gene and how many such genes are present in any one species. The Genome Network Project Core Group[133] has reported data that redefine the landscape of the mammalian genome. By quantifying transcription start and termination sites and analyzing previously unidentified full-length cDNAs of the mouse, they have increased the potential number of genes by an order of magnitude in the mouse genome! The majority of these have alternative promoters and polyadenylation sites. It has also been reported that transcripts are produced from both DNA strands from much of the genome,[134] leading the authors to conclude (p. 1564) "that antisense transcripts link neighboring 'genes' in complex loci into chains of linked transcriptional units." They also suggest from perturbation experiments that "antisense transcription contributes to control of transcriptional outputs in mammals." Thus, genes are being redefined at the molecular level as I write and the numbers of genes in any genome have yet to be counted.

NONCODING REGULATORY GENES

Another area of complexity is transcriptional units producing noncoding RNAs.[135,136,1123] These are not limited to the well-known transfer RNAs and ribosomal RNAs required for protein synthesis. Numerous small (19–25 nucleotides long) double-stranded, noncoding RNAs control protein synthesis, transcriptional processes, and alternative RNA splicing.[137–140] Originally these were called RNAi, for RNA interference, because they were associated with the repression of gene expression. RNAi was discovered as antisense translational repressors in nematode worms.[141] A functional RNAi is short because its initial transcript is broken down by an enzyme, Dicer, into small pieces.[141a] RNAi is double-stranded at first because it results from hybridization of two RNAs transcribed from complementary strands of DNA. The sense strand is destroyed before RNAi acts as a repressor. siRNAs (short interfering RNAs) silence genes by promoting cleavage of specific mRNAs and by interfering with transcription via recruitment of proteins that alter chromatin. siRNAs generally target genes from which they originated.

microRNAs (miRNAs) are another class of small RNAs that are present from plants to mammals.[142–144,144a,1124] These repress the translation of many mRNAs (not just individual mRNAs) and may cause mRNA degradation by binding to multiple sites by imperfect base pairing.[144b,1125] miRNAs also have other functions such as, for example, facilitating replication of viral RNA.[145] miRNAs appear to repress translation by tethering Argonaute proteins to reporter mRNAs.[146,147] miRNAs are expressed in a tissue-specific pattern[148] and are associated with various cancers in humans.[149] Many miRNAs are present in humans, and these probably control several thousand different mRNAs.[1124]

Still another interesting class of regulatory RNAs transcribed upon DNA that is interspersed between known protein-coding genes is conserved even more highly than coding sequences.[150] Their distribution within the genome suggests long-distance *cis* or *trans* chromosomal interactions.[150] Another type of noncoding RNA contributes to transcriptional activation by binding to DNA and recruiting epigenetic regulators.[151a] Two other regulatory RNAs—tiny noncoding RNAs (tncRNAs) and small modulatory RNA (smRNA)—are known. The former were discovered in *C. elegans* and are expressed tissue specifically, but their function remains obscure; and the latter were discovered in mice and repress specific genes in all tissues but neurons.[139]

Regulatory RNAs are not just byproducts of complex genomes of metazoa, but are also present in bacteria.[151,152] Bacterial small RNAs (sRNAs) have complex regulatory roles in transcription, translation, and RNA stability. Some 50 distinct sRNAs have been found in bacteria. They interact with other proteins in order to accomplish their functions and play crucial roles in diverse physiological processes such as iron metabolism and defense against oxidative stress.

Genomes are clearly filled with functional DNA previously considered as "junk." It turns out, however, that much of the junk is intimately connected with the expression of protein-coding genes. Because the concept of a "gene" is connected to its function and expression, should scattered DNA templates of RNA products regulating gene expression be incorporated as part of a protein-producing "gene"? Regulatory RNAs increase the complexity of the genome and the difficulty of establishing neat boundaries between a protein-coding gene and the extensive supporting cast of associated regulatory genes that do not encode proteins but are connected with the ultimate expression of the protein-coding gene. It makes common sense within the context of gene sharing that the amount of space and infrastructure devoted to the control of expression of protein-coding genes reflects their repeated use and reuse under different circumstances and for different functions.

PROTEIN DIVERSITY

As introduced in Chapter 1 (see Figure 1.4), protein diversity can be achieved by a single gene by different mechanisms. Although gene sharing does not apply when distinct polypeptides with different amino acid sequence have different molecular functions, the mechanisms generating protein diversity can in some instances lead to gene sharing.

Alternative pre-mRNA splicing: Many transcriptional units are alternatively spliced, producing multiple mRNAs generating different polypeptides.[124,133,153] In one extraordinary example, the *Drosophila Dscam* gene can potentially produce more than 38,000 protein isoforms by alternative RNA splicing![154] In humans, up to half of all the identified genes may undergo alternative RNA splicing,[34,155] and more than 30,000 alternative splice relationships have been reported.[156] Of these, greater than 600 splice forms are tissue-specific.[157] Alternative RNA splicing plays important roles in evolution and is associated with the creation and loss of exons. Interestingly, exons that are alternatively spliced in a species-specific manner are poorly conserved.[155] Alternative splicing of a small proportion of transcripts derived from a gene serves to probe for new gene functions without losing the original function because the wild type RNA and protein continue to be manufactured.[155] This provides a more compact mechanism than gene duplication for the evolution of novel functions for a given gene/protein.

Alternative RNA splicing is generally not in the domain of gene sharing even if it causes a single gene to have more than one function when it produces proteins with different amino acid sequences, which is usually the case. However, alternative RNA splicing can also generate multiple mRNAs with identical coding sequences but diverse 5′ or 3′ untranslated regions. In theory, these situations could affect gene sharing because the untranslated regions of mRNA can affect mRNA stability, intracellular location, and/or translational efficiency; any or all of these variables could affect the function of the polypeptide.

Eukaryotic mRNA stability and translation are subject to extensive post-transcriptional control by poly(A)-binding proteins[158] as well as a host of additional enzymes and proteins.[159] AU-rich motifs (that could be placed or removed by alternative RNA splicing) in the 3′-untranslated region of mRNAs represent a primary source of control elements for mRNA degradation.[160,161] The androgen receptor mRNA is an example of 5′ (CAG repeats) and 3′ (UC-rich protein-binding element) sequence motifs interacting with cytoplasmic proteins regulating mRNA stability and, probably, translation.[162] In addition, studies on nonsense-mediated mRNA decay (a process triggered by premature termination of translation) showed that translation termination and mRNA stability are dependent on the configuration of the 3′ untranslated region.[163] Clearly, alternative RNA splicing affecting the noncoding 5′and 3′ sequences of mRNAs can control tissue-specific expression of a gene and ultimately influence how the encoded protein is utilized as a consequence of gene sharing.

Bacterial mRNAs are controlled by sequence motifs called riboswitches; these control mechanisms monitor metabolic states and alter gene expression by changes in secondary structure via interaction with small metabolites.[164–166,1126–1128] Riboswitching involves two RNA domains: an aptamer domain (the name is derived from a specific adaptive binding causing a concomitant conformational change in the RNA) and a separate expression platform whose conformation regulates gene expression. Aptamer domains are generally in the 5′ untranslated sequence and can affect transcription before the mRNA is released from the DNA template or during translation. In general, aptamer binding represses mRNA translation by causing a conformational shift that buries the Shine-Delgarno site, which is the bacterial ribosome entry site initiating protein synthesis; more rarely, aptamer binding stimulates mRNA translation. Riboswitches may also be operative in eukaryotes, although further studies are necessary for confirmation.[167] Putative metabolite-binding domains are embedded in both the 5′ and 3′ regions of various RNAs of eukaryotes and reports on their functional significance should be forthcoming. It is interesting to note that riboswitches are believed to have originated 4.2 to 3.6 billion years ago at the dawn of life preceding DNA and proteins.[168] This RNA world depended on the self-replication and catalytic activities of RNA, properties that still exist at the present time. Thus, multitasking of RNA is as ancient as life itself and may be considered the first manifestation of gene sharing.

Alternative promoter utilization: Another mechanism by which a single gene generates mRNA and protein diversity is by initiating transcription at more than one site via alternative promoters.[133,169] This process may also lead to alternative RNA splicing. Alternative promoters will always generate different mRNAs, but these will only produce different polypeptides if the coding regions are affected.

Alternative transcription initiation sites can result in the transformation of noncoding introns into protein-coding exons and create proteins with different structures and functions. An example is the transcriptional repressor (*CtBP2*)/ribbon synapse (*RIBEYE*) gene[170,171] (Figure 3.1A). Use of an upstream promoter produces CtBP2, a ubiquitously expressed transcriptional repressor. Use of a different promoter within intron 1 produces RIBEYE, a protein whose N-terminal amino acid sequence is encoded by part of intron 1 of the *CtBP2* gene. RIBEYE is a structural component of the ribbon synapse that carries vesicles undergoing exocytosis at the synapse under the influence of calcium. These two proteins—CtBP2 and RIBEYE—with different structures and functions result from the alternative use of two transcription initiation sites of the same gene; in addition, the RNA transcript from the upstream site utilized for *CtBP2* expression is alternatively spliced to skip exon 2.

The cytoplasmic protein RIBEYE has nothing to do with transcription, and the nuclear protein CtBP2 has nothing to do with synaptic ribbons. Of particular importance, the *CtBP2* gene and the *RIBEYE* gene are regulated differently; the former is ubiquitously expressed and the latter is tissue-specific. Given these distinctions, I would consider the RNA transcripts generating CtBP2 and RIBEYE derived from two overlapping genes and outside the domain of gene sharing.

There are cases in which two transcription sites directed by alternative promoters result in a different expression pattern for an identical polypeptide. An example is guinea pig NADPH/quinone oxidoreductase that is expressed in the liver as an enzyme by using one promoter and more extensively in the eye lens as ζ-crystallin by use of a different, lens-specific promoter[172,173] (Figure 3.1B). The liver expression of the gene results in additional 5' untranslated sequence in the mRNA. Use of the alternative promoters results in gene sharing with the identical polypeptide being expressed in the liver at a relatively low concentration as an enzyme (quinone oxidoreductase) for metabolic purposes and in the lens at a higher concentration as a crystallin (ζ-crystallin) where it contributes to refraction. The same arguments suggesting a potential relationship between alternative promoter utilization and gene sharing apply as with alternative RNA splicing when the two resulting mRNAs differ in the untranslated regions but not in the coding region.

mRNA editing: Processing events at the level of mature mRNA further complicate the relationship between DNA sequence, protein structure, and function. On occasion a protein sequence will differ from that dictated by its genes as a consequence of mRNA editing.[34,174–179] Editing results in modifications of individual mRNA nucleotides so that the DNA gene and the edited mRNA transcript encode distinct proteins. An example of mRNA editing is the gen-

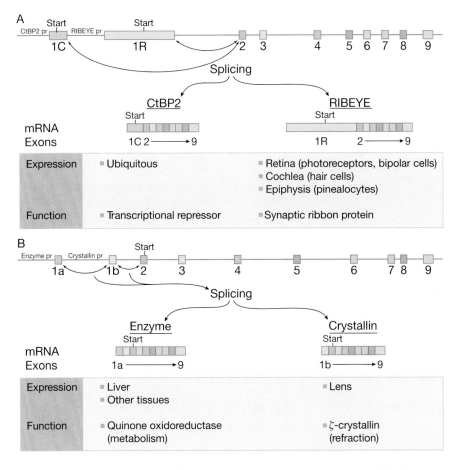

Figure 3.1. Overlapping genes and gene sharing by alternative promoter utilization. A. Two different proteins (CtBP2 and RIBEYE) with different expression patterns and functions are produced by alternative promoter utilization coupled with alternative RNA splicing. Adapted from Piatigorsky.[171] B. Alternate promoter utilization generates the same enzyme (quinone oxidoreductase), which functions as a crystallin (ζ-crystallin) only when targeted to the lens by the downstream promoter (crystallin pr) between exon 1a and 1b of the gene.

eration of protein diversity for electrical and chemical neurotransmission.[180] mRNA editing may also be associated with disease, such as in the case of fatal sporadic amyotrophic lateral sclerosis, where there is a defect in the editing of the mRNA encoding a subunit of glutamate receptors.[181] This posttranscriptional process does not encompass gene sharing when it results in a protein with a different amino acid sequence than that derived from the unedited mRNA. However, it would be interesting if an example of editing were found that involved changes confined to the untranslated regions of the mRNA, where it might become a mechanism leading to gene sharing by affecting the tissue-specific expression of the protein or its relative abundance. It is also possible that an edited mRNA generates a protein with more than one function, while the unedited mRNA does not, which would make this a derivative mechanism leading to gene sharing.

Posttranslational modifications: Posttranslational modifications of proteins (i.e., the covalent attachment of small molecules to proteins, such as phosphorylation, acetylation, methylation, glycosylation) are a major mechanism diversifying protein function (Bray[34]; see www.abrf.org). Gene sharing would apply to unmodified and modified proteins that have different molecular functions but the same linear continuity between amino acid sequence and nucleotide gene sequence.

THE AMBIGUOUS GENE

Despite the enormous amounts of detailed facts about genomic organization at our disposal, the early difficulties of conceptualizing a gene as both putative structure and undefined process to explain biological traits (phenotype) and inheritance (genotype) still haunt us in molecular terms (see references[76,77,79–85,114,182–184]). To define a gene strictly by the polypeptide(s) it encode(s) may potentially sharpen the boundaries and delimit exons interrupted by noncoding introns; however, alternative RNA processing, alternative promoters, and other molecular mechanisms generating protein diversity make genomes more like a buffet than a collection of discrete courses of a formal meal (read "genes"). To define a gene by the function(s) of its encoded protein(s), which depends on gene expression regulated by dispersed DNA sequence motifs and templates for regulatory RNAs, is like grasping mercury. If one considers regulation, the structure and boundaries of a "gene" will change during development or in response to the environment and biological needs. This leaves us once again with the nagging question of how to conceptualize a gene with respect to structure and function.

Keller describes in detail the difficulties in attempting to define a gene in light of molecular data[79] (p. 70).

> Ever since the term *gene* was first introduced, confidence in the physical reality of the gene has always been accompanied by the assumption that structure, material composition, and function were all properties of one and the same object, be it a bead on a string or a stretch of DNA. Today, it is precisely that self-identity which has been disrupted. We have learned not only that function does not map neatly onto structure but also that function must be distinguished from a particular and prespecified locus of the chromosome. To the extent that we can still think of the gene as a unit of function, that gene (we might call it the *functional gene*) can no longer be taken to be identical with the unit of transmission, that is, with the entity responsible for (or at least associated with) intergenerational memory. Indeed, the functional gene may have no fixity at all: its existence is often both transitory and contingent, depending critically on the functional dynamics of the entire organism.

She concludes (p. 72) that "perhaps it is time we invented some new words."

Keller is joined by many other investigators of similar opinion. Gelbart[183] (p. 660) wrote that "we may well have come to the point where the use of the term gene is of limited value and might in fact be a hindrance to our understanding of the genome." He goes on to consider that "we are entering a period in which we must shift to the view that the genome largely encodes a series of functional RNAs and polypeptides that are expressed in characteristic spatial, temporal, and quantitative patterns. The classical concept of the gene ultimately forms a barrier to trying to understand phenotypes in terms of encoded functional products." Burian[85] (p. 175) states that

> the task of delimiting genes contains an inextricable mixture of conceptual and factual elements. To be sure, the "lowest level" (i.e., molecular level), though it is most distant from naïve observations, brings the argument closer to a context-fixed factual basis than the others. But the price for this is that one must deal with the interactions of all of the relevant macromolecules within their physiological setting. This has the consequence that precise definitions of genes must be abandoned, for there are simply too many kinds of genes, delimited into many ways. Taken in combination, these arguments

combine to provide powerful support for the principal contention . . . that when we reach full molecular detail, we are better off to abandon specific gene concepts and to adopt, instead, a molecular biology of the genetic material.

THE "MOLECULAR GENE" CONCEPT

Waters[76] considered that molecular biologists might dispense with the gene altogether and refer instead to specific genomic regions such as introns, exons, promoters, and enhancers. However, he is skeptical (as am I) that continuing use of the term "gene," so deeply entrenched in our vocabulary, will fade away. Waters has made a heroic effort to define a gene by melding both classical and molecular views. He linked the classical gene concept with phenotypic differences; classical genes were identified by quirks of mutation, not by molecular structures. In attempting to define molecular genes, Waters recognized the ambiguities of combining different entities (alternative RNAs, multiple proteins, multiple functions) with one structure, and the difficulty of considering regulation with a particular genetic structure. He developed the molecular gene concept, which is an extension of the earlier classical molecular gene concept defining a gene as a stretch of DNA coding for a single polypeptide chain.[184] The molecular gene concept broadens a gene to "a linear sequence in a product at some stage of genetic expression." A molecular gene includes introns if the unit of gene expression under consideration is the primary transcript, but the "same" gene excludes introns if the unit under consideration is the encoded protein. A gene may be an intron alone, if that should contain a separate coding sequence expressed under a particular circumstance. Ambiguity is removed by identifying the precise linear sequence or stage of genetic expression being discussed. Alternatively spliced RNAs producing two proteins would be encoded by one gene for protein A and another gene for protein B. An unspliced primary transcript on-route to making a protein would be encoded by a gene that is context-free, while the protein that appears would be encoded by a context-dependent gene—in other words, its function would depend upon circumstances. Invoking regulation to unite the classical and molecular gene concepts, Waters notes that differences affecting transcription and further processing of the transcript of molecular genes always explain the phenotypic mutations identified by classical genes. He concludes by stating that "differences in classical genes produce differences in phenotypes because they affect the action of molecular genes. So, what really happened to classical genetics? It went molecular."

Waters has imaginatively confronted the inherent ambiguities of defining a modern gene by creating a movie from a photograph so that he could focus on different frames as needed. In Waters' scheme, the gene is dynamic, yet it can be viewed as rigidly encased depending upon circumstance. Sometimes it is free of context, and at other times it is bound by context. In a sense, the molecular gene concept may suffer from attempting to save the gene by changing its definition depending upon need. Nonetheless, by including regulation as the bridge uniting the phenotypic function of the classical gene and the linear information content of the molecular gene, the molecular gene concept embraces the idea of gene sharing.

THE "MOLECULAR PROCESS GENE" CONCEPT

Griffiths and Neumann-Held[184] have modified Waters' "molecular gene" concept to the "molecular process gene" concept. They have fused the epigenetic aspects of the gene in question to link regulation and developmental use with gene structure. Like Waters' view, the precise structural properties of a gene will vary with the function that the gene is performing. This acknowledges that DNA sequence can be used differentially under different conditions and in different tissues.

Griffiths and Neumann-Held[184] provide a dynamic view of genes and write (p. 659) that it is "misleading to think of functional descriptions of DNA, such as 'promoter region,' as explicable solely in terms of structural descriptions of DNA, such as sequence. The structural description is, at best, a necessary condition for the functional description to apply." The molecular process gene concept of Griffiths and Neumann-Held[184] (p. 659) includes, in addition to the DNA coding region, the whole range of processes "in whose context these sequences take on a definite meaning . . . it is the molecular process that produces the product rather than a sequence of nucleotides." In a fascinating way, the molecular process gene concept resonates in molecular detail with the original gene concepts of a process defining traits without a structural basis.

THE "EVOLUTIONARY GENE" CONCEPT

Griffiths and Neumann-Held contrast their "molecular process gene" concept to the "evolutionary gene" concept.[184] The concept of an "evolutionary gene" initially referred to a DNA stretch in competition with other DNA stretches for transmission to progeny.[52] An evolutionary gene did not correspond to a particular sequence, but for survival it required selection of a fit individual. This was the forerunner to the selfish gene of Richard Dawkins,[75,185] which is a DNA stretch that could be replaced by an alternative sequence in future generations (see Moss[82] for criticism of the selfish gene concept). Evolutionary genes do not necessarily encode polypeptides but are associated with phenotypic differences—particularly in rates of reproduction—within a population. This idea returns genes to inheritable units of potential for selecting phenotypic traits and, indeed, Griffiths and Neumann-Held consider an evolutionary gene as a theoretical entity with a role in the selection of phenotypic traits.[184] Because such traits may depend on complex developmental modules that function as single units, an evolutionary gene need not correspond to a particular DNA sequence.

Beurton has provided fascinating new twists on the evolutionary gene concept (in Beurton et al.[77]). He defines a gene (p. 286) "as the genetic underpinning of the smallest possible difference in adaptation that may be detected by natural selection." Beurton suggests (p. 299) that "an array of nonlocalized DNA variations, whose reproduction comes to be controlled by some such adaptive difference large enough for natural selection to detect, begins to qualify as a gene." He essentially turns the world upside down by proposing (p. 301) that the gene, or "unit of selection," "is not a unit *encountered* by natural selection," but "an *emergent* unit or one generated in the process of natural selection from a background of never-ceasing variation contained in the genome." Beurton's gene emerges and becomes defined during the process of natural selection, may be scattered throughout the genome, and may or may not encode a protein or incorporate DNA regulatory sequences. Beurton has made a gene a product of population genetics. In single individuals, he writes (p. 306) that "functional interdependencies fade away into the genomic horizon, and there is simply no yardstick by which to define the limits of a gene."

It is of historical interest that a broad concept labeled "reaktionsnorm" (also known as norm of reaction, or NoR) was proposed in 1909 by a German investigator, Richard Woltereck, to explain the varied phenotypes obtained in response to the same environmental change in pure lines of crustaceans (*Daphnia* and *Hyalodaphnia*) (see Sarkar[186]). A hereditary change in the reaktionsnorm led to modification in the range of phenotypic responses. Woltereck considered genotype an inherited enabling agent in phenogenesis rather than a deterministic force. The concept of the norm of reaction stimulated many ideas, such as phenotypic plasticity, expressivity (extent of manifestation of a trait), and penetrance (proportion of individuals manifesting a trait). Dobzhansky and Spassky[187] developed the idea of the norm of reaction to include the array of genotypes constituting the adaptive norm of a species of population, extending the concept to the genotypes themselves rather than phenotypic manifestations. While all this precedes the molecular evolutionary gene concepts, the norm of reaction concept foreshadows notions of selection on the entire genome rather than compartmentalized sequences affecting individual proteins or traits.

TWO CONCEPTS FOR ONE GENE: GENE-P/GENE-D

Moss has developed new terminology to account for the differing informational content of a gene[82] (see Burian,[85] Chapter 9). The more classical, Mendelian-like information is identified by a gene's relationship to phenotype. Moss points out that defining a gene by its relationship with phenotype is a type of preformationism used for instrumental utility, because the gene in question does not necessarily have a direct role in creating the normal phenotype; rather, lesions (mutations or deletions) connected with the gene lead to phenotypic changes. He identifies this indirect phenotypic information content by calling it Gene-P, for preformationist. Gene-P also contains a molecular entity that is revealed when the particular mutation associated with the phenotype is described at the DNA level. In other words, the sequence of a Gene-P is that which contains the mutation resulting in the phenotype associated with the gene.

Moss considers the normal, wild type DNA sequence as another form of information that he calls Gene-D for developmental resource. Gene-D is the transcriptional unit on a chromosome that contains all the sequences (resources) to give rise to all the protein products of the gene. In contrast to Gene-P, Gene-D is indeterminate with respect to phenotype and plays roles in epigenesis as it participates in its various functions. In Moss' words[82] (p. 46), "where Gene-P is defined strictly on the basis of its instrumental utility in predicting a phenotypic outcome and is most often based on the absence of some normal sequence, a Gene-D is a specific developmental resource defined by its specific molecular sequence and thereby by its functional template capacity; yet, it is indeterminate with respect to ultimate phenotypic outcomes."

Gene-D resembles the situation in gene sharing in that it can encode a protein with multiple molecular functions; however, the concept of Gene-D is more inclusive than that of gene sharing by being a resource that includes all the diverse proteins that arise during its expression, such as the multiple proteins that are generated from a single Gene-D by alternative RNA splicing. Gene-D and Gene-P are mutually exclusive concepts using different physical considerations for defining the same gene. The Gene-P/Gene-D dichotomy addresses the difficult issue of reconciling structure and function by creating two concepts to define one gene. A gene participating in gene sharing, on the other hand, requires at least two molecular functions for a single encoded protein.

GENE SHARING: A CONCEPT INCORPORATING AN "OPEN GENE"

Although there are a number of concepts of a modern gene, they generally acknowledge the need to recognize function (molecular, phenotypic or adaptive), encoded protein (or RNA), and expression. Without expression, a gene is reduced to a simple sequence of inert nucleotides with an unrecognizable potential of one sort or another, perhaps appropriately considered a "nongene gene." Singer and Berg[107] (p. 440) defined a eukaryotic gene as "a combination of DNA segments that together constitute an expressible unit, expression leading to the formation of one or more specific functional gene products that may be either RNA molecules or polypeptides." They chose to include the entire primary transcript (exons and introns), the promoter needed to initiate correct transcription, and all the sequence elements regulating rate and tissue-specificity of transcription. Portin[81] (p. 264) is basically in agreement, if less specific, and states that we need "to adopt a new, open, general and abstract concept of the gene." Burian[85] (p. 175) concludes "that even when one works at the molecular level, what counts as a gene is thoroughly context-dependent . . . what counts as a gene depends on what one chooses as a phenotype." The phenotype for gene sharing is the molecular/biochemical/biophysical function of a protein with a given amino acid sequence. Burian states poetically (p. 172) that "the term 'gene' in molecular biology is a genuine accordion term—its expansion and contraction make for a lot of semantic music and allied quibbling. But the arguments involved are not always empty semantic functions performed by nucleic acids that do not fit any of the standard structural constraints on genes. Underlying the different terminologies are serious disagreements about the status of parts of nucleic-acid molecules that behave or are treated in different ways in different cellular contexts and at different phases of ontogeny."

Thus, gene sharing—whereby two or more molecular functions are performed by a single polypeptide—is compatible with an open gene concept. The open gene includes its scattered and distant DNA regulatory sequences that are crucial to governing the multiple functions of the encoded polypeptide. Even the DNA motifs acting as *cis*-regulatory elements do not need to be on the same chromosome as the genes they regulate.[188] Indeed, mutations in plants and animals can operate via epigenetic modifications (called paramutations) invoking non-coding long RNAs, microRNAs, and protein complexes.[1129–1132] The open gene will thus have accordion-like boundaries, as envisaged by Burian. The entire range of influences on the expression of the gene engaged in gene sharing, including environmental factors, extragenic mechanisms, and DNA templates for microRNAs controlling expression of the gene, remain loosely associated with the shared gene, depending on circumstance. The gene concept is not abandoned in gene sharing, nor is the gene devoid of its encoded polypeptide if one considers the gene open in this sense.

The rationale for an open gene seems straightforward, but the specifics are still murky. Perhaps incorporating the distinction between a polypeptide chain encoded by a nucleotide sequence and the folded, functional protein performing various functions is needed to ground the open gene to biological realities. This idea suggests that there are as many genes, or perhaps subgenes, for each polypeptide as there are combinatorial arrangements of regulatory elements controlling the expression associated with distinct functions. Because the totality of regulatory control elements is not known yet for even one situation, it is premature to provide the structural details for any gene or to count the number of open genes that exist in a species.

TAKE-HOME MESSAGE

The concept and precise boundaries of a gene remain ambiguous even in the present molecular era. The concept of gene sharing is best served by considering the gene open so as to include its many scattered regulatory components, which play a crucial role in determining the various functions of the encoded polypeptide. Although gene sharing refers specifically to multiple molecular uses of an individual polypeptide, the modern "gene" in gene sharing is an interactive, differentially expressed gene that challenges the investigator to see in how many ways it can be enlarged through the functions of its polypeptide rather than how it might be subdivided into an elementary unit, as was the goal of investigators in the classical and neoclassical concepts of the gene.

4

Eyes and Lenses: Gene Sharing by Crystallins

The gene-sharing concept originated from investigations on the major water-soluble proteins—the crystallins—of the cellular eye lens (see Wistow and Piatigorsky,[22] Piatigorsky and Wistow,[23] and de Jong et al.[24] for reviews). A typical vertebrate eye and a diagrammatic representation of a lens are shown in Figure 4.1. Crystallins contribute to the optical properties of the transparent lens optimizing refractive index for image formation on the retinal photoreceptors. For lack of better criteria, crystallins are defined simply as abundant water-soluble lens proteins because protein concentration affects refractive index. The connection between the concentration of a given protein and refractive index of a tissue is complicated by the fact that the refractive power of any region of the transparent lens is determined by the total concentration of macromolecules, not that of an individual protein, thereby obscuring a concentration threshold required for a lens protein to be listed among its crystallins. Because the concentration of any one protein, including a so-called crystallin, may differ substantially in different regions of the lens, theoretically a protein could be considered a crystallin in one spatial compartment of the lens but not in another compartment.

Historically, crystallins have been thought of as specialized for their optical functions. It was assumed that crystallins play strictly structural roles in the lens inasmuch as transparency and refractive index are physical properties. But, as described in this chapter, crystallins are neither restricted to the lens nor confined to structural roles, indicating that they have varied biological roles. It turns out that many crystallins are common metabolic enzymes or stress proteins that are "overexpressed" in the lens and are also present at lower levels in many other tissues where they have nonrefractive functions. The diverse lens crystallins are also often taxon-specific—that is, they may differ in species belonging to different taxonomic groups. This diversity contrasts with the more commonly observed use of homologous proteins for similar functions, such as hemoglobin for oxygen transport.

The term "gene sharing" was used initially to generalize the finding that crystallins with a structural, optical function in the lens may also be expressed in other tissues, where they have a metabolic, nonrefractive function.[20] Thus, the protein encoded in the *identical* gene may perform entirely different functions depending on its expression pattern. This chapter summarizes our knowledge of the lens crystallins, and I stress two points in the discussion. First, the lens crystallins are exemplary proteins that have at least dual natures: a refractive function in the lens and a catalytic or stress function elsewhere as well as in the lens. Second, *crystallin* gene regulation is a critical factor for gene sharing, leading to the use of a metabolic or stress protein as a structural crystallin. It is ironic that the abundantly expressed crystallins, long

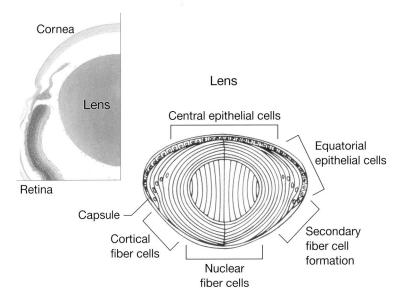

Figure 4.1. The vertebrate lens. Left: Histological section of half of a mouse eye. Adapted from Piatigorsky.[1110] Right: Diagram of a sagittal section of a human lens showing the anterior epithelial cells, cortical and nuclear fiber cells, and nuclear core. Note that the cell nuclei (as well as other organelles) disappear from the innermost cortical fiber cells and the nuclear fiber cells. Adapted from Piatigorsky.[1111]

considered inert bricks as it were, squirreled away in a tiny transparent tissue specialized for focusing an image on photoreceptors, have become landmarks for multifunctional uses of widely expressed proteins that illustrate the dynamism of evolutionary change!

EYE DIVERSITY: MANY FORMS TO PERFORM A FUNCTION

To give gene sharing by crystallins some biological perspective, we need to briefly review eye evolution. Tracing the history of eye diversification reveals strands that both weave together and go in their separate directions during evolution. The emerging story portrays the biological context of one system—the eye—in which gene sharing operates.

The eye has challenged evolutionary biologists due to its complexity and requirement for interdependent functions. Charles Darwin[44] (p. 186) stated famously that "to suppose that the eye, with all its inimitable contrivances for adjusting the focus for different distances, for admitting different amounts of light, and for the correction of spherical and chromatic aberration, could have been formed by natural selection, seems, I freely confess, absurd in the highest possible degree."

Not all animals have eyes; eyes of various design and complexity are found in about a third of the (approximate) thirty-three phyla comprising invertebrates and vertebrates.[189–198] One-third of the phyla lack organs for light detection and the remaining third have some form of light detection apparatus. Although image-forming eyes exist in only six of the extant thirty-three phyla, these comprise approximately 96 percent of the living species. The diverse eyes of animals come in many shapes and sizes, including those with pinholes for focusing (some mollusks), mirrors that reflect images on the retina (scallops), and multiple types of compound eyes organized into virtually hundreds of independent facets (best known in, but not confined to, insects).[196,199] Eyes are also located in many body locations, not just heads:[200] in the rear of some polychaete worms, on the edge of the mantle in scallops and some clams, on the tips of arms in starfish, on the genitals in a butterfly *(Papilio xuthus)*, and on the dangling rhopalia of cubozoan jellyfish. Even unicellular algae[201] and dinoflagellates[202,203] have organelle "eyes" with lenses. The significant anatomical and developmental differences among eyes indicate that eyes evolved in parallel in the different species during evolution.

Whether an eye originated once (monophyletic) or numerous times (polyphyletic) is contentious.[204–210] Differences in developmental, anatomical, and organizational details are consistent with convergent evolution—namely, that eyes originated a number of times.[193,196,197,211,212] It has been proposed that photoreceptors arose forty to sixty times independently during evolution,[213] an idea that fits with structural and biochemical differences between many invertebrate and vertebrate eyes. In general—though there are exceptions—rhabdomeric (microvillar) photoreceptor cells predominate in invertebrates and ciliary photoreceptors in vertebrates. In addition, the biochemistry of phototransduction differs in invertebrate and vertebrate eyes: Inositol triphosphate is used in the former and cyclic guanosine triphosphate in the latter as second messengers. A case for independent origins of compound eyes was also made within a single group, the crustaceans.[214] A pair of lateral compound eyes exists in myodocopids (ancient bivalves) even though these animals are nested within several groups of related species that do not have eyes; these observations are consistent with the possibility that the myodocopid compound eyes arose independently (although of course the same distribution of the compound eyes could have arisen by multiple losses). Computer-assisted estimations[215] that a light-sensitive skin patch could evolve into a complex, camera-type eye by incremental changes over 300,000–400,000 years, a mere "geological blink,"[216] support the possibility of multiple independent eye inventions during evolution. Nilsson and Pelger[215] pointed out that different eyes could have evolved independently more than 1,500 times since their original invention, thought to be during the Cambrian explosion 530 million years ago.

It is also possible that eyes originated only once and subsequently diverged, undergoing diverse individual modifications during extensive periods of parallel evolution. This is supported by various observations. The occurrence of rhabdomeric and ciliary photoreceptors are not strictly separated by species and both can be present in invertebrates, sometimes even in the same species (for example, the marine annelid *Platynereis dumerilii*); moreover, melanopsin, an invertebrate-like opsin (visual pigment used for phototransduction) is present in retinal ganglion (nerve) cells of vertebrates.[205,208,217] Gehring[209] considered this consistent with a single pre-bilaterian photoreceptor cell precursor diverging into rhabdomeric and ciliary cell types already present in Urbilateria (the last common ancestor of bilaterally symmetrical organisms). The similarity in molecular genetic aspects of development has been used to support the idea that the eye originated only once. The story began when the *eyeless* gene in *Drosophila* was identified as *Pax6*,[218] a transcription factor[219,220] responsible for eye defects in mammals, including humans.[221–224] What caught everyone's attention was that redirecting the expression of the *eyeless/Pax6* gene to noneye imaginal discs (embryonic precursors for different adult organs) in the fruitfly *Drosophila* induced compound eyes in the leg or antennae of the mature fly.[225] These amazing eyes in abnormal body locations were also created in the fly by misexpression of the mouse[225] and squid[226] *Pax6* gene. *Pax6* expression was subsequently correlated with eye development in other invertebrates—the planarian *Dugesia tigrina*[227] and the ribbon worm *Lineus sanguineus*,[228] protochordates (the ascidian *Phallusia mammillata*[229] and the cephalochordate *Amphioxus*[230]), and a larval vertebrate, the African frog *Xenopus laevis*.[231] *Pax6* was coined a "master control gene" for eye development, a term that lost some punch when it was found that many of the transcription factors used in eye development could induce ectopic eyes (see Fernald[198]), suggesting common ancestry for all eyes.[232,233] The prototypic eye, according to some researchers, might even have been acquired by ingestion of a unicellular symbiont.[234]

The idea of a monophyletic (single lineage) eye gained momentum as studies uncovered a conserved cascade of transcription factors in addition to Pax6 that directed development of the diverse eyes. The concern arose, however, that reutilization of transcription factors and similar developmental pathways may not prove homology, or evolution from a single source.[212,220,235–239] In other words, implementation of a similar set of tools (transcription factors) to build a structure (an eye) that can be used for a similar function (vision) may not prove that each type of eye was derived ultimately from the same original one (see Conway Morris[70,240] and Kozmik[240a]). The difficulty of establishing homology among structures that have analogous functions is not confined to eyes and has been of concern for many different tissues (see Zuckerkandl[241] and Arthur[242]).

The fact that sets of transcription factors within the same families can be redeployed to direct development of tissues with entirely different functions is a central issue of contention at present for linking homology to developmental processes. A striking example is the use of the *Pax/Six/Eya/Dach* network of transcription factors for development of the eye, ear, and skeletal muscle.[243–245] Indeed, there are many suites of interacting transcription factors that are used conservatively (repeatedly for the same structure) as well as innovatively (for different structures) during evolution in a modular fashion (see Kardon et al.[245] and Fernald[198]).

In my opinion, to reach consensus regarding whether eyes originated once or multiple times, we must first agree on a suitable definition for an "eye" (see Land[199]) and focus on individual components and issues one at a time: molecule (photopigment, crystallin), cell (photoreceptor, lens cell, pigment cell), function (vision, nervous excitation by photons, behavioral response to light), development (transcription factors, morphology). The most satisfying explanation seems to be that eyes in different species display different levels of homology or partial homology.[196,198,206,246] According to Fernald[198] (pp. 704–705), "as different eye types evolved, there was probably repeated recruitment of particular gene groups, not unlike improvisational groups of actors, interacting to produce candidates for selection. Trying out various routines could have led to numerous parallel evolutionary paths for eyes as we now envisage . . . So the answer to the question of whether eyes evolved from a single prototypical eye (monophyletic) or if they evolved repeatedly (polyphyletic) may be the wrong question since it depends on the level of comparison."

Nilsson[200] (p. 39) speculates that "eyes have evolved numerous times, in each case using whatever tissues and materials that were at hand." He goes on to write (p. 40) that "at all levels of organization in the visual system, there is either more than one functional solution, or the single solution is shared with functions other than vision." Sharing functions again! If eyes do have a common ancestral origin, we might expect that their specialized proteins, such as opsins in the retina and crystallins in the lens, would show corresponding homology in different species. While this is the case for the opsins, it is not true for the crystallins, which I explore next. Biology is never simple!

THE LENS

To appreciate crystallins, we must understand lenses. Image-forming mechanisms vary among the diverse eyes of different species.[196] A smattering of examples follows. Light is bent by hardened corneal sheaths on the facet surfaces of ommatidia in most compound eyes of invertebrates. In some cases (butterflies) the image is focused on cone cells in front of the retina of the compound eye and then sent to the retina via a powerful lens cylinder. There are complex mirrored eyes (scallops) in which the light passes through a transparent cornea, lens, and retina before being reflected back by guanidine plates (the argenteum) to form an image on retinal photoreceptors. Some marine annelid worms have jelly-like (meaning uncharacterized!) lenses to direct light into the eye, while some gastropod snails use secreted proteins as a lens beneath the transparent cornea. There are even some protozoa (especially dinoflagellates) that have surface "lenses" that refract light on "retinal-like" structures carrying photopigments.

As with image formation, there are also diverse mechanisms for focusing, a process called accommodation.[189,196,247] The relative position of a hard spherical lens within the eye has a different resting point depending on the species. Fish are atypical in having muscles to draw the spherical lens toward the cornea (away from the retina) for focusing distant objects. As abounds in nature, there are many exceptions. The fascinating four-eyed fish *(Anableps anableps)*, which has its eye bisected by the water surface, receives light from air and water simultaneously and has its single ovoid lens positioned to refract focused images from the two sources[248] (see Figure 4.3). Nocturnal rodents have hard, round lenses that can move back and forth for accommodation, while birds, reptiles, and diurnal mammals (including humans) have soft lenses that can be squeezed or released to induce shape changes that result in different focal points of the emerging light. Diving ducks squeeze the anterior region of the lens to create a bulge of high refractive power when they enter the water, giving them the ability to see focused images above and below the water. In some birds the cornea can also change shape for accommodation and cooperates with the lens to cast a proper image on the retina.

Nonaccommodative eyes also exist. The cellular lens in ocelli (described below) of jellyfish or in the median ocelli of *Drosophila* apparently cannot focus; it may serve to both gather light and keep a fixed focal distance that is sufficient for the particular biological needs of the eye. The optics of the lenses of the complex jellyfish ocelli show long focal lengths that converge behind the retinal photoreceptor cells, suggesting that the jellyfish may be able to detect large objects (such as mangrove roots) and avoid obstacles but cannot see focused images.[249,250]

CRYSTALLINS AND THE OPTICAL PROPERTIES OF THE LENS

Crystallins are the major, cytoplasmic water-soluble proteins conferring refractive properties on lenses of invertebrates and vertebrates.[22,251–253] The name "crystallins," first used by Berzelius (1830) for bovine lens proteins, comes from their abundance in the crystal-clear or crystalline lens (see McDevitt and Brahma[254] and Robinson and Lovicu[255] for historical reviews). However, crystallins are neither crystals nor arranged in a type of regular lattice-like structure in the lens. They are globular proteins present as single polypeptide monomers (γ-crystallins in vertebrates, J-crystallins in jellyfish), some as dimers (some β-crystallins in vertebrates, S-crystallins in cephalopods) or tetramers (δ-crystallins in birds and reptiles, Ω-crystallins in mollusks), and some as aggregates containing numerous polypeptides (α-crystallins in vertebrates). The aggregation state of the crystallins (especially the α-crystallins) increases with age, contributing to the unfortunate time-dependent loss of accommodative power and cataract.

The crystallins account for as much as 80–90 percent of the water-soluble proteins of the lens cells and are closely packed in order to minimize concentration fluctuations that scatter light.[256–259] Their absolute concentrations vary with species (higher in aquatic than terrestrial animals) and spatial location in the lens (higher in the center than periphery). Crystallin concentration can be as high as 70 percent protein in the center of squid lenses, but as low as 10 percent protein in the periphery of chicken lenses. Indeed, image formation lacking spherical aberration is dependent on crystallin concentration declining smoothly from the center to the periphery of the lens creating a gradient in refractive index that bends the transmitted light as it passes through the lens.[196] Remarkably, even cellular lenses of cubomedusan jellyfish[260] have precise refractive index gradients, even though in that particular case the focal point of transmitted light resides behind the retina and thus would not lead to a focused image.[250]

The crystallin concentration gradient in vertebrates is achieved by developmental modulation of gene expression in the closely packed fiber cells comprising the lens (see Piatigorsky[261] and Duncan et al.[262] for reviews). It is the short-range order of crystallins and their carefully controlled concentrations that determine the refractive properties of the transparent lens. Despite their high concentration in the lens, crystallins remain water-soluble. There is no protein turnover in the center of the lens because the cell nuclei and other organelles are degraded already in the embryo. Moreover, cells cannot enter or leave the lens, which is enclosed by a collagenous capsule. Consequently, the identical crystallin proteins are physically present in the lens throughout life, making them among the most long-lived intracellular proteins in the body. Crystallins undergo numerous posttranslational modifications (includ-

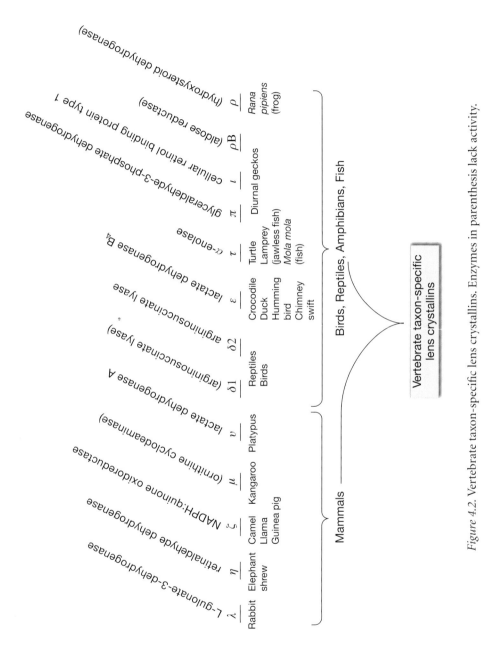

Figure 4.2. Vertebrate taxon-specific lens crystallins. Enzymes in parenthesis lack activity.

ing cleavages, glycosylations, phosphorylations, aggregations) and these may differ with species (see Zigler et al.[263] and Wistow[264] for reviews). For the most part, the functions of the posttranslational modifications of crystallins are not known.

DIVERSITY AND TAXON-SPECIFICITY OF LENS CRYSTALLINS

In accordance with the diversity of eyes, lenses, and focusing mechanisms, there is a diverse array of crystallins. Different crystallins present in lenses of vertebrates are shown in Figure 4.2. Mörner[265] first separated bovine lens proteins into the insoluble albuminoid fraction and three groups of water-soluble proteins called α-crystallin, β-crystallin, and albumin. A plethora of studies, many immunological in nature, established the conservation of α-, β- and γ-crystallins in vertebrate lenses (see Harding and Dilley[251] and Bloemendal[252,266] for reviews). However, it was noted first that bird lenses deviated in their crystallin composition (see Clayton[266]). Initially an electrophoretic variant called "typical songbird crystallin" was discovered in a number of passiforme bird lenses; Rabaey later found a major new crystallin in chicken and duck lenses.[267,268] Because this novel protein was the first crystallin to appear in the embryonic chicken lens, it was called "first important soluble crystallin" or FISC. Final dignity came to this deviant when it was renamed δ-crystallin, bestowing it with enough crystallin status to join the elite α-, β-, and γ-crystallins.[36,269]

δ-crystallin is confined to bird and reptile lenses, while the α-, β-, and γ-crystallins are present in the lenses of all known vertebrates, although their relative abundance and developmental appearance differ among species (see Piatigorsky[261]). α-, β-, and γ-crystallins were considered ubiquitous crystallins, while δ-crystallin became known as a taxon-specific crystallin due to its appearance only in birds and reptiles. It turns out that α-, β-, and γ-crystallins are not ubiquitous if we consider invertebrates, which lack these crystallins and have their own sets of taxon-specific crystallins (Figure 4.3). Early immunological studies showed that squid and lobster crystallins do not cross-react with vertebrate crystallins,[270–272] and other investigations established that cephalopod,[19,273–275] scallop,[276] fly,[277] and jellyfish[260,278,279] lenses have distinct crystallins (see Tomarev and Piatigorsky[194]). Comparative studies have established that entirely different proteins are used as lens crystallins in different species or taxonomic groups. Despite their taxon-specific crystallin compositions, lenses in different species are transparent and, with rare exception (i.e., jellyfish),[250] focus appropriately upon the retinal photoreceptors without spherical aberration (Figure 4.4).

The taxon-specificity of crystallins might be explained, at least in part, by multiple independent derivations of eyes or lenses during evolution. In view of the fact that there are numerous structural and developmental differences among lenses of different species, and that there are eyes lacking lenses altogether, the independent origin of lenses between some species is a reasonable (but not the only) hypothesis. Independent lens acquisitions remain possible even if all lens-containing eyes had a common ancestry. An alternative hypothesis is that taxon-specific crystallin diversity resulted from species-specific variations in the expression of different genes. In any case, taxon-specific crystallins raise a red flag about evolutionary ancestry of proteins

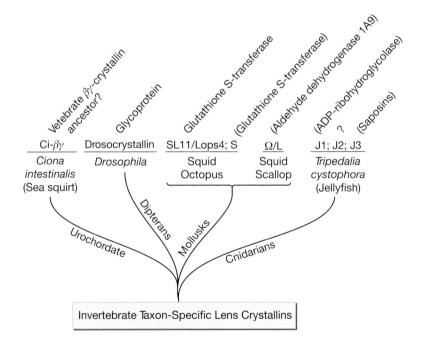

Figure 4.3. Invertebrate taxon-specific lens crystallins. Ci-βγ-crystallin in urochordates is thought to be ancestral to the vertebrate βγ-crystallins although this organism does not have a lens in the larval eye. Enzymes in parenthesis lack activity.

for specialized roles. Taken together, the diversity and taxon-specificity of lens crystallins throughout the animal kingdom are consistent with overlapping functions by different proteins, indicative of convergent evolution for crystallin roles.

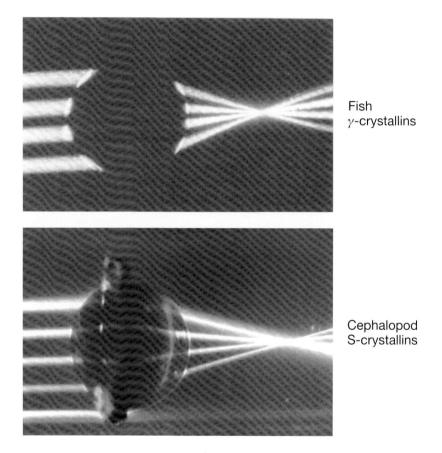

Fish
γ-crystallins

Cephalopod
S-crystallins

Figure 4.4. Refraction by a fish (African cichlid; *Haplochromis burtoni*) and squid (*Sepioteuthis lessoniana*) lens. The lens was removed from the eye. Laser beams focus without spherical aberration in both cases despite the fact that the fish lens contains mainly γ-crystallins and the cephalopod lens mainly S-crystallins. The top image was supplied by Dr. Robert Fernald (Stanford University, California) and is reproduced from Fernald and Wright[1112] with permission from the publisher; the bottom image was supplied by Dr. Jacob Sivak (University of Waterloo, Ontario, Canada) and is reproduced with permission.

CRYSTALLINS AND GENE SHARING

There were early immunological[266,280,281] and nucleic acid hybridization[282] signals, sometimes doubted (see Zwaan[283]), that crystallins are not confined to the lens but are expressed in other ocular and nonocular tissues. Clayton and colleagues[280] (p. 11) concluded after reexamining organ specificity of lens antigens that "much tissue specificity is the result of a unique selection of antigens genetically available, any one of which may be found in other tissues, rather than being due exclusively to the possession of some specific protein restricted to that tissue." The idea was developed that low levels of crystallin gene expression were a precondition for transdifferentiation of a heterologous tissue (i.e., retina) into lens.[284]

The wake-up call that crystallins are not restricted to a refractive lens function did not come from the reports that they are diverse and occasionally taxon-specific—which was accepted without fanfare since the discovery of FISC (δ-crystallin) in chickens[267]—or from the evidence that they are expressed weakly outside of the lens, but from the announcement by Ingolia and Craig[285] that α-crystallins are homologous to small heat shock proteins (shsps) of *Drosophila*. Even though homology does not necessarily mean identity but simply places the proteins into a family, α-crystallin/shsp homology shattered the belief that the specialized characteristics of the transparent lens could be achieved without correspondingly specialized proteins. Almost a decade passed before it was established that αB-crystallin is a *bona fide*, stress-inducible,[286] and functional[287] shsp that is expressed in many different tissues.[288–290]

In retrospect, nature's selection of a shsp for refraction in the lens makes sense because the lens cannot replace damaged cells or proteins and therefore must protect what it has. High expression of shsp/αB-crystallin in the lens thus serves the dual purposes of (1) conserving the soluble properties of irreplaceable lens proteins and (2) refraction. If it were not so abundant in the lens, αB-crystallin would be known simply as one of the members of the shsps. The main point is this: A shsp gained a crystallin, or refractive, function in the lens of vertebrates during evolution by virtue of its lens-preferred accumulation, and it did not lose its old function(s) in other tissues, or even in the lens, by acquiring a new role. The innovation of a refractive function depended upon a modification in the expression of its gene although it no doubt also underwent some structural changes in amino acid sequence as it adapted for its dual roles. There is no evidence that acquisition of a lens refractive role by the shsp/αB-crystallin was associated with a loss of its nonrefractive functions outside of the lens. It turns out that the α-crystallin functions are not even limited to stress protection and refraction (Figure 4.5).

There are in fact two *α-crystallin* genes (αA and αB) that originated by duplication of an ancestral *shsp/αB-crystallin* gene at least 500 million years ago[291,292] (see Chapter 9). Both α-crystallins are abundant in vertebrate lenses, although αA-crystallin is more specialized for lens than is αB-crystallin. The lens specialization of αA-crystallin resulted in restricted stress inducibility;[1133] nonetheless it is still able to act as a chaperone, meaning that it binds to and protects partially denatured proteins from aggregation.[293] The greater lens specialization of αA- than αB-crystallin is reflected in the fact that elimination of the *αA-crystallin* gene in mice by homologous recombination leads to cataract,[294] while elimination of the *αB-crystallin* gene does not.[295]

Despite their prevalence in the lens, both *α-crystallin* genes are differentially expressed outside of the lens.[288–290] The number of different functions that each crystallin polypeptide has is not known. Because the chaperone ability of the α-crystallins prevents thermal aggregation,[287] it has been assumed that they also have a stress-protective and/or stabilizing function in nonlens tissues. αB-crystallin can modulate the assembly of intermediate filaments[296] and may be involved in the stabilization or morphological reorganization of the developing notochord, lens, and myotomes in chicken embryos[297] and of the kidney in rat embryos.[297a] Early studies found αB-crystallin in the Z-bands of the myofibrillar structure of the heart[298] and associated with the cytoskeletal components of skeletal muscle.[299] Cytoplasmic αB-crystallin translocates to the Z-I-region of myofibrils and binds to desmin and titin filaments in the heart during ischemia.[299a,299b] Evidence that αB-crystallin protects these myocardial proteins was obtained in *αB-crystallin*-overexpressing transgenic[299c] and *αB-crystallin*-deficient[299d] mouse hearts recovering from ischemia. Recent experiments using isolated papillary muscles from *αB-crystallin*-deficient mice provided evidence that αB-crystallin may have less to do with protecting the contractile ability of muscle than with maintaining muscular elasticity.[299c] Numerous muscle phenotypes are observed in αB-crystallin-deficient mice, the most dramatic being a severe spinal curvature known as kyphosis followed by death at approximately eight months of age[295] (see Figure 4.5). A caveat to the interpretation of the late

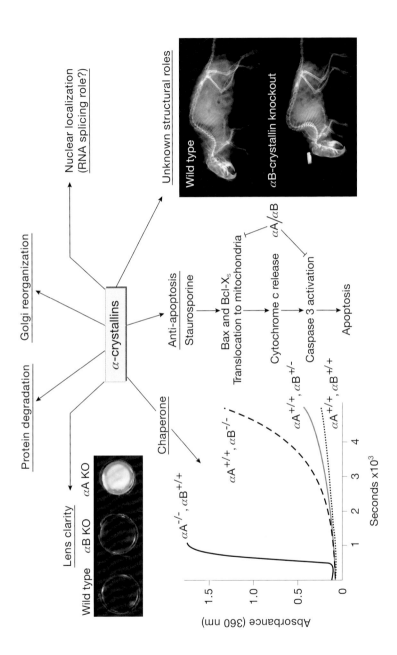

death is that these αB-crystallin-deficient mice also lack the linked αB-*crystallin*-related gene encoding the heat shock myokinase-binding protein.

αA-crystallin is expressed at a low level in other tissues, including the spleen and thymus[300] and teeth.[301] In contrast to mice lacking the αB-*crystallin* gene, αA-*crystallin*-deficient mice have smaller, opaque lenses.[294] The lens opacity is associated with inclusion bodies comprising αB- and γ-crystallins, suggesting that one lens function of αA-crystallin is to maintain solubility of the other abundant crystallins. The smaller sizes of the αA-crystallin-deficient lenses suggest a role in growth control for this shsp-derived crystallin that is consistent with the known involvement of other heat shock proteins in cellular growth.[302] Indeed, αB-crystallin is commonly expressed in basal-like breast tumors of humans, and overexpression of αB-crystallin transforms immortalized human mammary epithelial cells, induces luminal filling, and produces neoplastic-like changes in mammary acini.[302a] Together, this supports the connection between αB-crystallin and cellular growth and indicates that this shsp can act as a novel oncoprotein.

An antiapoptotic role for αA-crystallin is suggested by an increase in apoptosis and disorganization of F-actin (bundles around the nucleus) in UVA-irradiated cultured αA-*crystallin*-deficient lens cells.[303] By contrast, while cultured αB-*crystallin*-deficient lens cells are not prone to increased apoptosis,[304] they are predisposed to genomic instability and hyperproliferation.[305] These features may be caused by an impairment of a p53 checkpoint function.[306] An antiapoptotic role was ascribed for αB-crystallin in a rabbit lens epithelial cell line (N/N1003A) transfected with human *BCL2*.[307] Unexpectedly, the

Figure 4.5. Many roles of the α-crystallins. The lens clarity insert (upper left) shows lenses taken from mature mice that lack the αA-*crystallin* or the αB-*crystallin* gene (KO, knockout). The graph insert (lower left) shows aggregation (noted as increasing absorbance) of the water-soluble fractions of proteins in extracts of mouse lenses from knockout mice with the indicated α-crystallin composition. The extracts were heated to 63° C. Proteins aggregate faster in the extracts lacking αA-crystallin than αB-crystallin because there is more of the former than the latter in normal lenses. The insert showing spinal curvature (lower right) shows approximately one-year-old mice that lack the αB-crystallin gene. See text for details and additional α-crystallin functions. The inserted figures are courtesy of Eric F. Wawrousek (National Eye Institute, NIH). The antiapoptotic activity is adapted from Mao et al.[310]

transfected cells containing the antiapoptotic BCL2 protein were more, rather than less, susceptible to hydrogen peroxide–induced apoptosis. Due to a decrease in αB-crystallin gene expression in the hydrogen peroxide–treated cells, this can be counteracted by restoring the level of αB-crystallin in the BCL2-transfected cells. An antiapoptotic activity by αB-crystallin was also noted in cardiomyocytes during myocardial ischemia followed by reperfusion in transgenic mice overexpressing αB-crystallin[308] and in hydrogen peroxide–treated human retinal pigmented epithelial cells transfected with αB-crystallin cDNA.[309] In addition to their chaperone function protecting against aggregation of partially denatured proteins, then, αA- and αB-crystallins have different roles in the growth and stability of lens cells.

Several mechanisms have been proposed for the antiapoptotic function of the α-crystallins. Apoptosis is regulated by multiple pathways involving the cytoplasmic Bcl-2 family of proteins, cytochrome c release from the mitochondria, and cascades of calcium-activated serine proteases called caspases (see Chapter 6 and Mao et al.[310]). Released cytochrome c ultimately leads to the activation of caspase 3, an executioner of apoptosis. Detailed studies using carcinoma[311] and the C_2C_{12} myoblast[312] cell lines have addressed the regulation of apoptosis by shsp/αB-crystallin. The proposed scheme converges on both the mitochondrial (cytochrome c/caspase-9 mediated) and death (TNF-α/caspase-8 mediated) receptor pathways by repressing the autoproteolysis component of caspase-3 activation from its procaspase. Additional studies on lenses from mice lacking both the αA-crystallin and αB-crystallin genes related fiber cell formation to an attenuated apoptotic mechanism and indicated that αA-crystallin interacts with caspase 6, while αB-crystallin interacts with caspase 3.[312a] This is distinct from other known apoptosis repressive mechanisms and from that employed by shsp 27 or Hsp 70; these other heat shock proteins repress caspase-9 activation at different positions in the mitochondrial receptor pathway. Human αA- and shsp/αB-crystallins also prevent apoptosis by interacting with Bax and Bcl-X$_s$ in cells treated with the apoptosis-promoter staurosporine[310] (see Figure 4.5); both of these Bcl-2 protein family members promote apoptosis by releasing cytochrome c from the mitochondria. The α-crystallin-bound Bax and Bcl-X$_s$ proteins are sequestered in the cytoplasm and are unable to release cytochrome c from the mitochondria. Finally, overexpression of αA- and αB-crystallins in transfected cells appears to repress an increase in Bak (another Bcl-2 family member) by staurosporine by an unknown mechanism,[310] suggesting still another anti-apoptotic mechanism employed by the α-crystallins.

In addition to regulation of cell growth and apoptosis, αB-crystallin modulates protein degradation[313] and Golgi reorganization.[314] αB-crystallin is also associated with centrosomes of dividing cells.[315] Interestingly, αB-crystallin has been identified in the cell nucleus in the form of speckles known as interchromatin granule clusters of cultured cells.[316–318] αB-crystallin colocalizes with the splicing factor SC35 in the nuclear speckles, suggesting a role in RNA splicing or in protection of splicing factors. The control of membrane fluidity, especially during periods of stress, is still another proposed function of the α-crystallins.[319] Artificial model membranes were used to show that α-crystallin can stabilize the liquid-crystalline state and increase the molecular order in the fluid-like state of the membranes, with the nature of the interaction depending on the lipid composition and extent of lipid unsaturation.

The α-crystallin polypeptides are phosphorylated by cyclic-AMP-dependent kinases.[320–322] The biological roles of the phosphorylations are being revealed slowly. The nuclear import of αB-crystallin is phosphorylation-dependent.[318,323] Phosphorylation of αB-crystallin also regulates its interaction with cell membranes, where it has a role in cell migration.[324] In addition to being phosphorylated by kinases, α-crystallins have a capacity for autophosphorylation.[325,326] This suggests that they may have a regulatory role in signal transduction pathways, although there is no direct evidence for that yet. Much research remains to establish whether these different biological functions of the α-crystallins are performed by the same or different inherent properties of these versatile proteins.

THE $\beta\gamma$-CRYSTALLINS: A SUPERFAMILY WITH DISTANT STRESS CONNECTIONS

Some connection with stress proteins continues with the β- and γ-crystallins. Although the monomeric γ-crystallins and the oligomeric, complex β-crystallins form separate classes, they fall into a related $\beta\gamma$-crystallin superfamily with internal symmetry.[327–330] The family relationship between the β- and γ-crystallins was established by the three-dimensional crystallographic structures of the proteins[331–333] and the exon/intron structures of the genes.[328,334,335] In brief, the γ-crystallins are symmetrical proteins comprising two linked domains; each domain has a signature folded structure composed of two Greek key motifs. The exons encoding the two motif-domains of the γ-crystallin polypeptides are separated by an intron.[336,337] By contrast, the genes for the β-crystallins differ by having separate exons coding for each of the four Greek key motifs.[334] These gene and protein structures suggest that the $\beta\gamma$-crystallin superfamily grew by motif and domain duplications, fusions, and rearrangements of their exon/intron organizations (see D'Alessio[329]).

Two γ-crystallins—γS and γN—are present in vertebrates and may link the β- and γ-crystallins.[330] γS-crystallin, formerly called βS-crystallin, has β-crystallin-like protein properties but a γ-crystallin-like gene structure (exons encode two motifs, or one domain). The gene structure for γN-crystallin is a hybrid between the γ- and β-crystallin genes. The first exon encodes a short N-terminal arm and the second exon encodes the first domain (which comprises two Greek key motifs), as in the γ-crystallin genes; the next two exons each encode one of the two Greek key motifs of the second domain, as in the β-crystallin genes. Assuming that β-crystallins are the more ancient members of the $\beta\gamma$-crystallin family, the gene structure of γN-crystallin raises the possibility that the γ-crystallin family was amplified in vertebrates by a complex processes of gene duplication and intron loss at the two-domain stage. The separate γS- and γN-crystallins appear as intermediates between the β- and γ-crystallin families.

There is no known function for the vertebrate $\beta\gamma$-crystallins aside from their refractive roles in the lens. However, they are expressed outside of the lens, which suggests additional functions.[330,338–345] The nonlens expression of the β-crystallins is most evident in the retina; however, rat βB2 is expressed in the brain and testis as well as the lens and retina.[344] The nonrefractive functions of these crystallins in the lens or other tissues are unknown. βA3/A1-, γB-, γC-, γE-, and γF-crystallins have also been observed in the teeth of mice, where it has been proposed they have a stress-related role.[301]

Close progenitors of the $\beta\gamma$-crystallins of the vertebrate lens are found in urochordates[346] (see Piatigorsky[346a]). Urochordates, the closest living relatives of vertebrates,[346b] have larvae with vertebrate-like characteristics (cf. notochord, bilateral symmetry). The urochordate larva contains simple eyes (ocelli) lacking a cellular lens; the eyes are lost after metamorphosis into sedentary adults. A $\beta\gamma$-crystallin-like gene encoding a one-domain (two motifs) protein called Ci-$\beta\gamma$-crystallin was cloned from *Ciona intestinalis* (a urochordate with the undignified nickname of sea squirt). Each Greek key motif of Ci-$\beta\gamma$-crystallin is encoded by a separate exon, as in β-crystallin genes. The sequence similarity between vertebrate $\beta\gamma$-crystallins and Ci-$\beta\gamma$-crystallin is weak; but their crystal structures are well conserved (Figure 4.6), raising the possibility that the urochordate Ci-$\beta\gamma$-crystallin is the evolutionary precursor to the vertebrate $\beta\gamma$-crystallins. Ci-$\beta\gamma$-crystallin does not perform a refractive function because sea squirts lack lenses. Ci-$\beta\gamma$-crystallin is expressed in the otolith and the palps of the sea squirt larva, but not in the ocellus, and thus clearly must have a nonrefractive function. The otolith is a sensory, opsin-containing structure in the brain that shares a common developmental origin with the ocellus; the palps are structures extending from the head for adhering to a substratum during metamorphosis.

Differences in structure and expression among the many $\beta\gamma$-crystallin family members in invertebrates and vertebrates provide evidence that these proteins have a variety of functions. However, while the extended members of the $\beta\gamma$-crystallin superfamily are all defined by having similar domain structures, the proteins and the $\beta\gamma$-like Greek key motifs comprising the domains show

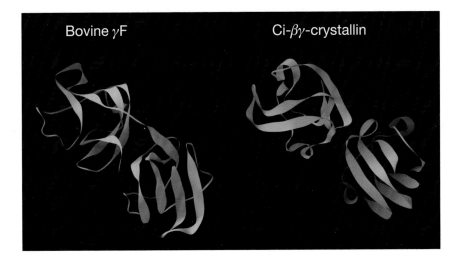

Figure 4.6. "Greek key" structural motifs in the $\beta\gamma$-crystallin protein family. Ribbon models, derived from the RCSB Protein Data Bank, of a $\beta\gamma$-crystallin "Greek key" fold in γF-crystallin from the bovine lens[331] and from Ci-$\beta\gamma$-crystallin from the urochordate *Ciona intestinalis*.[346]

numerous differences. Some proteins contain only a single domain (two motifs), while others have two or more domains. There are also two distinct Greek key motifs (A and B), which allows further structural discrimination among the proteins. The arrangement of the motifs within the two domains of the lens $\beta\gamma$-crystallins is AB AB in the two domains of the lens crystallins. This is not universally true for all members of this family; moreover, the motifs show stability, packing, and calcium-binding differences in the various family members.

Protein S, a two-domain spore coat protein of a soil bacterium *(Myxococcus xanthus)*,[333,347] and Spherulin 3a, a one-domain abundant protein in spherulating plasmodia of a slime mold *(Physarium polycephalum)*,[348–350] are abundant in dormant microbes. Protein S and Spherulin 3a accumulate during stressful conditions (starvation and darkness), leading to spores or cysts that are resistant to desiccation, ultraviolet rays, and heat. These $\beta\gamma$-crystallin family members may thus be involved in stress protection as are the α-crystallins. However, differences among Protein S and Spherulin 3a and the $\beta\gamma$-crystallins suggest that they evolved separately.[351] The motif arrangement of Protein S is BA BA (rather than AB AB), and the Spherulin one domain is a homodimer whose stability requires calcium. The lens $\beta\gamma$-crystallin domains are stable without calcium and their modes of pairing differ from that of Protein S.

Another related $\beta\gamma$-crystallin family member—Yersinia crystallin—was isolated from the gram-negative bacterium *Yersinia pestis*.[352] This pathogenic bacterium is responsible for the dreaded plague that has caused three human pandemics.[353] Yersinia crystallin is a putative exported protein with three calcium-binding domains comprising an AA or BB Greek key motif arrangement and might be involved in calcium-dependent stress responses.[352]

A distantly related two-domain epidermis differentiation-specific protein (EDSP) is present in the amphibious newt *(Cynops pyrrhogaster)*.[354] Homology with the lens $\beta\gamma$-crystallins is weak. Nonetheless, computer-assisted modeling indicates that EDSP has an AB AB motif arrangement in its domains, as do the $\beta\gamma$-crystallins of the lens. These proteins are expressed in ectodermal structures and have an unknown function.[355,356] Membrane-associated proteins, called PCM1–4, present in the protozoan *Paramecium tetraurelia* contain $\beta\gamma$-crystallin domains.[357] The PCM1–4 proteins bind calcium and calmodulin and are suspected to mediate protein:protein interactions that are important for signal transduction events. The possibility of γ-crystallins having a membrane-associated function in the vertebrate lens is supported by their ability to bind MIP/aquaporin-0, the main intrinsic water-channel protein of the mammalian lens.[358,359]

Another $\beta\gamma$-crystallin family member is present in the marine sponge *Geodia cydonium*.[360] The motifs comprising this two-domain protein are similar but nonidentical, with an apparent AB AB pattern as in the lens $\beta\gamma$-crystallins. Unlike the lens crystallins, however, the sponge protein is encoded in an intronless gene.[361] The function of this $\beta\gamma$-crystallin family member is not known.

There are several known yeast proteins with $\beta\gamma$-crystallin-like motifs that have apparently evolved convergently. One is WmKT, a killer toxin in *Williopsis mrakii*.[362] Another is a one-domain metalloproteinase in *Streptomyces nigrescens*.[363] A DNA sequence potentially encoding a similar domain is found in a pathogenicity island of the bacterium *Vibrio cholerae*.[351,364]

Finally, a trimeric six-domain (twelve-motif) human melanoma suppression protein called AIM1 (for absence in melanoma) binds calcium and structurally resembles the β-crystallins.[365,366] Human AIM1 and its mouse counterpart[367] are expressed in many tissues and have weak similarity to filament or actin-binding proteins, suggesting a possible cytoskeletal function. The calcium-binding property and its absence in melanoma cells suggest a role in growth control.

Whatever their many functions, the $\beta\gamma$-crystallin superfamily of proteins have clearly exploited this versatile structural domain made up of two Greek key motifs. Themes for the function of these proteins relate to protein stability and stress protection. Learning the multiple functions of the lens $\beta\gamma$-crystallins remains as an important challenge for the future.

THE ENZYME-CRYSTALLINS OF VERTEBRATES

The structural and functional connections between shsps and the α-crystallins, and the evolutionary relationships between physiological stress-protective proteins and the $\beta\gamma$-crystallins, resonate with the biological needs of the long-lived, stressful existence of anucleated lens cells that are constantly exposed to light and oxidative stress. But the stress connection of the classical crystallins is just the beginning, not the end, of the surprising story. Crystallins are also related or identical to different metabolic enzymes.[9,22-24,368] These include lactate dehydrogenase B_4/ε-crystallin,[18] argininosuccinate lyase/

δ2-crystallin,[19] and α-enolase/τ-crystallin,[369] among others. These diverse enzymes that appear as lens crystallins in a taxon-specific manner are called enzyme-crystallins (see Figure 4.2).

Many of the enzyme-crystallins in vertebrate lenses are metabolically active proteins. While some of the enzyme-crystallins are linked to redox and energy-producing pathways (e.g., lactate dehydrogenase/ε-crystallin, α-enolase/τ-crystallin, quinone oxidoreductase/ζ-crystallin) that could protect stressed, long-lived lens cells, other enzyme-crystallins belong to different pathways (e.g., argininosuccinate lyase/δ-crystallin) that are not obviously connected to stress protection. Some of the taxon-specific crystallins are unquestionably related to enzymes, but they lack enzymatic activity. This suggests that their lens function is structural, although it is possible that some may possess unknown substrate specificities. As a group, the enzyme-crystallins do not fall into a consistent class of metabolic activities.

After the discovery that some crystallins are related to enzymes, even to the extent of having catalytic activity, it remained to be established that the *identical* gene encodes the abundant lens crystallin and the metabolic enzyme: Similarity does not mean identity. The first cases to link identity between crystallin and enzyme by analyses of genes and cDNAs were argininosuccinate lyase/δ2-crystallin,[20,370] lactate dehydrogenase B$_4$/ε-crystallin,[21] and α-enolase/τ-crystallin[369] in the chicken and duck. This established that the same gene encodes a crystallin and an enzyme, and that its differential, tissue-specific expression has a controlling role in directing the function of the protein.

The phylogenetic distributions of the enzyme-crystallins do not fit neatly into different niches. Although the argininosuccinate lyase/δ-crystallins are restricted to birds and reptiles,[36] consistent with phylogeny being important, the appearance of enzyme-crystallins among species resembles a mosaic in search of a pattern (see Figures 4.2): Lactate dehydrogenase B$_4$/ε-crystallin is present in lenses of ducks, hummingbirds, chimney swifts, and crocodiles;[18] lactate dehydrogenase A/υ-crystallin is present in lenses of the platypus; α-enolase/τ-crystallin is present in lenses of lampreys,[371] turtles,[372] and the ocean sunfish;[373] L-gulonate-3-dehydrogenase/λ-crystallin,[374] a relative of dehydroascorbate reductase[375] and acetyl CoA dehydrogenase,[376] is confined to rabbit and hare lenses; oxidoreductase/ζ-crystallin is found in lenses of guinea pigs,[377] some hystrichomorph mammals,[173] camels,[378] llamas (see [264]), and the

Japanese tree frog;[379] retinaldehyde dehydrogenase/η-crystallin (an aldehyde dehydrogenase class 1) is present in elephant shrews;[380,381] an ornithine cyclodeaminase homologue/μ-crystallin is found in kangaroos;[382,383] aldose-reductase homologue/ρB-crystallin[384,385] and GAPDH/π-crystallin[386] are present in lenses of diurnal geckos; and a hydroxysteroid dehydrogenase relative (another aldo-keto superfamily protein)/ρ-crystallin is found in lenses of different frogs.[387,388] The potpourri of taxon-specific enzyme-crystallins shows no overall phylogenetic or metabolic consistency. It is likely that selection of the different enzyme-crystallins was governed by a combination of neutral increases in lens-specific gene expression and by specific advantages that each protein had for the lifestyle of the particular species involved. Interesting cases for the latter have been made.

Wistow[264,368] speculated that enzyme-crystallins came about to soften hard, γ-crystallin-enriched lenses when vertebrates took to land during evolution. The idea was that softer lenses containing hydrated enzyme-crystallins are easily deformed to allow rapid accommodation, a process that may be relatively unimportant for aquatic species but critical for terrestrial species. Mammals, especially poorly accommodating nocturnal ones that have hard spherical lenses, would have reacquired γ-crystallins. This hypothesis has merit but is inconsistent with the prevalence of lens enzyme-crystallins in aquatic invertebrates with eyes that predate terrestrial species and with the imperfect correlation between lens hardness and the use of enzyme crystallins. For example, aldehyde dehydrogenase/Ω-crystallin is the sole crystallin in the soft scallop lens[276] and glutathione S-transferase derivatives are the major crystallins in the hard cephalopod lens.[273] Moreover, the highly accommodating, relatively soft primate lenses are filled with α-, β-, and γ-crystallins and lack enzyme-crystallins. While these facts do not eliminate the possibility that one or more enzyme-crystallins were selected for a lens-softening role in terrestrial vertebrates, they suggest that lens softening, like other possible adaptive uses of enzyme-crystallins, may have been a secondary exploitation of another primary cause.

That many, but not all, of the enzyme-crystallins are catalytically active raises the question as to whether both the enzymatic and structural roles are used in the lens, or which of the metabolic roles were important in their selection as crystallins. If lenses in general require high levels of certain metabolic activities, why are the distinct enzyme-crystallins so scattered among different species? Because enzymes employed for catalytic purposes should be adequate at relatively low concentrations (more enzyme than substrate makes little biochemical sense), their abundance in the lens suggests a structural role in that tissue. The adaptive advantage of enzyme activity is called into question in some cases by the vastly different activity levels of the same enzyme-crystallin found in different species. For example, argininosuccinate lyase/δ-crystallin has poor activity in chickens and pigeons yet this enzyme-crystallin is active in ducks, swans, geese, and ostriches.[20,389–391] Differences in relative gene expression of the inactive δ1- and active δ2-crystallin polypeptides account for the differences in argininosuccinate lyase activity in the chicken and duck lens.[20] The question concerning the need for enzyme activity could also be reversed: How does the lens *protect* itself from an active enzyme-crystallin whose activity is under stringent control in other tissues?

The full biological uses of taxon-specific enzyme-crystallins remain unclear. Metabolic activity may have provided a selective advantage for one enzyme-crystallin but not for another, or for a given enzyme-crystallin in one species but not for the same enzyme-crystallin in another species. Could the metabolic strains of deep-water diving and long-distance flying have added catalytic requirements to the structural needs for both lactate dehydrogenase B$_4$/ε-crystallin and argininosuccinate lyase/δ2-crystallin in their role as enzyme-crystallins in a duck lens, while only their structural properties were useful in the lens of a land-locked, nonmigratory bird such as a chicken?

Considerations of possible adaptive values of enzyme-crystallins have not been limited to catalysis. One possibility is that enzyme-crystallins that bind nicotinamide adenine dinucleotide cofactors (NAD$^+$, NADH, and NADPH) can serve as UV filters to reduce glare[18,264] and to protect against oxidative stress.[392] These ideas are consistent with the positive correlation between NAD and NAD-binding enzyme-crystallins in lenses of different species.[393,394,1134] Comparisons of the time-dependent loss of tryptophan fluorescence upon photodamage for lenses with high and low concentrations of the NAD cofactor suggest that high levels of the reduced cofactors protect against oxidative damage.[395] Another correlation consistent with a redox protective role for high levels of enzyme-crystallin/NADH in the lens was made for geckos.[386] Diurnal geckos exposed to high illumination accumulate GAPDH/π-crystallin and NADH in the lens, while nocturnal geckos do not. In addition, a positive correlation exists between species (especially some birds and reptiles) with NAD-binding enzyme-crystallins and the degree to which they are exposed to glare:[264] Birds feeding off the water (herons, gannets, and gulls) and in bright light (swifts and hummingbirds) as well as amphibious crocodiles accumulate the NADH-binding lactate dehydrogenase/ε-crystallin, while ground (sparrows, chickens, and owls) or underwater (penguins) feeders not exposed to glare do not. A case for protection against excess ultraviolet light by NADPH absorption has also been made for the use of an enzymatically active oxidoreductase/ζ-crystallin in the Japanese tree frog instead of a completely different inactive aldo-keto reductase family protein/ρ-crystallin in many other frogs.[379] The different frogs all live by the waterside, with high exposure to ultraviolet light; and the common trait of the enzymatically active ζ- and inactive ρ-crystallins, which were derived from different members of the aldo-keto family of enzymes, is their ability to bind NADPH. These are interesting correlations worthy of further study.

The use of crystallin-bound retinoids for protection against incident light was also proposed for ι-crystallin in geckos.[396] ι-crystallin binds 3,4-didehydroretinol (vitamin A_2), yellows the lens, and presumably protects the retina by absorbing short-wave radiation coming from the environment. ι-crystallin is identical to cellular retinol-binding protein type 1, which is involved in retinoid storage and transport, and binds vitamin A_2 and retinol with equal affinity *in vitro*. However, only vitamin A_2 is found in the gecko lens; retinol is excluded because it fluoresces in visible light and is not photostable. The ι-crystallin/vitamin A_2 complex absorbs UV and short-wave blue light in the intense ambient light environment of diurnal geckos. The events to achieve this evolutionary adaptation are beautifully concerted, because vitamin A_2 is synthesized in the pigmented retinal epithelial cells and must be transported into the lens, while the cellular retinol-binding protein accumulates as a crystallin. As in the case of glyceraldehyde-3-phosphate-dehydrogenase/π crystallin, an argument favoring the use of ι-crystallin as a filter is that it is only present as a crystallin in diurnal geckos exposed to desert sands; nocturnal geckos lack this crystallin. Similarly, μ-crystallin, an ornithine cyclodeaminase homologue, binds NADPH and is present in lenses of diurnal and not nocturnal marsupials.[383] Taken together, it is possible that the common property of binding NAD as light filters in the lens may have been a selective feature that contributed to the recruitment and rerecruitment of such different proteins as crystallins in a species-specific manner.

CRYSTALLINS OF INVERTEBRATES

Crystallins of cellular lenses of invertebrates are also diverse and taxon-specific (see Figure 4.3). This description extends even to the secreted corneal lens of compound ommatidial eyes of insects. Invertebrate crystallins have been reviewed by Tomarev and Piatigorsky[194] and are described briefly here. The varied expression patterns of invertebrate crystallins suggest that they are multifunctional, as are the crystallins of invertebrates.

Ommatidial eyes: Drosocrystallin, a glycoprotein secreted from pigment cells, is one of three proteins comprising the refractive corneal sheaths within the facets of *Drosophila* compound eyes.[277,397] Two other dipteran flies lack this protein in their corneal lenses, suggesting species specificity. Drosocrystallin appears to have another function because it is a good substrate for cholera- and pertussis-dependent ADP-ribosylation. Similarly, antigen 3G6, a glycoprotein in glial and retinal cells, is present in structurally different (secreted and cellular) crystalline cones of compound eyes of different arthropods.[398] Due to its appearance in nonrefractive cells and refractive cones, it is another candidate for the use of gene sharing to obtain a protein used for refraction.

Camera-type eyes: The lens-containing camera-type eyes of invertebrates have also derived crystallins from enzymes or other proteins with nonrefractive functions. S-crystallins of squid and octopus lenses are related to glutathione-S-transferase.[19,274] The *S-crystallin* genes have duplicated many times, forming a family comprising over 20 members (Tomarev and Piatigorsky[194] and Tomarev, Zinovieva, and Piatigorsky[399]; see Chapter 9). All but one of the lens-specific S-crystallins (SL11/Lops4) has lost enzymatic activity by a combination of sequence changes and acquisition of an internal peptide. SL11/Lops4 is also lens-specific but has low enzymatic activity when tested *in vitro*. Because another gene encoding the authentic glutathione S-transferase enzyme is present and expressed principally in the digestive gland,[400] it appears as if the S-crystallins were derived by multiple gene duplications followed by specialization for crystallin function.[401]

Cephalopods[275,402] and scallops[276] have also utilized an aldehyde dehydrogenase as a crystallin (Ω-crystallin).[275] These are designated ALDH1C1 (octopus), ALDH1C2 (squid), and ALDH1A9 (scallop) (www.ALDH.org). Ω-crystallin is equally related by sequence to ALDH1 and ALDH2. Model building of scallop Ω-crystallin[276] indicates a very similar structure to vertebrate ALDH2 (Figure 4.7). Although the active site is conserved, Ω-crystallin has been found to be enzymatically inactive with all the substrates tested. This appears to be due to poor or no binding of NAD or NADP; poor cofactor binding may result from a single change (from isoleucine to valine) in the binding pocket. In contrast to Ω-crystallin, ALDH/η-crystallin of the elephant shrew lens has retinaldehyde dehydrogenase activity.[381] Ω-crystallin appears to have evolved by duplication of an ancestral gene encoding an active enzyme and subsequently specialized for refraction in the transparent lens.[275] Despite the inactivity of scallop Ω-crystallin, its mRNA is expressed weakly outside of the eye, suggesting that Ω-crystallin has an enzymatic activity that has not been identified yet or has a nonenzymatic function outside of the lens.[276] It also remains possible that the extralenticular Ω-crystallin-like mRNA is derived from a closely related gene duplicate (although there is no evidence for this) encoding an enzymatically active protein.

The muscle-derived cellular lens of the squid *(Euprymna scolopes)* photophore is filled with Ω-crystallin, also called L-crystallin for its presence in the light organ.[403] Photophores emit light produced by resident bioluminescent bacteria. One proposed function for photophores is to camouflage the shadow of a squid as it cruises on the water surface seeking prey in the moonlight. Because photophores and eyes are not homologous organs, selection of the same protein as a crystallin in their respective transparent lens provides a compelling case for independent innovation of a new function for an old protein.

Borrowing proteins for an optical function in transparent lenses is a very old custom indeed.[194] This happened already in cubomedusan jellyfish, ancient metazoans with sophisticated lens-containing eyes. The eyes of the jellyfish *Tripedalia cystophora* exist within four rhopalia dangling from a stalk embedded in a notch in the surface bell.[249,250,260] The jellyfish crystallins represent collectively at least 90 percent of the lens proteins. *Tripedalia* has three crystallin families, J1-, J2-, and J3-crystallin.[260] The J1-crystallin family resolves into three novel 35 kDa polypeptides that share 95–98 percent identity in amino acid sequence[278] and display sequence similarity to ADP-ribosylation enzymes[404,405] and group with the selenoprotein family called SelJ.[405a] *SelJ,* a J1-crystallin homologue in zebrafish, is preferentially expressed

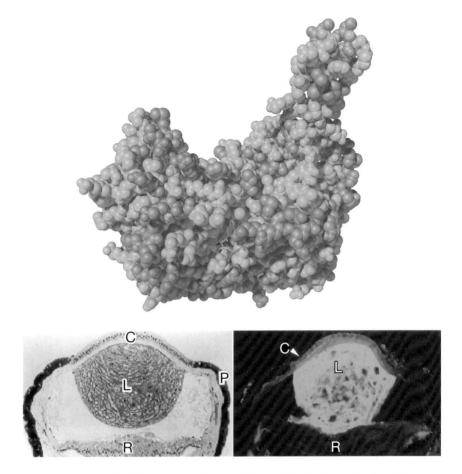

Figure 4.7. Aldehyde dehydrogenase-derived Ω-crystallin in the scallop lens. Upper panel: homology model of scallop Ω-crystallin colored by sequence identity to the human ALDH2 template structure. The light green amino acids have identical nearest neighbors in the template structure; the other amino acids are colored light blue. The NAD cofactor, drawn as ball-and-stick, was inserted in the model by copying its coordinates from the superimposed ALDH2 complex. Adapted from Piatigorsky et al.[276] Lower left panel: light micrograph of a transected eye from a sea scallop *(Placopecten magellanicus)*. Lower right panel: immunofluorescence of the sea scallop eye using a primary polyclonal antibody made against squid L-crystallin/ALDH of the light organ lens. Photomicrographs from Piatigorsky et al.[276] C, cornea; L, lens; R, retina; P, pigmented epithelium.

in the zebrafish lens during early development,[405a] although it does not accumulate as a crystallin in this species. A J1-crystallin homologue is present in the zebrafish gene database, but its function is not known.

More information is available for J3-crystallin, a 19 kDa polypeptide encoded in a single copy gene.[279] Sequence analysis indicates that J3-crystallin is similar to the saposin family of proteins present in vertebrates.[406–408] Saposins act as chaperones in that they physically bridge membrane lipids and lysosomal hydrolases, aiding membrane turnover. In some cases the 6-cysteine saposin motif activates the enzyme as well as bringing it into proximity to its substrates. In general, several distinct vertebrate saposin polypeptides are encoded in a single transcript and are released by proteolysis of the prosaposin protein. While not proved, it appears that a similar situation may exist for J3-crystallin, whose mRNA potentially codes for a protein larger than the mature J3-crystallin in the lens. Vertebrate saposins also give rise to a secreted peptide that acts as a neuronal survival factor. Thus, saposins are multifunctional proteins that encompass chaperone and enzyme characteristics, two properties of vertebrate crystallins.

The *J3-crystallin* gene is expressed outside of the lens, suggesting that its protein has a noncrystallin function. *In situ* hybridization has revealed J3-crytallin RNA in the statocyst (another rhopalial structure believed to be used for orientation), in outward radiations from the retina, in the tip of tentacle, and in larvae as well as in the lens, indicating nonoptical roles.[279] The *J1A-crystallin* gene is also expressed in the statocyst of the rhopalia and in the larva, which lacks an eye, suggesting that it too has a nonoptical role. Each crystallin has its own tortuous history, finding cracks and crevices of opportunities by a gene-sharing mechanism whenever possible and whenever it was lucky enough to be selected and survive.

CRYSTALLIN GENE REGULATION IN VERTEBRATES: A SIMILAR CAST OF TRANSCRIPTION FACTORS

Abundance is the most important, if not sole, criterion that is used for a water-soluble lens protein to be considered a crystallin. The protein must be at a sufficient concentration to affect the refractive properties of the transparent lens in a biologically meaningful way. The minimum concentration of 5 percent of the water-soluble protein has been suggested for a lens crystallin.[409] Accumulated data suggest that a fundamental consideration for unifying the diverse lens crystallins is the presence of similar mechanisms for high lens-preferred expression of their genes. While crystallin concentration in the lens can be controlled theoretically at multiple levels, including protein and/or mRNA stability, posttranscriptional processing, and translational efficiency, tissue-specific transcriptional control is a critical feature defining the crystallin function of a multifunctional protein (see Piatigorsky and Zelenka[410]).

Crystallin gene regulation in vertebrates has been reviewed extensively.[262,290,411–415] Not surprisingly, the details of lens-specific expression differ among *crystallin* genes. However, the similarities of the mechanisms of high lens expression of nonhomologous genes in different species are impressive. Two features emerge: First, the transcription factors controlling gene expression are generally similar for different crystallins within and between species and second, many of the factors also direct eye and lens development.

Early experiments showed preferential lens expression of the chicken *δ1-crystallin* gene (that encodes an inactive taxon-specific enzyme-crystallin) in microinjected mouse cells[416] and transgenic mice.[417] These investigations suggested that lens specificity is programmed into the *δ1-crystallin* gene regulatory apparatus and is recognized even in a species (mouse) that does not use this protein (argininosuccinate lyase) as a taxon-specific enzyme-crystallin in its lens. The DNA sequence responsible for lens-preferred expression exists in the third intron of the *δ1-crystallin* gene;[418] a lens enhancer is also present in the chicken *δ2-crystallin* gene that encodes an enzymatically active argininosuccinate lyase polypeptide.[419] Despite the presence of similar enhancer sequences in both the chicken *δ1-* and *argininosuccinate lyase/δ2-crystallin* genes, additional regulatory mechanisms are superimposed because there is much greater expression of the former than the latter in the lens of this species.[20,420,421] This is not true in the duck, in which both the enzymatically inactive *δ1-crystallin* and active *argininosuccinate lyase/δ2-crystallin* genes are highly and equally expressed in the lens.[20,370]

A similar story unfolds with other vertebrate *crystallin* genes. Lens-preferred enhancers (often in the 5′ flanking region) and proximal promoter fragments display lens-specific activity in transgenic mice in a cross-species fashion, with fragments derived from the chicken *αA-crystallin*[422] and *βB1-crystallin*[423] genes serving as two striking examples.[262,410] A similar group of transcription factors interact to promote lens-preferred expression of unrelated *crystallin* genes in different species. Pax6, c-Maf/Nrl (mammals), L-Maf (chicken), Prox1, RAR/RXR, and Sox1/2 have been found to be particularly important and have been used consistently to regulate expression of the diverse *crystallin* genes throughout the animal kingdom, as illustrated in simplified fashion in Figure 4.8. A group of similar, more general factors (for

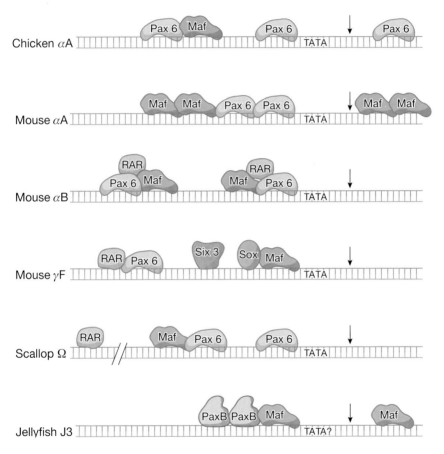

Figure 4.8. Critical transcription factors for lens promoter activity of diverse *crystallin* genes. While many more factors have been implicated, this provides a simplified view that suggests convergent evolution of high lens expression of different genes in a taxon-specific manner.

example, AP-1, CRE, CREB/CREM, and USF)[410] and transcriptional co-activators (for example, CBP/p300)[424] are also utilized for lens promoter activity of *crystallin* genes.

Because transcription factors regulating *crystallin* genes in the lens are also expressed in nonocular tissues as well as in species lacking eyes, lens-preferred promoter activity must require combinatorial, context-dependent parameters. This fits with the fact that the precise combinations of similar transcription factors as well as the relative positions of the DNA motifs to which they bind (the *cis*-control elements) differ in the diverse *crystallin* genes. There are even situations in which a transcription factor that activates one *crystallin* gene represses another. For example, Pax6 activates most *crystallin* genes but represses the chicken βB1-crystallin gene.[425] The present model suggests that Pax6 occupies the βB1-crystallin promoter in lens epithelial cells, where Pax6 concentrations are high, inhibiting access of the activating c-Mafs and Prox1 to the promoter.[426] A change in the relative levels of these transcription factors results in the substitution of c-Mafs and Prox1 for Pax6 on the regulatory motifs of the chicken βB1-crystallin promoter, with consequent activation of gene expression in the lens fibers. In mammalian lenses, Pax6 activates the αB-crystallin promoter but represses the γF-crystallin promoter.[427] This activating or repressing role of Pax6 on different crystallin promoters has been interpreted to be due to a combination of differences in the architecture of the αB-crystallin and γF-crystallin promoters and in their specific interactions with the developmentally regulated transcription factors and other members of the machinery used for gene expression. Despite these finely tuned differences, the similarity in the cast of transcription factors controlling *crystallin* gene expression is striking in view of the diversity of proteins comprising the crystallins.

CONVERGENT EVOLUTION OF *CRYSTALLIN* GENE EXPRESSION

The simplest hypothesis for the commonality of lens-preferred expression of diverse *crystallin* genes is that it results from similarities in their *cis*-control elements. The acquisition of *cis*-regulatory elements used for high promoter activity of *crystallin* genes in the lens probably resulted from site-specific nucleotide changes and rearrangements, deletions, and/or insertions of DNA. It is unlikely that the common use of similar transcription factors for diverse *crystallin* genes was accompanied by mutations affecting the transcription factors themselves and resulting in new DNA binding specificities. Mutations changing the binding specificity of transcription factors involved with developmental processes and many genes (as is the case for the transcription factors regulating crystallin genes) would have far-reaching impact much beyond *crystallin* gene expression *per se*. We may conclude that the similarities of *cis*-control elements regulating the expression of diverse *crystallin* genes were acquired independently during evolution by convergent evolution because many of the genes are unrelated (nonhomologous).

The similarities in *cis*-control elements and transcription factors used for lens-preferred expression of *crystallin* genes were not obvious initially for several reasons. First, the consensus binding motifs for transcription factors allow variations in sequence. Moreover, the transcription factor binding sites often overlap[426,428] and on occasion differ spatially in different *crystallin* genes.[262] Pax6 binding is an example of spatial differences of comparable regulatory motifs in different *crystallin* genes. While generally in the 5′-flanking sequence of *crystallin* genes, Pax6 binding is in the third intron of the $\delta 1$-*crystallin* gene.[418]

The orthologous *αA-crystallin* genes of the chicken and mouse illustrate the evolutionary dynamism of *cis*-control elements.[429] These small heat shock–related genes are structurally similar in both species and are expressed highly and almost exclusively in the lens, resulting in extreme specialization for lens-associated functions. The mouse *αA-crystallin* gene has a lens-specific enhancer at sequence positions $-111/-18$, and the promoter fragment $-111/+46$ shows lens-specific activity when tested in transgenic mice.[430] The DNA sequence of the analogous sequence in the chicken *αA-crystallin* gene is similar but not identical to that in the mouse, yet a chicken promoter fragment must contain a longer sequence ($-162/+77$) than the mouse fragment to achieve lens-specific activity when introduced as a transgene in transgenic mice.[422] This has been explained by differences in the relative positions of similar *cis*-control elements in the promoter fragments of the mouse and chicken genes. The mouse gene has a sequence motif at position $+24/+43$ (called PE2) in exon 1, which is transcribed and precedes the coding sequence, while a similar sequence is found at positions $-138/-130$ in the nontranscribed 5′ flanking region of the chicken gene. The mouse PE2 and the chicken $-138/-130$ sequence both bind AP-1 related transcription factors. Ilagan and colleagues[429] proposed that the juxtaposition of similar but

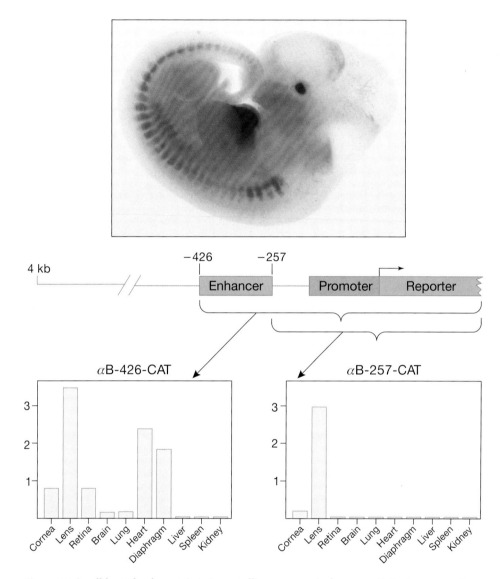

Figure 4.9. Small heat shock protein/αB-crystallin promoter/enhancer activity in transgenic mice. The 5′ flanking region of the mouse *αB-crystallin* gene was fused to the *β-galactosidase* reporter gene and used as a transgene in transgenic mice. The top photomicrograph shows a thirteen-day-old embryonic transgenic mouse; blue demonstrates *β*-galactosidase activity in certain tissues that resulted from activity of the reporter gene driven by the αB-crystallin promoter/enhancer. Note the strong blue in the eye (lens) as well as in the heart and multiple somites. Adapted from Haynes et al.[1113] The bottom graphs show the expression profiles obtained in different experiments when the enhancer was present (αB-426-CAT) or deleted (αB-257-CAT). In these cases the reporter gene was the *chloramphenicol acetyltransferase* gene. Adapted from Gopal-Srivastava et al.[506]

nonidentical *cis*-control motifs that bind either the same or family members of the same transcription factors contributes to the differences in intensity as well as spatial and temporal features of *αA-crystallin* gene expression in the chicken and mouse lens.

While the shsp-related *αA-crystallin* gene is expressed highly in the lens and is not heat inducible, as are other shsps, the αA-crystallin protein itself has maintained chaperone function for protecting proteins from thermal and other stresses. By contrast, the *shsp/αB-crystallin* gene illustrates how acquiring promoter elements for high lens expression and modulating enhancer function led to specialized crystallin function of its protein without loss of the complex expression pattern and heat inducibility (Figure 4.9). The *shsp/αB-crystallin* gene is expressed highly in the lens, heart, and skeletal muscle, and to a lesser extent in most other tissues in mice and other mammals;[288–290] it also retains stress-inducibility.[286] Lens-preferred expression of the *shsp/αB-crystallin* gene requires at least two control elements called LSR1 and LSR2 (for lens-specific regions 1 and 2) in the proximal promoter. These promoter elements are synergistically activated in the lens by binding Pax6, the alternatively spliced isoform Pax6(5a), retinoic acid receptors (RARβ/RARβ) and Maf family members (MafA, c-Maf).[427,431–433] In addition, a muscle-preferred enhancer in front of the promoter boosts lens activity of this promoter approximately seven-fold.[434] The enhancer is required for expression of the *shsp/αB-crystallin* gene in nonlens tissues; by contrast, the proximal promoter alone (LSR1 and LSR2) is sufficient to drive lower level lens expression in the absence of the enhancer.

EVOLUTION OF GENE SHARING BY SHSP/αB-CRYSTALLIN

The molecular details of *shsp/αB-crystallin* gene expression provide a glimpse of the evolutionary dynamism of the gene- sharing process. In mice and other mammals, the *shsp/αB-crystallin* gene is upstream of a related gene called *myokinase binding protein/heat shock protein B2* (*Mkbp/HspB2*), which encodes a protein-binding and activating myotonic dystrophy protein kinase.[435–437] The *Mkbp/HspB2* gene situated 5′ to the *shsp/αB-crystallin* gene is transcribed in the opposite direction. Thus, the relatively short intergenic sequence (<1000 bp) between the *Mkbp/HspB2* and *shsp/αB-crystallin* genes contains regulatory sequences that we expect might be shared by the two genes. It has been suggested that bidirectional arrangement of genes, especially prevalent in the human genome (~10 percent of genes), may have its root in the selective pressures imposed by shared *cis*-control elements.[438–440] However, the expression patterns of the *shsp/αB-crystallin* and *Mkbp/HspB2* genes are entirely different. In contrast to the stress-inducible *shsp/αB-crystallin* gene, which is expressed highly in lens and widely in other tissues, the *Mkbp/HspB2* gene is not stress-inducible, is not expressed in the lens, and is expressed at low levels in heart, skeletal muscle, and intestine. What has happened during evolution to achieve such different expression patterns of adjacent sibling genes that have acquired such different functions?

First, the shsp/αB-crystallin enhancer is orientation-dependent (apparently not generally true of other enhancers), which restricts its ability to direct gene expression to many tissues to the *shsp/αB-crystallin* gene.[441] The intergenic region also contains a sequence close to the *Mkbp/HspB2* gene that exerts a negative effect on gene expression. This putative silencer may further limit activity of the Mkbp/HspB2 promoter. Thus, in addition to acquiring the ability to bind transcription factors that regulate essentially all *crystallin* genes, the αB-crystallin intergenic region has undergone numerous specializations that separated the regulation of the closely linked, divergently arranged *shsp/αB-crystallin* and *Mkbp/HspB2* genes. The result was the acquisition of a refractive function for shsp/αB-crystallin in the lens by a gene-sharing process.

Evolutionary changes of *shsp/αB-crystallin* gene expression can be seen in connection with lens degeneration in the subterranean blind mole rat *(Spalax ehrenbergi).* The blind mole rat begins developing eyes early in embryogenesis, but these regress into subcutaneous, atrophied structures with a retina relegated to photoperiodicity.[442,443] As expected for the degenerate lens of the blind mole rat, the highly lens-preferred *αA-crystallin* gene has accumulated numerous mutations in its coding region.[444] Not so for the *shsp/αB-crystallin* gene, inasmuch as the encoded protein is an authentic shsp expressed in many tissues of the blind mole rat.[445] The αB-crystallin promoter/enhancer of the mole rat has a similar, but not identical, DNA sequence to that of the orthologous mouse gene.[446] Experiments testing the activity of equal-sized (~600 nucleotides) promoter/enhancer fragments of the mouse and the blind mole rat *shsp/αB-crystallin* gene revealed interesting differences in their ability to direct gene expression. After initiating lens activity in the embryonic transgenic mouse lens, the mole rat shsp/αB-crystallin promoter/enhancer abruptly ceased functioning in the lens at midgestation but continued directing gene expression in the heart and lens, even in the adult mouse. The mole rat promoter/enhancer activity was approximately ten-fold higher in the skeletal muscle of the transgenic mouse than was the comparable promoter/enhancer fragment of the mouse gene. These data suggest that the promoter/enhancer of the mole rat *shsp/αB-crystallin* gene underwent adaptive changes associated with the subterranean evolution and burrowing needs of the species. It remains possible that the blind mole rat promoter/enhancer retains some lens-specific activity because another study analyzing the entire kilobase intergenic region indicated lens activity in transgenic *Xenopus* tadpoles.[437] Indeed, rearrangement of DNA regulatory elements of *crystallin* genes is known to occur.[429] In any event, these results indicate molecular changes affecting *shsp/αB-crystallin* gene expression that are consistent with eye degeneration and the need for powerful digging muscles in the nonvisual blind mole rat. These experiments reveal the dynamic state of regulatory regions of orthologous, conserved genes during evolution.

CONVERGENT EVOLUTION OF INVERTEBRATE AND VERTEBRATE CRYSTALLIN PROMOTERS

Comparisons of invertebrate and vertebrate *crystallin* gene expression suggest convergent evolution for lens promoter activity in these distantly related groups of animals. The first inkling of convergence for the mechanism of expression of *crystallin* genes between invertebrates and vertebrates came from studies on the glutathione S-transferase-derived *S-crystallin* genes of squid.[194] The minimal promoters of the lens-specific *SL20–1-* and *SL11-crystallin* genes contain an overlapping, protein-binding AP-1/antioxidant responsive (ARE) element that is required for promoter activity in transfected vertebrate cells.[447] These *S-crystallin* gene control elements resemble the PL-1 and PL-2 *cis*-control elements of the lens-specific chicken *βB1*-crystallin promoter.[448] AP-1 and other stress-responsive regulatory elements are prevalent in vertebrate crystallin promoters, consistent with the stress connection of the *crystallin* genes. Pax6 may also be connected to the regulation of squid *crystallin* genes as it is in vertebrate *crystallin* genes,[411] although a direct connection has not been shown yet. The *Pax6* gene is expressed in the developing squid eye and can induce ectopic eyes in *Drosophila*.[226] These facts suggest resemblances between *crystallin* gene expression in squids and vertebrates.

The promoter of the scallop aldehyde dehydrogenase (ALDH)-related *Ω-crystallin* gene also shows similarities in the mechanism of lens-specific gene expression in vertebrates.[449] ALDH is unique in being used as an enzyme-crystallin both in vertebrates (*η*-crystallin in elephant shrews) and invertebrates (cephalopods, scallops). The scallop Ω-crystallin promoter has putative binding sites for a similar set of transcription factors used for mouse and chicken αA-crystallin promoter activity. These include members of oxidative stress response factors (CREB/Jun and AP-1) and developmental transcription factors (Pax6 and Maf) as well as other factors (αA-CRYBP1, RAR/RXR, and USF) that have been implicated in the activity of many crystallin promoters (see Duncan et al.[262] and Cvekl and Piatigorsky[411]). In addition to the structural similarities of the scallop Ω-crystallin and vertebrate αA-crystallin promoters, site-specific mutagenesis experiments suggest that overlapping CREB/Jun and Pax6 sites are functional in the Ω-crystallin promoter.[449] Because the vertebrate shsp-derived αA-crystallins and the scallop ALDH-derived Ω-crystallin have no homology to each other, the structural and functional similarities of the promoters of their genes indicate evolutionary convergence.

The *J-crystallin* genes provide evidence of convergence in the mechanism of lens-preferred crystallin promoter activity between the distant cubomedusan jellyfish *(Tripedalia cystophora)* and vertebrates, although here we see an intriguing variation on the theme. The promoters of the jellyfish *J1A-*, J1B, and *J1C-crystallin* genes contain several potential AP-1 sites[278] and can also bind a jellyfish retinoic acid receptor, jRXR[450]—both characteristics of other crystallin promoters. While the *Pax6* gene appears to be missing from jellyfish, they have a *PaxB* gene[451] (see Figure 4.8). The PaxB protein contains a Pax2-like paired domain (which selectively binds a Pax2 DNA-binding regulatory motif, not a Pax6-DNA binding motif) and a Pax6-like homeodomain (which can bind both DNA and other proteins). The chimera-like jellyfish *PaxB* gene is believed to have given rise to both by gene duplication during evolution. PaxB binds two contiguous potential binding motifs (equivalent to Pax2 binding sites) in the J3-crystallin promoter, strongly activating it in experimental transfection tests using mammalian cells. The J3-crystallin promoter is not activated by Pax6 or other members of the Pax family of transcription factors.

Taken together, Pax6 contributes, either positively or negatively, to the regulation of the diverse, nonhomologous vertebrate and invertebrate *crystallin* genes. The single known exception is the jellyfish *J3-crystallin* gene, where Pax6 is replaced by its evolutionary precursor PaxB (a Pax2-like binding protein). However, it has been shown that the chicken δ1-crystallin enhancer, which binds and is activated cooperatively by Pax6 and Sox2 in vertebrates,[452] is preferentially activated in the lens-secreting cone cells of *Drosophila* embryos by Pax2 in combination with SoxN (a Sox2 homologue) in experimental situations.[453] Thus, the substitution of PaxB in jellyfish for Pax6 that is used for *crystallin* gene expression in other species remains consistent with a general convergence for the mechanism of lens-preferred *crystallin* gene expression in diverse *crystallin* genes from jellyfish to mammals.

POTENTIAL FOR LENS-SPECIFIC PROMOTER ACTIVITY

While lens specificity appears to have been acquired convergently in many crystallin promoters, studies on the tunicate *Ciona intestinalis* revealed a potential for lens-specific activity for the Ci-$\beta\gamma$-crystallin promoter[346] (see Piatigorsky[346a]). This is diagrammed in Figure 4.10. The endogenous *Ci-$\beta\gamma$-*

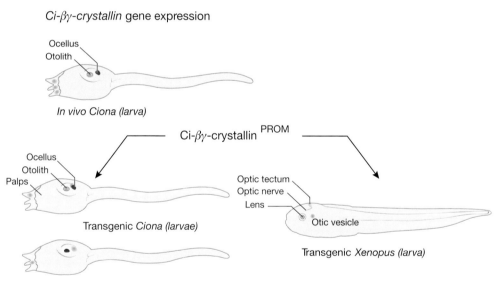

Figure 4.10. Expression and promotor activity of the *Ci-$\beta\gamma$-crystallin* gene. Expression of *Ci-$\beta\gamma$-crystallin* in *Ciona* and of *Ci-$\beta\gamma$-crystallin*[PROM] in transgenic larvae of *Ciona* and *Xenopus*. Expression of the endogenous gene in *Ciona* (top left) was identified by immunofluorescence using a rabbit polyclonal anti-Ci-$\beta\gamma$-crystallin antibody. The transgene, *Ci-$\beta\gamma$-crystallin*[PROM], has one kilobase of 5' flanking region of the *Ci-$\beta\gamma$-crystallin* gene as a promoter fragment fused to the *green fluorescent protein* reporter gene. The green color represents expression of the endogenous gene (top) or the transgene (bottom); the black color indicates pigmented tissue without expression of the endogenous transgene. Taken from Piatigorsky.[346]

crystallin gene is not expressed in the lens-less ocellus of the sea squirt; rather, this gene is expressed in the sensory, opsin-containing otolith (derived from a pigmented sister cell of the light sensitive ocellus) and adhesive, nonvisual palps. A similar pattern of activity exists for the cloned Ci-$\beta\gamma$-crystallin promoter fragment in transgenic *Ciona*. Remarkably, the Ci-$\beta\gamma$-crystallin promoter fragment is active in the lens and other structures of the visual system when tested in transgenic *Xenopus* tadpoles, a vertebrate with eyes containing cellular lenses. The structural *Ci-$\beta\gamma$-crystallin* gene itself is believed to have given rise to the *$\beta\gamma$-crystallin* genes of vertebrates. Thus, Shimeld and colleagues[346] (p. 1688) conclude "that the evolution of the lens did not derive from a new association between a visual system regulatory circuit and co-opted lens structural genes but from the reuse of a pre-existing regulatory interaction linking these components in the central nervous system of a primitive chordate." While this may be applicable to the *$\beta\gamma$-crystallin* genes, which are expressed ubiquitously in vertebrates, it does not explain the high lens expression of the *α-crystallin* or taxon-specific *crystallin* genes, whose promoters are unique among other family members in being adapted for high lens expression. Convergent acquisition of lens specificity seems to be the simplest explanation for high lens expression for most of the *crystallin* genes.

CONVERGENT EVOLUTION AND RELAXED STRINGENCY FOR CRYSTALLINS

The diversity of crystallins throughout the animal kingdom gives a rare opportunity to detect convergent evolution of tissue-preferred gene regulation connected to a specific function—namely, lens transparency and refraction. It is likely that all genes have selective pressures on independently derived modifications in gene regulation that are appropriate for the biological roles of the encoded proteins. In many cases, however, protein functions—such as oxygen transport and release by hemoglobin—are so vital that selective pressures on protein phenotype override those detectable at the level of gene expression, limiting the use of entirely different proteins for the same function. This is apparently not the case for crystallins. Transparency is achieved when the lens lacks discontinuities of refractive index greater than half the wavelength of the transmitted light; this can be realized by macromolecular packing in the lens cells. Because different refractive indices can be established by varying the concentration of any number of different macromolecules, various (not all) proteins can be used as lens crystallins as long as they remain water-soluble in the lens, they do not interfere with other proteins and processes that are needed for lens functions, and, especially, their genes are highly expressed in the lens.

The hypothesis is as follows: DNA modifications independently generating *cis*-control elements responsive to the transcription factors necessary for development and maintenance of the lens make the affected genes candidate *crystallin* genes. Subsequently, nonhomologous genes expressed highly in the lens may be selected as *crystallin* genes, leading to the basketful of taxon-specific crystallins that exist. Regulatory DNA motifs are known to change convergently in many cases.[1135–1137]

TAKE-HOME MESSAGE

The gene-sharing concept arose from the twin observations (1) that lens crystallins responsible for specialized optical functions (transparency and refraction) are surprisingly diverse water-soluble proteins; and (2) that lens crystallins are multifunctional proteins that are expressed in tissues where they have nonrefractive roles (often stress-related or enzymatic). The diversity and taxon-specificity of crystallins indicate that a surprising number of proteins can fulfill their optical role in the lens. We might say that the multifunctional crystallins are a specialist's nightmare!

The required high lens expression of the nonhomologous crystallin genes in invertebrates and vertebrates is regulated by similar *cis*-elements and developmental transcription factors. This indicates that convergent changes in DNA regulatory motifs must have been a major mechanism for creating a wide array of candidate crystallin genes.

The evolutionary scenario of gene sharing emergent from the multifunctional crystallins indicates that independently derived modifications in gene regulation provide a source for functional innovation by placing a protein in another tissue or at different concentration in the same tissue where it can acquire another molecular role. In short: Changes in gene regulation serve to explore new molecular functions for old proteins. Gene sharing ensues when a differentially expressed protein keeps its original function and takes on one or more new molecular functions. Crystallins exemplify the powerful role of gene regulation for deriving new protein functions.

5

The Enigmatic "Corneal Crystallins": Putative Cases of Gene Sharing

Chapter 4 summarized gene sharing by the taxon-specific lens crystallins—abundant cytoplasmic proteins contributing to lens optics as well as having nonrefractive functions in other tissues. This chapter explores the intriguing parallel between a group of taxon-specific corneal-enriched proteins and the diverse lens crystallins.[454,454a] The abundance of these selected corneal proteins has led to their being called "corneal crystallins." However, unlike the lens crystallins, the optical function(s) of the corneal crystallins are not established, making it premature to claim that they are multifunctional proteins engaged in gene sharing. The corneal crystallins are considered here because the combination of their taxon-specificity and relative abundance in the cornea suggest an analogy to lens crystallins in terms of having a specialized optical role in the cornea as well as metabolic roles in the cornea and elsewhere.[455]

THE CORNEA

For light to reach the photoreceptor cells of complex eyes, it must pass through the cornea as well as the lens and, in most cases, the extracellular vitreous body. Thus, the transparent cornea, like the lens, is a critical component of the refractive machinery of eyes.[196]

The cornea of vertebrates differs from the lens developmentally, structurally, and functionally. Although both lens and cornea are derived from the surface ectoderm of the embryonic head, the thin cornea is more complex than the thick cellular lens (Figure 5.1). The mature cornea consists of an extracellular stroma containing keratocytes (fibroblasts) scattered among regularly arranged, alternating orthogonal arrays of collagen fibrils with equal diameters. The stroma is situated between an anterior layer of a relatively thin (depending on the species; the layer is thicker in fish than mammals)

96

stratified epithelial cell layer and a monolayer of posterior endothelial cells. While the lens is surrounded by a collagenous capsule that prevents cellular turnover, the corneal epithelial cells slough off at the corneal surface and are ultimately replaced by the influx of limbal stem cells and the progeny of basal epithelial cells that migrate anteriorly. Another major difference between lens and cornea is that the former has negligible extracellular space while the latter contains a thick extracellular stroma comprising the bulk of the corneal mass in most vertebrates.

The lens and cornea act together as the "window" of the eye and cooperate to cast a focused image onto the retina. Nonetheless, studies have concentrated on their differences rather than their similarities due to the surface lo-

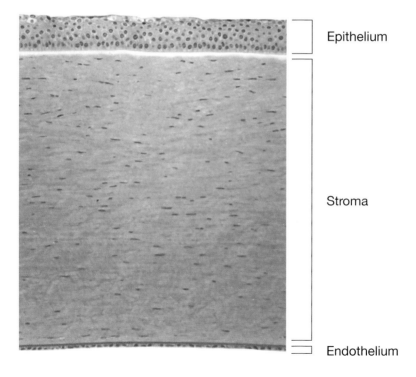

Epithelium

Stroma

Endothelium

Figure 5.1. Photomicrograph of a transected central region of a mouse cornea. See text for description. Photomicrograph courtesy of the late Dr. R. Gerald Robison, Jr. (National Eye Institute, NIH, Bethesda, Maryland).

cation of the cornea, compared to the interior placement of the lens, and their marked differences in structure. The precise collagen organization in the extracellular stroma has been emphasized in considerations of corneal transparency;[456] lens transparency, by contrast, is due largely to the close packing of the cytoplasmic crystallins that minimizes concentration fluctuations.[256–259] The focusing power of terrestrial corneas is due to a combination of the difference in refractive index between air and cytoplasm and to the curvature of the cornea; the optical power of the lens is dictated by the smooth concentra-

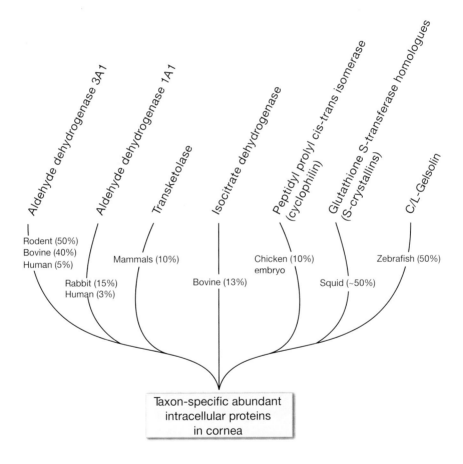

Figure 5.2. Taxon-specific corneal crystallins. Enzymes in parenthesis lack activity. Numbers in parenthesis indicate the percentage of water-soluble protein in lens. See Jester et al. [454] for additional abundant corneal proteins.

tion gradient of the packed, cytoplasmic crystallins. The cornea of aquatic species is more protective than refractive because of the similarity in refractive index between water and cytoplasm, resulting in dependence on the abundant crystallins in the prominent lens for image formation on the retina. Despite the many differences between lens and cornea, both of these transparent tissues accumulate selected widely expressed proteins in a taxon-specific manner[454,454a] (Figure 5.2). Does this signify gene sharing?

ALDEHYDE DEHYDROGENASE: A CANDIDATE CORNEAL CRYSTALLIN

The first corneal, taxon-specific, water-soluble, abundant cytoplasmic protein reported was in cattle (BCP54 for bovine corneal protein of 54 kiloDaltons in size). This undefined protein, except for size, accounts for about 40 percent of the corneal protein in this species.[457,458] Appreciable amounts (5–50 percent) of BCP54 exist in most mammalian corneas; but little, if any, is present in the corneas of other vertebrates, despite the similarities in the structures of vertebrate corneas.[459,460] Silverman and colleagues[459] proposed that BCP54 has a structural, crystallin-like role in the cornea. This led to BCP54 being referred to as "transparentin,"[461] but no mechanism was provided for how this protein might contribute to corneal transparency.

Subsequent investigations showed that BCP54 is an active aldehyde dehydrogenase class 3 (ALDH3A1) enzyme.[462–464] The accumulation of a detoxification enzyme in the cornea made sense due to its surface location, requiring protection from exposure to environmental insults. One culprit is UV light, because it generates toxic oxidative products in cells such as malondialdehyde and 4-hydroxy-nonenal. UV exposure is considered a risk factor for lens cataract as well, even though the lens is partially protected by being buried within the eye. ALDH3A1 is also prevalent in skin, tongue, and stomach, where it can serve detoxification functions. Thus, it was generally accepted that the high ALDH3A1 in mammalian corneal epithelial cells exists strictly to eliminate toxic aldehydes.

The importance of stress protection by corneal ALDH3A1 is supported by the SWR/J mutant mouse.[465] In this mouse strain, ALDH3A1 carries a four–amino acid replacement, making the enzyme unstable. The SWR/J mouse corneas have trace amounts of ALDH3A1 activity at best and become opaque more readily than wild type corneas when experimentally exposed to UV light. *In vitro* experiments have indicated that the enzymatic activity of ALDH3A1 protects corneal epithelial cells against oxidative damage induced by either UV radiation or the highly reactive product of lipid peroxidation, 4-hydroxy-nonenal,[466] and these results are consistent with experimental results using SWR/J mice.

However, the abundance of corneal ALDH3A1 remains an enigma. While it must play enzymatic roles in the cornea for detoxification and redox control, there appears to be simply too much of it in some species (rodents, cattle, pig) to serve a strictly catalytic function (see Piatigorsky[454,455] and Jester et al.[454a]). There are also reasonable questions that may be asked concerning why ALDH3A1 would be the enzyme of choice if detoxification were the only reason for its abundance in the cornea. For example, even though one of the main products of UV-induced oxidation, 4-hydroxy-nonenal, is a substrate for ALDH3A1, the other major toxic product, malondialdehyde, is not a favorable substrate for this enzyme. The idea that ALDH3A1 is used for detoxification of UV-derived aldehydes also does not mesh well with its abundance in the corneas of rodents (mice and rats), which are nocturnal creatures, and its low level (if present at all) in corneas of birds, which fly and are exposed to high levels of UV in the sky.[460] The rabbit also presents a taxon-specific enigma: Rabbit corneas accumulate ALDH1A1 rather than ALDH3A1.[467] Unlike ALDH3A1, ALDH1A1 can utilize malondialdehyde as a substrate, but why should this be necessary only in rabbits? There is an additional difference between rabbit and rodent corneas. Rabbits have higher levels of ALDH1A1 in the stromal keratocytes than in the epithelium, although the enzyme occurs there as well, along with surprisingly high amounts of lactate dehydrogenase (LDH), which does not accumulate in rodent corneas.

Revealing nonenzymatic functions of the overly abundant corneal ALDH3A1 is a challenge. Even elimination of the mouse *Aldh3a1* gene by homologous recombination gives no gross corneal phenotype.[468] Nonetheless, a number of possible nonenzymatic functions have been proposed for corneal ALDH3A1. One is that it physically absorbs much of the incident UV radiation, thereby protecting the corneal cells by its mass rather than its enzymatic activity.[462] The high relative absorption of UV (290–300 nm) light by the water-soluble component of bovine corneal extracts prompted Mitchell and Cenedella to call these proteins "absorbins."[469] There is also evidence for a chaperone-like function preventing aggregation of partially denatured proteins in the cornea by ALDH3A1.[470] Another possible nonenzymatic function for corneal ALDH3A1 is to bind NADP to create a filter for glare, as was suggested for LDH/ε-crystallin in the duck lens.[18] A fourth possibility derived from experiments using cultured cells is that ALDH3A1 can act as a negative regulator of the cell cycle.[471] Highly expressed ALDH3A1 in transfected cells retards growth. In addition, ALDH3A1 is present in the nucleus and cytoplasm, suggesting a connection between this enzyme, the cell cycle, and gene expression.

By analogy with lens crystallins, a logical function for corneal crystallins would be to homogenize discontinuities in refractive index to optimize transparency and minimize light scattering in the cytoplasm.[257,259] Creation of a gradient of refractive index by corneal crystallins, which is one of functions of lens crystallins, is a less likely requirement due to the structure and thinness of the cornea and its surface location where it interfaces with air of low refractive index (at least in terrestrial species). Initial evidence that corneal crystallins minimize light scattering was obtained in freeze-injured rabbit corneas.[467] Specific reductions in corneal crystallins (ALDH1A1 and transketolase) were accompanied by corneal haze attributed to light reflection by the keratocytes. More recently, Jester and colleagues[454a] linked corneal crystallins with cellular transparency and haze in cultured keratocytes. Crystallin expression patterns were phenotypically modulated by culture conditions and light scattering was assayed by *in vivo* confocal reflectance microscopy. In general, light scattering increased as crystallin concentration decreased in the cultured keratocytes. These experiments strongly support the idea that the abundant corneal crystallins participate in a gene-sharing process and contribute physically (nonenzymatically) to the optical properties of this transparent tissue.

OTHER CANDIDATE CORNEAL CRYSTALLINS: TRANSKETOLASE, ISOCITRATE DEHYDROGENASE, AND CYCLOPHILIN

Transketolase (TKT), a member of the pentose phosphate pathway, is another abundant enzyme in the cornea of mice[472] and other mammals (see Piatigorsky[454]). At a minimum, TKT represents 10 percent of the water-soluble proteins of mouse corneal epithelial cells.[468,473,474] This is higher than expected for a strictly catalytic role, although the amount is not as glaringly out of line for an enzyme as the 40–50 percent levels of corneal ALDH3A1 in this species. While more concentrated in the corneal epithelial cells, TKT is also found in appreciable amounts in other cells of the cornea.[475] Unlike the more tissue-restricted ALDH3A1 (cornea, skin, stomach, testes, tongue), TKT is ubiquitous due to its importance for energy metabolism (especially NADPH production) and for generating pentoses used in many biochemical pathways, including nucleic acid synthesis. As with corneal ALDH3A1 and lens enzyme-crystallins, the abundance of TKT in the cornea is taxon-specific, being high in mammals but very low in chicken and zebrafish.[476] Complete elimination of TKT by homologous recombination in mice has not been possible, because TKT-deficient mice die early in development.[473] The corneas of TKT-heterozygous mice appear normal. Again we wonder, why does a key metabolic enzyme like TKT exist at such different concentrations in structurally similar corneas of different species?

Cytosolic isocitrate dehydrogenase (ICDH) is another unexpectedly abundant, taxon-specific corneal enzyme. It represents approximately 13 percent of the water-soluble protein of the bovine cornea, and is 31 and 929 times higher in the cornea than in the liver and heart, respectively.[477] Because ICDH concentrates in the corneal epithelium, this enzyme represents an even higher proportion of water-soluble protein in the epithelium than the overall 13 percent with respect to the total bovine cornea. Human, mouse, rat, and rabbit corneas contain much lower quantities of ICDH, which is consistent with its use as an enzyme. ICDH is a mitochondrial enzyme that reduces isocitrate to α-ketoglutarate, the first oxidative conversion in the tricarboxylic acid cycle. Due to its central position in this metabolic pathway, it would not be expected to regulate the entry of carbon into the cycle. Like TKT, ICDH supplies NADPH, which link its activity to the redox state of the cell.[477a-d] It is unlikely that ICDH simply substitutes for TKT in the cornea for this function, because bovine cornea have significant amounts of TKT as well as ICDH.[474,476] Nonetheless, ICDH has been shown by transfection experiments to protect cells against UV irradiation and oxidative damage.[477a-d] ICDH is also involved in glutamate synthesis. The high ICDH levels in the cornea do not fit with the need for accelerated amino acid synthesis because glutamate dehydrogenase and α-ketoglutarate dehydrogenase activities, both downstream enzymes in the biosynthetic pathway for glutamate, are no higher in the cornea than in the bladder, heart, or liver.

The abundance of ICDH in the bovine cornea, especially in view of the high level of ALDH3A1, provokes the same question that I asked for rodent corneas: Why would such massive protection against oxidative insult be bestowed upon a grazing animal that would seem to have no greater exposure to UV light than do other animals? Detoxification, scavenging dangerous free radicals, and redox control via NADH may not be sufficient answers. According to the gene-sharing idea, ICDH may be performing any number of nonenzymatic functions in the bovine cornea. For example, yeast ICDH, like other enzymes that use NAD as a co-factor, binds the 5'-untranslated region of mitochondrial mRNAs.[478] The functional significance of this binding is not known, but it has been suggested[478] that it is part of a general regulatory circuit connecting mitochondrial biogenesis to function. Whatever the function(s) of the abundant ICDH in the bovine cornea, it has been called a taxon-specific corneal crystallin.[477]

Cyclophilin, also known as peptidyl-prolyl *cis-trans* isomerase, is another taxon-specific, abundant corneal protein that comprises approximately 10 percent of the water-soluble protein of the embryonic chicken cornea.[460] Whether this quantity of cyclophilin in the embryonic chicken cornea reflects a huge need for its enzyme activity, perhaps associated with the deposition of stromal collagen, or whether it plays another optical role remains to be determined.

C/L-GELSOLIN: A CORNEAL CRYSTALLIN IN ZEBRAFISH

Approximately 50 percent of the water-soluble protein in the corneal epithelium of the zebrafish *(Danio rerio)* is a gelsolin-like protein that contains six repetitive, characteristic amino acid sequence motifs.[479] This abundant corneal protein is also expressed in lower amounts in the lens and thus has been called C/L-gelsolin. ALDH and TKT are negligible in the zebrafish cornea. A similar gelsolin-like protein is also the major cytoplasmic corneal protein of the rosey barb *(Puntius conconius)* and tricolor shark *(Balantiocheibus melanopterus)*[479] as well as being the dominant protein in the corneas of the remarkable "four-eyed" fish *(Anableps anableps)*.[248] The four-eyed fish cruises at the surface of brackish water where its eye is bisected by the meniscus, resulting in the dorsal half being exposed to air (so the fish can presumably search for insects) and the ventral half being submerged (Figure 5.3). Thus, the mysterious taxon specificity of abundant corneal proteins extends to fish.

Gelsolin is one of several members of the gelsolin superfamily.[480–482] Although gelsolin is best known as a cytoskeletal regulator that severs actin filaments, caps their ends, and nucleates actin assembly, it also participates in a number of other functions, including gene expression, signal transduction,

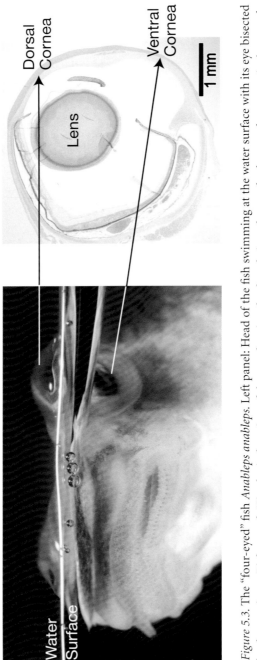

Figure 5.3. The "four-eyed" fish *Anableps anableps.* Left panel: Head of the fish swimming at the water surface with its eye bisected by the meniscus. Right panel: Histological section of the eye showing the dorsal air and ventral submerged corneas. A pigmented iris flap separates the two corneas. Light transmitted by either cornea passes through a different aspect of the same ovoid lens; images coming from the air are received by a specialized region in the ventral part of the retina, while images coming from the water are received by a region in the dorsal part of the retina. I am grateful to Dr. Shivalingappa Swamynathan (National Eye Institute, NIH, Bethesda, Maryland) for the preparation of this figure. Left panel photograph reproduced with permission from David Denning: BioMEDIA ASSOCIATES and adapted from Kanungo et al.[483]; right panel adapted from Swamynathan et al.[248]

and apoptosis (see Chapter 6). The gelsolin superfamily consists of at least seven members (gelsolin, adseverin, villin, capG, advillin, supervillin, and flightless I). Adseverin is the closest homologue of gelsolin and severs actin.

The corneal function(s) of the gelsolin-like protein is not known. One possibility, by analogy with lens crystallins, is that it performs an optical role.[483] It is interesting that C/L-gelsolin is the dominant protein in the cornea exposed to air as well as the one continually submerged in water in the four-eyed fish. Thus, high C/L-gelsolin does not seem incompatible with vision in air. In addition to being transparent, fish corneas must be resistant to the harsh environment because they do not have eyelids for protection. Cytochemical staining with phalloidin suggests that most of the actin of fish corneas is not fibrillar, presumably due to the high content of C/L-gelsolin, which can sever the actin filaments. What this means for cytoskeletal structure or "toughness" of the corneal epithelial cells is not known.

Perhaps corneal C/L-gelsolin accelerates wound healing in fish, which involves epithelial cell migration followed by stratification. Gelsolin is known to affect cell migration and wound healing. It seems reasonable that a corneal scrape in fish lacking eyelids needs to be healed rapidly to avoid infection, and the sculpin cornea heals at least seven times faster than the mouse cornea.[484] The riddle is not easily solved, however, because a similar gelsolin is not abundant in the sculpin cornea.[479] At present, overexpression of C/L-gelsolin in the corneal epithelial cells of many fish adds to the enigma of the taxon-specific abundant corneal proteins and suggests, but does not establish, a novel, tissue-specific role for this ubiquitous protein.

A SIGNALING ROLE FOR C/L-GELSOLIN

Zebrafish C/L-gelsolin has a signaling role in early development.[485] Before being concentrated in the cornea, C/L-gelsolin mRNA is expressed during early cleavage of zebrafish embryos. It subsequently becomes enriched in the spinal cord and rostral brain region, and finally the cornea. Reduction of embryonic expression of C/L-gelsolin by injecting fertilized eggs with a specific morpholino oligonucleotide ventralizes the embryos and severely impairs brain and eye development (Figure 5.4A and B). Conversely, microinjection of zebrafish C/L-gelsolin mRNA (or of human gelsolin protein) into fertilized eggs dorsalizes the zebrafish embryos and causes partial duplication of the body axis. Although the molecular role of C/L-gelsolin in dorsal-ventral signaling

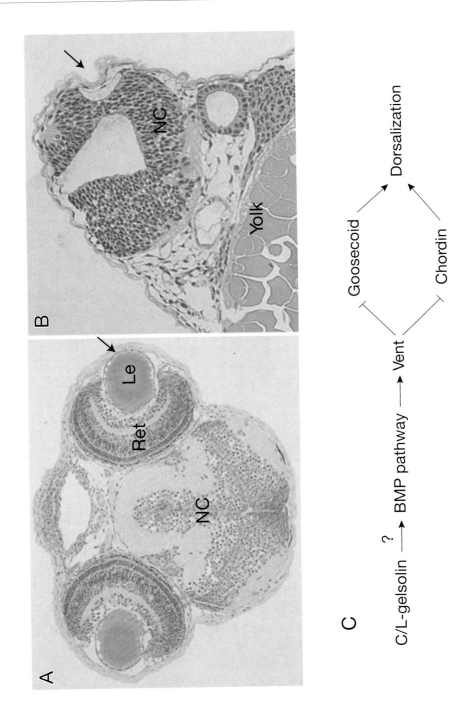

in zebrafish development is not known, it may involve the bone morphogenetic pathway (BMP)[486–488] (Figure 5.4C).

Ventralization due to inhibition of *C/L-gelsolin* gene expression can be rescued by coinjection of mRNA chordin (a dorsalizing factor), which operates via the BMP pathway.[485] Moreover, microinjected C/L-gelsolin mRNA increases chordin and goosecoid gene expression, both characteristics of dorsalization by the BMP pathway. Interference of C/L-gelsolin synthesis by microinjection of a specific morpholino oligonucleotide into fertilized zebrafish eggs reduces chordin and goosecoid mRNAs and increases Vent mRNA, a marker for ventralization, consistent with the BMP pathway.

Possibly C/L-gelsolin's signaling role during early development involves interaction with the actin cytoskeleton. Studies in *Drosophila* have implicated three actin-binding proteins that nucleate actin filament formation (Spir, Cappucino, and Profilin) in axis specification during development.[489–491] Zebrafish C/L-gelsolin may also have acquired secretory functions similar to adseverin (its closest homologue) by affecting exocytosis.[1138–1141] Further investigations are necessary to establish whether C/L-gelsolin has a different molecular function in the cornea and in its role as a regulator of dorsal-ventral patterning in the zebrafish embryo.

CORNEAL GENE EXPRESSION

Promoter fragments of mouse *Aldh3a1*[492] and *keratocan*[493,494] and rabbit *Aldh1a1*[495] show corneal-preferred activity in transgenic mice. Environmental stresses influence high gene expression in the corneal epithelium. In contrast, crystallin gene expression in the lens appears unaffected by environmental cues such as light.[496] The taxon-specific corneal crystallins do not accumulate until eye opening in the mouse, when the cornea is directly exposed to light and other oxidative stresses.[472] The environmental effect on corneal gene expression is limited, however. When mice are reared in total darkness, expres-

Figure 5.4. Possible role of C/L-gelsolin in zebrafish eye development. A. Histological section of the normal eyes of a twenty-four-hour-old zebrafish embryo. B. Histological section of the head and eye of a twenty-four-hour-old zebrafish embryo after microinjection of a morpholino oligonucleotide specifically interrupted expression of C/L-gelsolin. Note the underdeveloped eye. Panels A and B display minor adaptations from Kanungo et al.[483] C. Possible signaling pathway affected by C/L-gelsolin. Le, lens; Ret, retina; NC, notochord.

sion of the *Aldh3a1*[497] and *Tkt*[476] genes is delayed but not eliminated, suggesting that both environmental and developmentally programmed mechanisms are operative. Many stress-inducible *cis*-regulatory elements are associated with the abundantly expressed corneal genes, further implicating environmental induction.

Hypoxia-related pathways appear to be one of the mechanisms used for corneal-preferred gene expression of *Aldh*. In rats, a xenobiotic response element (XRE) 3 kb upstream of the *Aldh3a1* gene has been implicated in high expression of corneal genes.[498] XREs are DNA *cis*-control elements that are used for regulation of numerous genes encoding enzymes, including ALDHs that detoxify xenobiotics. The rat XRE that contributes to the control of the *Aldh3a1* gene binds the aryl hydrocarbon nuclear translocator (ARNT) as well as other factors.[498] In general, ARNT is present in most cells and regulates gene expression by forming heterodimers with other PAS-bHLH proteins, such as the hypoxia-inducible proteins (HIF) and the aryl hydrocarbon receptor (AhR).[499] PAS proteins contain a domain homologous to that in *PER* (a ciradian rhythm protein), *ARNT*, and *SIM* (a fly regulatory protein); bHLH denotes a structural motif of proteins comprising a basic *He-lix-Loop-Helix*. These proteins are sensors of environmental conditions, such as light, oxygen concentration, and various toxic ligands. Control of the rat *Aldh3a1* gene by an XRE is consistent with its marked sensitivity to hypoxia.[500,501]

Studies on the rabbit provided additional evidence for the use of a hypoxia and/or detoxification pathway for high, cornea-preferred gene expression.[495] Rabbits accumulate ALDH1A1 in the cornea rather than ALDH3A1.[467] The rabbit *Aldh1a1* gene contains three XREs within its 5 kb upstream sequence as well as an E-box *cis*-element 3 kb upstream of the transcription initiation site. E-boxes also bind bHLH transcription factors, including homodimers of bHLH sequences of ARNT.[502,503] Gene expression and DNA:protein binding experiments suggested that HIF-3α/ARNT heterodimers activate the *Aldh1a1* gene by interacting with these gene control elements.[495] Use of the hypoxic/detoxification pathway employing PAS-bHLH proteins for gene regulation in the cornea fits reasonably with the facts that the oxygen levels vary considerably in the avascular cornea when the eyelids are open or closed and the need to protect the exposed cornea from environmental insults. Zebrafish C/L-gelsolin expression in the cornea may also involve a stress response because adseverin, a close paralogue, is inducible by the toxin dioxin in mice[1142,1143] and marmosets.[1144] The abundant corneal crystallin genes, like the lens crystallin genes, are regulated by Pax6, among other factors.[504–506] At present, knowledge of the molecular basis for corneal gene expression is virtually in its infancy.

THE REFRACTON HYPOTHESIS: IMPLICATIONS FOR GENE SHARING

The unexpectedly high expression of certain enzymes and C/L-gelsolin in a taxon-specific manner in the cornea suggests that they are multifunctional and participate in a gene-sharing process as do the lens crystallins (Figure 5.5).[454,455] The putative specialized corneal functions of these abundant proteins are not established, however. Because together the transparent lens and cornea form the window of the eye that is responsible for refraction and image formation on the retina, and both may have implemented a gene-sharing process during evolution, I have proposed that the two optical structures be incorporated under a single umbrella: the "refracton."[455] That six of fifteen putative corneal crystallins from diverse species (rabbit, human, bovine, pig, mouse, chicken, fish, and squid) have known counterparts as lens crystallins[454a] supports the contention that gene sharing coevolved in the lens and cornea. The extent of refraction performed by the lens/cornea—the re-

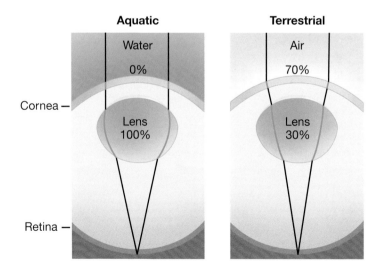

Figure 5.5. The "refracton" hypothesis. Left diagram: The lens is responsible for essentially all of the light refraction in aquatic eyes. Right diagram: The cornea is responsible for the greater share (~70 percent) of light refraction in terrestrial species.

fracton—depends on niche: Terrestrial species use mostly cornea and aquatic species almost entirely lens for focusing an image on the retinal photoreceptors. Amphibious species must vacillate back and forth. One might consider the cornea the anterior aspect and the lens the posterior aspect of the refracton.

The cornea uses environmental cues for high tissue-preferred gene expression, while the lens does not. Years of evolution appear to have internalized the mechanisms for high gene expression during lens development, resulting in noninductive events, despite the continued use of a shsp as a crystallin. During evolution, it is likely that high *crystallin* gene expression and gene sharing occurred first in the cornea and later in the lens; it makes little sense to evolve a transparent, refractive lens dependent upon crystallins without first having evolved a transparent cornea. Considering gene expression in the cornea and lens as an evolutionary continuum may facilitate the dissection of the mechanics that resulted in gene sharing in these transparent tissues. Detailed comparisons between high tissue-specific expression of the crystallins of lens and cornea should provide new clues into how novel protein functions arise by changes in the expression of their genes.

TAKE-HOME MESSAGE

Corneas of different species accumulate certain ubiquitous, water-soluble, cytoplasmic proteins (enzymes, gelsolin) in a taxon-specific manner. The reason for their extreme abundance in the cornea remains unexplained but suggests that they have additional specific corneal functions that differ from their functions in other tissues. By analogy with lens crystallins, the abundant corneal proteins have been considered corneal crystallins, thereby extending the gene-sharing concept from lens to cornea.

In contrast to the lens-preferred expression of crystallins, the corneal-preferred expression of the putative corneal crystallins appears to be influenced, at least partially, by inductive, environmental parameters such as light or oxygen levels. A challenging task for the future is to establish the range of mechanisms, including both internally (developmental) and externally (environmental) directed processes, for establishing gene sharing during evolution. The adjacent cornea and lens, which may be considered a single unit—the refracton—provide an opportunity to delineate an array of mechanisms used for gene sharing during the coevolution of two tissues for their combined optical role of image formation.

6

Gene Sharing As a Common Event: Many Multifunctional Proteins

Crystallins are one example of proteins having different, tissue-specific functions or serving multiple roles in the same tissue. Review of the scientific literature makes it clear that protein multifunctionality is more the rule than the exception for protein behavior. Most, if not all, polypeptides are probably involved in multiple tasks, exploiting both similar and different regions to perform various functions.

Enzymes exemplify specialization. They show stereo and substrate specificity as they direct metabolism by their catalytic activity at precise positions in biochemical pathways. We do not think of enzymes as structural building blocks. Nonetheless, specific enzymes accumulate in the lens (Chapter 4), where they are used to affect optical properties, and in the cornea (Chapter 5), where their relative abundance suggests that they have structural as well as metabolic roles. Opportunistic use of the different capacities of individual proteins creates an interlocking network of structural and metabolic functions.

This chapter reviews selected proteins that participate in multiple functions. Some enzyme-crystallins are included in this chapter along with other proteins in order to emphasize that their refractive role in the lens is only one of many functions outside of their typical catalytic roles. Other examples I present illustrate the stark functional differences that can exist for individual polypeptides, such as nucleotide metabolism and lipid secretion, redox control and eyespot morphogenesis, and circulating structural carrier protein and liver-detoxifying enzyme, to name a few.

GLYCOLYTIC ENZYMES AND THE VERSATILE HEXOKINASES

A number of enzymes used for glycolysis (some of which are also enzyme-crystallins) bind cytoskeletal proteins, especially actin and tubulin (also mi-

111

crotubules), as well as cellular membranes (see Masters,[507,508] Uyeda,[509] and Knull and Walsh[510] for discussion and references). There has been speculation that this binding micro compartmentalizes the energy-producing pathway within the cytoplasm to deliver products where they are needed, such as during the dynamic actions of the cytoskeleton during cell motility and shape changes. While simple energy distribution could explain cellular compartmentalization of glycolytic enzymes in some instances, the variable binding:activity relationships indicate additional complexity. Binding of some glycolytic enzymes (hexokinase, phosphofructokinase) increases enzyme activity, while binding of others (aldolase, GAPDH) decreases activity, linking compartmentalization with enzyme regulation. Moreover, hormones and substrates increase binding of some of the glycolytic pathway enzymes and decrease binding of others.

The gene-sharing concept suggests that different cellular locations of an enzyme may reflect functions beyond catalysis. Considerable amounts of a bound glycolytic enzyme to a membrane, suggesting a structural rather than metabolic role, suggests gene sharing. Although not established,[509] GAPDH may comprise as much as 5–7 percent of the human red cell membrane protein. Different nonenzyme roles may be due to physical properties critical for its enzyme function (such as substrate or cofactor binding site), or may be due to other properties of the protein that were selected for nonenzymatic functions. Gene sharing makes it necessary to examine each enzyme of a metabolic pathway separately.

Hexokinase is a glycolytic enzyme that has evolved multiple functions connected with membrane associations.[511] A hexokinase is even a hemoglobin receptor in a parasitic protozoan.[1145] This enzyme converts glucose to glucose-6-phosphate as a first step in the glycolytic pathway that generates ATP; glucose-6-phosphate can in turn be utilized in other biochemical pathways. There are four hexokinase isozymes with different intracellular distributions and gene expression patterns.[512] Type I hexokinase is ubiquitously expressed, enriched in brain that relies heavily on glycolysis, and binds mitochondria. Type II hexokinase is preferentially expressed in liver, skeletal muscle, mammary gland, and adipose tissue. It is generally distributed between a soluble and mitochondrially bound form. A hallmark of Type II hexokinase is that it is overexpressed in glycolytic tumor cells, where it is mostly bound to mitochondria.[513] Type III hexokinase is soluble and localized around the nuclear periphery.[514] Finally, Type IV hexokinase, also known as glucokinase, is expressed in many tissues, especially in liver and pancreas, and has several alternative forms.[515] The four hexokinase isozymes have different properties and biological roles. Type I favors catabolic functions related to using glucose-6-phosphate for energy production, while Types II and III favor anabolic functions, providing glucose-6-phosphate for glycogen synthesis or lipid synthesis via the pentose phosphate pathway. Type IV (glucokinase) is a glucose sensor that plays a key role in glucose metabolism.

Gene sharing appears to occur as a consequence of the binding of Types I and II hexokinase to the mitochondrial outer membrane protein VDAC (voltage-dependent anion channel, also known as porin).[511] Bound hexokinase can be very active because it is less sensitive to product inhibition by glucose-6-phosphate, it is protected against proteolysis, and it has preferential access to ATP within mitochondria. Linkage of hexokinase activity to mitochondrially derived ATP coordinates the enzyme reaction with ongoing aerobic metabolism. This serves to prevent the accumulation of lactic acid, a toxic product of glycolysis, in the cytoplasm. If little aerobic metabolism is taking place, the bound hexokinase uses extramitochondrial ATP as a source to phosphorylate glucose. This switch is governed by a conformational change in hexokinase.[516]

An entirely different function of Type II hexokinase is implicated by interaction with the VDAC protein of mitochondria. Hexokinase-bound VDAC cannot bind Bax, a pro-apoptotic protein that releases cytochrome c and other mitochondrial proteins.[517] Although some alternative models exist,[518] evidence indicates that mitochondrial proteins escape into the cytoplasm through VDAC, the pore regulating membrane potential. Assuming this mechanism is operative, Type II hexokinase promotes cell survival through a pathway that does not involve its enzymatic activity. This cell death protection might contribute to the malignancy of tumors that overexpress Type II hexokinase.[513] Finally, studies with Type I hexokinase showed that the bound enzyme closes the VDAC channel in the mitochondrial membrane.[519] Thus, interaction of hexokinase with mitochondrial VDAC affects both catalytic activity of the enzyme and VDAC association with Bax, leading to cytochrome c release. The latter phenomenon is a structural, nonenzymatic function of hexokinase that influences cell survival. Together, these interactions represent a balanced synchrony whereby distinct functions of hexokinase have fused to integrate metabolism, anion channel function, and cell survival.

Phosphoglucose isomerase (PGI) is another ubiquitous enzyme with a number of isoforms that participates in the glycolytic and the glucogenetic pathways. PGI catalyzes the reversible interconversion of glucose-6-phosphate to fructose-6-phosphate; however, the role of this enzyme is not restricted to these pathways.[520] PGI is also known as neuroleukin (NLK),[521] maturation factor (MF),[522] and autocrine motility factor (AMF).[523] NLK is present in numerous tissues and acts as a survival factor for embryonic spinal neurons, skeletal motor neurons, and sensory neurons. It can also induce immunoglobulin secretion by cultured human peripheral blood mononuclear cells. MF mediates the differentiation of human myeloid leukemic cells to terminal monocytic cells. AMF is secreted from some tumor cells or lectin-stimulated T cells. The secreted AMF binds a specific cell receptor (a 7 transmembrane 78 kDa glycoprotein) and stimulates cell motility, which depends on protein kinase C and tyrosine kinase activities.

CITRATE SYNTHASE: AN ENZYME AND A CYTOSKELETAL STRUCTURE

Citrate synthase of the protozoan *Tetrahymena thermophila* is a fascinating example of a metabolic enzyme that has a structural cytoskeletal role.[524] Citrate synthase catalyzes the stereospecific synthesis of citrate from acetyl coenzyme A and oxaloacetate during Krebs tricarboxylic acid cycle in the mitochondria. Citrate synthase also forms cytoplasmic 14 nm filaments that participate in oral morphogenesis during binary fission and during conjugation of the protozoan. In addition, these citrate synthase filaments control germ nuclear behavior (movements of macro and micronuclei and pronuclei) at fertilization. Studies with purified[525] and recombinant[526] protein have established that the 14 nm filaments have citrate synthase activity.

Posttranslational changes are associated with the functional role of citrate synthase: Phosphorylations promote enzyme activity and dephosphorylations lead to assembly of the 14 nm filaments.[527] The enzymatic and structural filament roles of citrate synthase have been joined in mitochondria with the discovery that citrate synthase activity decreases during filament formation, suggesting that monomer-polymer conversion modulates enzyme activity.[528] [524] This possibility fits with the accumulation of citrate synthase in rod-shaped structures in mitochondria during physiological states associated with low growth. The development of physiological systems dependent on the fusion of separate molecular functions of a protein is a beautiful aspect of gene sharing.

The complexity of gene sharing is seen not only with the multifunctional citrate synthase/14 nm cytoskeletal filaments but also with proteins that are associated with these filaments. Two examples are hsp60[529] and polypeptide elongation factor 1α (EF-1α).[524,530] hsp60, a mitochondrial chaperone known for its role in the folding of mitochondrial proteins, binds the citrate synthase/14 nm filaments of *Tetrahymena* and is believed to have a role in the formation of the oral apparatus of the protozoan. EF-1α catalyzes the GTP-dependent binding of aminoacyl-tRNA to ribosomes during protein synthesis. In addition, EF-1α binds the 14 nm filaments of citrate synthase and has a role in actin bundling and possibly in tubulin interactions (see Numata[524]). With respect to the latter, EF-1α is a calcium/calmodulin-binding protein associated with axonemal microtubules of ciliated protozoans and appears to be involved in ciliary motility.[530] The complexity of functions continues: EF-1α has been implicated in DNA replication in another protozoan, *Euplotes eurystomus* (see Numata[524]). The scenario is multifunctional proteins interacting with multifunctional proteins to govern a large network of functions.

LACTATE DEHYDROGENASE: AN ENZYME FOR ALL SEASONS

There are three separately encoded subunits (A, B, and C) of lactate dehydrogenase (LDH) isozymes in vertebrates.[531–533] LDH C gene expression is specialized for mammalian testes and spermatozoa.[534] Because LDH A and B are coexpressed in many tissues, they can interact to form five isozymes. LDH B/ε-crystallin is a homotetramer of the polypeptide that predominates in the heart and favors the conversion of lactate to pyruvate under aerobic conditions. LDH A_4 accumulates in skeletal muscle and favors the reverse reaction: conversion of pyruvate to lactate under anaerobic conditions. However, LDH functions are not limited to these enzyme reactions (Figure 6.1).

Quite separate from enzyme activity, LDH A_4 (also called M_4) binds AU-rich elements (AREs) that act as *cis*-stability determinants of RNA.[535] AREs are a class of sequences governing mRNA translation by controlling mRNA turnover, especially mRNAs encoding lymphokines, cytokines, and proto-oncogenes.[160,536,537] LDH binding to AREs has considerable specificity, with strong functional binding to granulocyte-macrophage colony stimulating factor (GM-CSF) mRNA. The binding occurs as a complex on active polysomes and increases mRNA turnover.

In the absence of RNA binding, LDH binds AUF1 (also called hnRNP D), itself a multifunctional protein, in a large cytoplasmic complex including Hsp 70, translation initiation factor eIF4G, and poly(A)-binding protein. LDH appears to remain associated with AUF1 during mRNA translation,[535] although this has been questioned.[538] The ARE sequences that bind the LDH complex also bind a complex (presumably lacking LDH) that includes another protein, HuR, which stabilizes the mRNA instead of promoting its degradation.[538] The mechanism by which the LDH complex increases RNA turnover is not established. One model proposes proteosomal targeting and degradation of the RNP complex[539] through an ubiquitin-dependent pathway.[540] Binding of LDH to the AU-rich RNA motifs is reduced by low concentrations of NAD+, indicating that this nonenzymatic function of the enzyme operates through the Rossman fold cofactor binding site in the protein. Thus, regulation of mRNA turnover by LDH is incompatible with its enzymatic activity, which also requires binding of the cofactor to the Rossman fold. Although overlapping regions of LDH are used for enzymatic activity and regu-

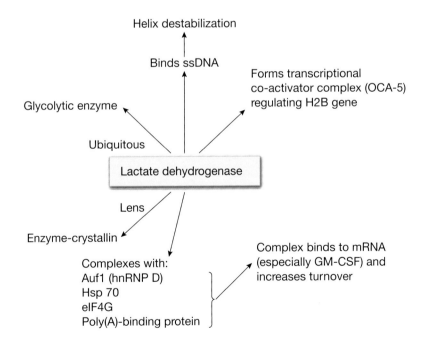

Figure 6.1. Multiple functions of lactate dehydrogenase.

lation of mRNA turnover, these are very different uses of the protein and examples of gene sharing.

AUF1 (hnRNP D) itself appears to be another elegant example of gene sharing. Because AUF1 interacts with LDH to regulate RNA turnover, it deserves further attention here. AUF1 has at least four isoforms (p37, p40, p42, and p45) derived by alternative RNA splicing.[541–543] Each of the four isoforms binds RNA, but they are associated with different levels of mRNA turnover; p37 and p42 are more powerful destabilizers than p42 and p45.[544] A heminregulated factor controls the function of the complex by sequestering AUF1.

As an aside, in addition to regulating mRNA turnover, the p45 AUF1 isoform forms a complex (called LR1) with nucleolin in B cells that binds G-rich DNA duplexes in a sequence-specific manner.[545] LR1 is believed to contribute to the coupled transcription and DNA recombination involved in immunoglobulin heavy chain switching.[546] The ability to influence transcription and DNA recombination is seen also in lymphomas, in which LR1 activates the *c-myc* promoter and contributes to deregulation of the gene by its translocation to the immunoglobulin locus.[547] In addition, LR1 contributes to B cell transformation by Epstein-Barr virus by activation of the *EBNA-1* gene that is essential for latent replication of the virus.[548] The story does not end here. AUF1 isoforms bind to G-G paired structures found in single-stranded DNA at the ends of telomeres and recruit telomerase to maintain the telomeres.[549] Multifunctionality is a way of life for these promiscuous proteins. It is not known to what extent or exactly how LDH may be involved in any of these processes.

LDH is one of several glycolytic enzymes (others include phosphoglycerate kinase, aldolase, and glyceralde-3-phosphate dehydrogenase) that have been observed in cell nuclei and shown to bind DNA.[550] The NADH binding site and the loop peptide of LDH A bind to single-stranded DNA *in vitro* and can act as a helix destabilizing protein.[551,552] Of particular interest is a remarkable report that LDH is one of seven enzymes comprising a multicomponent coactivator complex (called OCA-S) of the transcription factor Oct-1.[553,554] The redox-sensitive OCA-S complex is utilized to activate transcription of the histone 2B gene and functions only during S-phase of the cell cycle. This is described further in the section on glyceraldehyde-3-phosphate dehydrogenase, another member of the OCA-S transcription complex.

Still another point of interest is that LDH B_4 has been reported to represent as much as 5 percent of the total soluble protein of the mature oocyte in mice.[555,556] There is a sharp decrease in this enzyme after ovulation until blastulation; after implantation there is an increase in LDH A. This abundance of LDH suggests that it might have a nonenzymatic function in the oocyte. As with the abundant corneal proteins (see Chapter 5), the presence of higher levels of enzymes than we might expect for a strictly catalytic function is a frequent occurrence in tissues and may in many instances signify gene sharing.

REGULATION OF mRNA TRANSLATION BY ENZYME BINDING

LDH is not the only enzyme that binds and regulates mRNA translation.[557] This apparently nonenzymatic role often involves a (di)nucleotide-binding domain, as in the case with LDH. Thymidylate synthase, which converts deoxyuridine monophosphate to deoxyribosylthymine monophosphate, and dihydrofolate reductase, which converts dihydrofolate to tetrahydrofolate in the pathway to synthesis of thymidylate, purines, and amino acids, bind and repress their own mRNAs. Glutamate dehydrogenase binds cytochrome c oxidase mRNA, and yeast isocitrate dehydrogenase binds 5′ untranslated regions of yeast mitochondrial mRNAs. The antioxidant enzyme catalase, which converts hydrogen peroxide to water and oxygen, also binds the 3′ untranslated region of its own mRNA. All these cases appear to involve an RNA binding site that either overlaps with or resembles that binding to nucleotide substrates or cofactors.

Aconitase is one of the best-known cases of translational control by binding of an enzyme.[558–560] While aconitase activity converts citrate to isocitrate, the enzyme is also known as iron regulatory protein (IRP). Aconitase/IRP stimulates translation by interaction with a specific binding site in the 5′ untranslated sequence of ferritin mRNA in response to iron.

GLYDERALDEHYDE-3-PHOSPHATE DEHYDROGENASE: CONSTANT SURPRISES

Glyceraldehyde-3-phosphate dehydrogenase (GAPDH) is such a versatile, multifunctional protein (Figure 6.2) that I hardly know where to begin. The best-known role, of course, is as a NAD-dependent catalyst for the conversion of glyceraldehyde-3-phosphate to 1,3-bisphosphoglycerate during glycolysis. Apart from its role as a dehydrogenase and its abundance as an enzyme-crystallin (π-crystallin) and UV filter (via NADH binding) in diurnal geckos,[386] GAPDH also participates in membrane transport, membrane fusion and endocytosis, microtubule bundling, nuclear RNA export, ADP-ribosylation, protein phosphotransferase/kinase reactions, translational control of gene expression, DNA replication and repair, and transcriptional control.[553,561,562]

The multiple functions of GAPDH in eukaryotes extend to bacteria, where it is a surface protein in streptococci.[563] This streptococcal surface enzyme (SDH for surface dehydrogenase) has GAPDH activity and binds to fibronectin, lysozyme, myosin, actin, plasminogen, and a transferrin receptor (see Modun et al.[564] for review). Transferrin receptor binding by GAPDH is

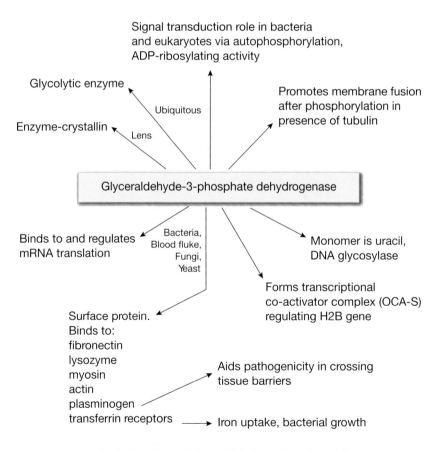

Figure 6.2. Multiple functions of glyceraldehyde-3-phosphate dehydrogenase.

important for iron uptake and consequently for bacterial growth. In addition to being a structural receptor protein, the surface GAPDH may enzymatically produce 1,2-diphosphoglycerate and 1,3-diphosphoglycerate, metabolites that are capable of mediating iron release from transferrin. Plasminogen binding by surface GAPDH may be directly related to the invasive pathogenicity of the bacteria in crossing tissue barriers because plasmin is capable of cleaving extracellular matrix proteins as well as dissolving blood clots. Its surface location and affinity for fibronectin and plasminogen, as well as the importance of lysozyme as a secretory component of the host's defense against infection by gram-positive bacteria, all suggest that SDH (GAPDH) plays a significant role in the infection process.

Bacterial SDH may participate in cell:cell communication by inducing phosphorylation of several human pharyngeal cell proteins, including nuclear core histone H3. This suggests that it engages a signal transduction pathway that activates histone H3-specific kinases.[565,566] A signal transduction role between bacteria and host is consistent with the ADP-ribosylating ability of GAPDH, which is known to be involved in transduction events.[567] GAPDH purified from rabbit skeletal muscle has been reported to be autophosphorylated in the presence of magnesium and ATP, further suggesting that this enzyme might be involved in signal transduction.[568] Cell surface GAPDH is also present in other microbes, including the blood fluke *Schistosoma mansonii*,[569] the fungus *Candida albicans*,[570] and the yeast *Saccharomyces cerevisiae*.[571] Temperature upshift and/or starvation of exponentially growing *C. albicans* causes an activation of cell wall GAPDH activity but not an increase in surface enzyme.[572]

Immunolocalization coupled with biochemical assays using cultured human fibroblasts indicates that intracellular perinuclear and nuclear regions concentrate GAPDH but show relatively little enzyme activity, consistent with major nonglycolytic functions for the enzyme.[573] Despite the large number of reports concerning the equally large number of putative functions of GAPDH, it has been pointed out that some of the findings are *in vitro* and, therefore, require further elaboration before their physiological relevance is proved.[574]

The mechanisms regulating the multiple functions of GAPDH are not known. The bulk of the GAPDH in the subcellular fractions of the cultured cells has been reported as tetrameric,[573] which is the oligomeric state for glycolytic activity and suggests that control of oligomerization is not a major mechanism governing different functions for the protein. Oligomerization, however, does affect the function of GAPDH. The monomeric GAPDH polypeptide is uracil DNA glycosylase, an enzyme utilized for DNA repair;[575,576] GAPDH dimers bound to microtubules (in the presence of ATP) have been implicated in the dissociation of microtubule bundles.[577]

At least some of the functions of GAPDH are regulated by protein modification. An example is its role in membrane fusion and secretion of vesicular clusters, which must pass through the endoplasmic reticulum to the Golgi apparatus. GAPDH is recruited to the vesicular clusters by small GTPase Rab2[578] and is subsequently phosphorylated by protein kinase Cι/λ.[579] The ability of GAPDH to promote membrane fusion is dependent upon its phosphorylation. Phosphorylated GAPDH is necessary to bind β-tubulin to the endoplasmic reticulum membrane. It has been proposed that a phosphorylation-induced conformational change allows membrane fusion to take place in the presence of tubulin. The basis for this idea is that tubulin specifically inhibits membrane fusion in the presence of nonphosphorylated GAPDH.[580]

Competition for NAD$^+$ binding in the Rossman fold, which is necessary for glycolytic activity, is another possible mechanism regulating different functions of GAPDH. For example, binding of GAPDH to AU-rich regions of cytokine mRNA during its presumed role in the regulation of mRNA translation (as for LDH) is inhibited by NAD$^+$, NADH, and ATP.[581] Similar inhibition occurs for GAPDH binding to single-stranded DNA,[582,583] which is responsible for a helix-destabilizing activity.[584] Still another mechanism for controlling the function of GAPDH is S-nitrosylation, which reduces enzymatic activity, modulates membrane interactions, and reversibly increases binding to actin (see[562,585–587]). Clearly, there are many opportunities for GAPDH to be shunted from one function to the next.

Let us return to the Oct-1 coactivator complex OCA-S, which is responsible for histone H2B promoter activity and includes LDH.[553] The OCA-S complex comprises seven polypeptides encompassing considerable enzymatic potential: p20 (nm23-H1) and p18 (nm23-H2) have been associated with nucleoside diphosphate kinase activity, p36 doublets have LDH and uracil-DNA glycosylase activities, and p38 has GAPDH activity. Nuclear p38/GAPDH interacts directly with the transcription factor Oct-1. p38/GAPDH is essential for H2B transcription *in vivo* and only occupies the H2B promoter when it is expressed during the S-phase of the cycle. Together, these experiments demonstrate convincingly that GAPDH is a specific transcriptional coactivator.

Although the transcriptional role of GAPDH differs from its enzymatic role in glycolysis, both its transcriptional coactivator and enzyme functions involve binding of the nicotinamide adenine diphosphate cofactor.[553] NAD+ enhances and NADH inhibits *in vitro* transcription via the ability of p38/ GAPDH to interact with Oct-1 and the H2B promoter. Potentially, then, p38/ GAPDH could link the OCA-S complex and specific gene expression to the redox state of the cell. This is not the first report linking intracellular redox to gene expression or employing GAPDH as a vehicle for such linkage. Yeast Sir2 is a gene-silencing protein with NAD$^+$-dependent histone deacetylase activity (see Denu[588]). Interestingly, Sir2-like enzymes generate the metabolite O-acetyl-ADP-ribose during the deacetylation reaction that may have a yet-to-be-discovered role as a second messenger. Another case involves the Clock:BMAL1 and NPAS2:BMAL1 heteromeric transcription factors that bind NAD cofactors and control gene expression during circadian light-dark cycles.[589,590] The reduced forms of NAD (NADH and NADPH) enhance while the oxidized forms inhibit DNA binding of the transcription factors, as is the situation with p38/GAPDH in the OCA-S complex.

Another protein connecting redox, transcription, and GAPDH is the transcriptional corepressor CtBP (carboxy-terminal binding protein).[591,592] While primarily known as a transcriptional corepressor, CtBP can act as a transcriptional coactivator and also has a role in the Golgi apparatus. CtBP was first discovered as a cellular phosphoprotein that represses oncogenic transformation by interacting with a carboxy-terminal binding domain of adenovirus 2 E1A protein.[593] It is now known to interact with many DNA-binding and other regulatory proteins. CtBP is homologous to 2-hydroxy acid dehydrogenases (see Turner and Crossley[591]) and has a NAD-regulated dehydrogenase activity.[594–597] It also has acyl transferase activity.[591] CtBP binding to its partners is modulated by NAD cofactors, with reduced NADH causing it to be a more effective repressor of promoter activity than oxidized NAD$^+$.[598] The relative importance of enzyme activity for the transcriptional function of CtBP has not been resolved. Enzyme activity *per se*, however, does not seem to be critical, inasmuch as mutagenesis of the active site does not prevent the repressive and biological functions of CtBP.[591,599]

The redox sensitivity and role of the NAD⁺/NADH ratio for controlling GAPDH as a transcriptional coactivator is a magnificent example of the integration of several completely different biological roles of a single protein. It also provides a mechanism for integrating cellular metabolism affected by environmental conditions with gene expression and cell growth.[1146] This aspect of gene sharing is discussed elegantly by McKnight,[554] who considers the linkage between metabolism and gene regulation "far more extensive, baroque and intricate than what can be expected from rational approaches evolved from classical studies going back to Jacob and Monod." To me, this is another way of expressing the pragmatic, unexpected interrelations and functional applications arising from gene sharing. McKnight ends his commentary by questioning whether GAPDH plays a surrogate role as a transcriptional coactivator in the OCA-S complex, such as refraction by the enzyme-crystallins in the lens, or whether its enzymatic and coactivator roles are integrated (which he favors). Whatever the interpretation, GAPDH is employing its physical properties for different tasks, all of which are adaptive, constrained by natural selection, and exemplify gene sharing.

ENOLASE: ANOTHER VERSATILE PROTEIN

Enolase is another glycolytic enzyme with an isozymic form (α-enolase) that is a taxon-specific crystallin (τ-crystallin). Like many of the other enzyme-crystallins, enolase has been associated with numerous functions unrelated to either enzyme activity or refraction (see Pancholi[600]) (Figure 6.3). For example, enolase is an actin-binding protein[508,510] and has been reported to be in centrosomes of HeLa cells, where it may have a microtubule-organizing role.[601] In yeast, α-enolase was identified as a heat shock protein induced at elevated temperatures (Iida and Yahara, 1985).[601a] Neuronal-specific enolase (γ-enolase) can act as a neuronal survival factor in the bovine central nervous system.[602] Experimentally, bovine γ-enolase has been shown to promote survival of primary cultured neurons from embryonic rats. The neurosurvival role of γ-enolase does not seem to be related to enzymatic activity because neither of its glycolytic reaction products, 2-phosphoglycerate and phosphoenolpyruvate, are effective neuroprotectors. Unexpectedly, neither ubiquitous

α-enolase nor muscle-specific β-enolase has neuronal survival activity, which suggests considerable specificity of γ-enolase for this function. While this observation needs confirmation, it is consistent with enolase functions beyond enzyme activity and lens refraction.

α-enolase is also a plasminogen-binding surface protein in microbial and eukaryotic cells.[600,603–606] Like GAPDH, enolase coats the surface of streptococci.[607–609] The plasminogen-binding ability of the surface enolase in the gram-positive streptococcal bacteria exceeds that of GAPDH. Because plasminogen is involved in the fibrinolytic process of the infected host, it may have a major role in the pathogenesis of streptococcal disease and in post-streptococcal autoimmune diseases.[607] Enolase is also a surface, plasminogen-binding protein in *Aeromonas hydrophila,* a bacterial pathogen that causes gastroenteritis.[610] It is up-regulated in the bacterium in a murine peritoneal culture model of gastroenteritis and is believed to contribute to pathogenesis.

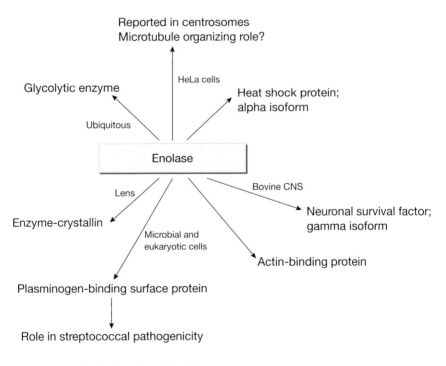

Figure 6.3. Multiple functions of enolase.

Finally, MBP-1, a nuclear protein comprising the C-terminal region of enolase that lacks enzyme activity, is derived from the human enolase gene by alternative RNA splicing (see Pancholi[600]). MBP-1 negatively regulates the c-myc-P2 promoter and thus is a tumor suppressor. While this is not gene sharing because it involves amino acid sequence changes by alternative RNA splicing (see Chapter 3), it does illustrate the versatility of sequences comprising the glycolytic enzyme.

BACTERIAL SURFACE ENZYMES

The surface location of metabolic enzymes in pathogenic bacteria is not restricted to GAPDH and enolase. Ornithine carbamoyltransferase, phosphoglycerate kinase, purine nucleoside phosphorylase, and glucose-6-phosphase isomerase have been identified on the surface of *Streptococcus agalactiae*.[609] Phosphoglycerate kinase has also been identified on the surface of *Schistosoma mansoni*,[611] and ornithine carbamoyltransferase is a possible cell surface protein in *Staphylococcus epidermidis*.[612] The surface location of these enzymes provides a potential host defense mechanism for infection. For example, antisera against ornithine carbamoyltransferase and phosphoglycerate kinase protect against infection in a neonatal murine model.[609] The roles of all these surface metabolic enzymes in bacteria are not established; nonetheless, it seems likely that these enzymes are not restricted to conventional metabolic activities.

XANTHINE OXIDOREDUCTASE: ENZYME AND ENVELOPE

Xanthine oxidoreductase (XOR) is a large (300 kDa) homodimeric housekeeping enzyme that catalyzes the oxidation of hypoxanthine to xanthine and then to uric acid in purine degradation. XOR exists *in vivo* more commonly as an NAD$^+$-dependent dehydrogenase, but it can be modified to an O$_2$-dependent oxidase depending on the oxidation state of its cysteine thiols.[613] While XOR is found in many tissues, large quantities are unexpectedly associated with milk fat droplets in many mammals (see Mather[614]). The NAD$^+$-dependent dehydrogenase form predominates in the mouse mammary gland; it is principally the O$_2$-dependent oxidase form of the enzyme that is found in the secreted milk.[615] There is little XOR in the mammary gland of virgin mice, but XOR begins to accumulate at midpregnancy and becomes prevalent during lactation.[616,617] Moreover, XOR becomes associated with adipophilin and butyrophilin in membranous material on the surface of lipid droplets.[618-620] Mather and Keenan[621] have proposed that XOR:butyrophilin:adipophilin interactions mediate the secretion of the lipid droplets at the apical plasma membrane of lactating mammary epithelial cells.

Convincing support for the structural role of XOR was obtained in mice that were made hemizygous for the *XOR* gene by homologous recombination.[622] The *XOR*-deficient mice died by six weeks of age; however, the *XOR*[+/−] female mice were healthy and fertile but could not maintain lactation due to premature involution of the mammary gland. Consequently, their pups starved by two weeks after birth. Direct electron microscopic evidence indicated that XOR is necessary to envelope milk fat droplets with the apical plasma membrane for secretion. Vorbach and colleagues[622] concluded that "XOR may have primarily a structural role, as membrane-associated protein, in milk fat droplet secretion" and then stated that "gene sharing of XOR in the lactating mammary gland, an evolutionarily young organ, provides a striking example of a gene acquiring multiple functions during its evolution."

THE THIOREDOXIN/RIBONUCLEOTIDE REDUCTASE SYSTEM AND THIOREDOXIN FAMILY MEMBERS: FROM REDOX TO MORPHOGENESIS

Thioredoxin, first purified from *Escherichia coli* as a hydrogen donor for ribonucleotide reductase in the synthesis of deoxyribonucleotides,[623] is a ubiquitous, relatively small (12 kDa) cellular redox regulating protein that works in conjunction with thioredoxin reductase.[624-626] All thioredoxins from archebacteria to man have a conserved active redox site comprising the amino acids cysteine-glycine-proline-cysteine. The oxidized sulfhydryl bridge (S-S) between these two cysteines is reduced to a dithiol (-SH HS-) by NADPH and the flavoprotein thioredoxin reductase. Catalytically active thioredoxin family members are found in the cytoplasm, associated with cellular surfaces, in nuclei and in extracellular regions, consistent with this metabolic regulator having several functions. Redoxins have an extensive role in redox regulation, influencing many physiological and cellular phenomena (see Hirota et al.[626]) (Figure 6.4). Redox regulation clearly has far-reaching and diverse effects on gene expression, signal transduction, cellular proliferation, and the immune response. Mechanisms of thioredoxin involvement in these processes include, but are not limited to, acting as an intracellular protein disul-

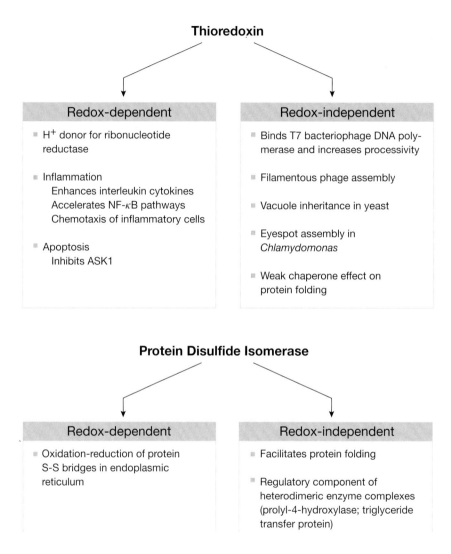

Figure 6.4. Redox-dependent and -independent functions for thioredoxin and protein disulfide isomerase.

fide oxidoreductase, scavenging reactive oxygen species, regulating the activity of transcription factors, and being a proinflammatory factor.

These diverse biological roles may be complex and dependent on redox-based catalysis involving thioredoxin. An example is the role of thioredoxin in inflammation. Thioredoxin is secreted (by an unknown mechanism because it lacks a signal peptide) by oxidatively stressed tissues and acts as a cytokine and chemokine-like factor (see Nakamura et al.[627]). For example, thioredoxin has a costimulatory role on numerous cells, enhancing the expression of the interleukin cytokines.[628] This involves acceleration of the transcription pathway utilizing NFκB, whose activity is known to be redox sensitive.[629] Secreted thioredoxin contributes further to inflammation by being a chemotactic agent for monocytes, polymorphonuclear leukocytes, T lymphocytes, and neutrophils.[630,631] The mechanism for chemotaxis appears to be linked to the catalytic function of thioredoxin because it operates in the nanomolar concentration range and mutants in the active site lose both redox-regulating and chemotactic activity.[630] Consequently, despite the spectrum of biological roles regulated by thioredoxin, these should not be attributed to gene sharing without more detailed information: They appear to be based on the same redox, catalytic property of the protein.

In addition to its major role in redox regulation, thioredoxin has entirely different functions that are in the domain of gene sharing. One of these, entangled with its participation in the redox regulation of apoptosis, is its ability to bind and inhibit apoptosis signal-regulating kinase (ASK) 1.[632] ASK1, which activates the c-Jun N-terminal kinase and p38 MAP kinase pathways leading to apoptosis, binds thioredoxin. This binding inactivates ASK1 and blocks apoptosis. Although thioredoxin binding to ASK1 does not regulate cellular redox, it connects redox with apoptosis because the interaction of these two proteins occurs only when thioredoxin is reduced. Oxidized and reduced thioredoxins have conformational differences that no doubt account for their difference in affinity for ASK1.[624,625] Thus, oxidative stress converts thioredoxin to its oxidized form, releases it from (or prevents its binding to) ASK1, and results in apoptosis of the cell. This is another elegant example of gene sharing whereby the role of thioredoxin in redox—in this case as a sensor—and its distinct ability to bind ASK1 fuse in its role governing apoptosis.

The versatility of thioredoxin extends further. E. coli thioredoxin stoichiometrically binds T7 phage gene 5 protein (DNA polymerase)[633] and enhances its DNA polymerase and double-stranded DNA exonuclease activities.[634] The C-terminal residue, histidine 704, of the T7 DNA polymerase is crucial for the polymerase activity and partitions the exonuclease and polymerase activities of the thioredoxin:T7 DNA polymerase complex.[635] Thioredoxin has no affinity for DNA so its stimulatory effect on T7 DNA polymerase activity depends entirely on its protein:protein interaction. This activation is due to increased processivity of the T7 phage polymerase—namely, the thioredoxin allows the enzyme to polymerize many more nucleotides upon a single-stranded DNA template without dissociation.[636] Thioredoxin acts as an accessory protein that stabilizes the bound polymerase complex to the primer, increasing its half-life from less than a second to approximately five minutes.[637] Complex formation requires reduced thioredoxin, although extensive DNA synthesis does not need further reductant, which is consistent with a stringent structural requirement for the protein:protein interaction but no significant dependence on catalysis for function of the complex.[638] Catalytically inactive thioredoxin, which lacks cysteines in the active site, has low affinity for T7 polymerase.[634] However, once the complex is formed, the catalytic properties of thioredoxin are not necessary for its ability to enhance polymerase or double-stranded DNA exonuclease activity of the T7 polymerase.

The thioredoxin binding domain (TBD) on the T7 polymerase has been identified in a stretch located in what is called the "thumb" of the protein.[639,640] Introduction of the homologous TBD from T3 DNA polymerase, which shows 96 percent identity in amino acid sequence with T7 DNA polymerase, into *Taq* DNA polymerase, which has only 25 percent amino acid sequence similarity with T3 DNA polymerase, leads to a thioredoxin requirement for high processivity of the Taq polymerase.[640] Thus, the TBD of T3 DNA polymerase appears sufficient to confer the thioredoxin-dependent stimulation of T7 DNA polymerase activity.

Thioredoxin is also required for filamentous phage assembly and extrusion in an ATP-dependent manner.[641–643] Active site mutants showed that phage assembly does not depend on the catalytic activity of thioredoxin.[644] Protein cross-linking studies suggest that pI and pXI, which are host inner membrane proteins, associate and then form a preinitiation complex with pIV, a host outer membrane protein; subsequently this preinitiation complex interacts with two minor coat proteins (pVII and pIX) and thioredoxin to form the initiation complex leading to phage assembly and export.[645] Thus, as in the case of T7 DNA polymerase, the accessory role of thioredoxin on filamentous phage assembly and export depends on structure, not catalysis.

Another, less defined role of thioredoxin is its participation in the "early pregnancy factor" system.[646] This phenomenon, the earliest sign that human fertilization has taken place, is noted in the rosette inhibition assay. Rosettes are formed by a subpopulation of lymphocytes that bind red blood cells in the presence of a complement source. Pregnancy serum contains multiple components that increase the titer of a rosette inhibition assay that depends on antilymphocyte serum. Thioredoxin plays a central permissive role as an early pregnancy factor. As in the case of bacterial thioredoxin in T7 polymerase processivity and in filamentous phage assembly, mutagenesis tests show that human thioredoxin's participation in the early pregnancy factor system does not depend on the two cysteines in the active site for redox catalysis. However, human thioredoxin has another cysteine pair in the C-terminal region of the protein, which is not present in bacterial thioredoxin and is essential for its role in rosette formation. It is interesting that only the reduced form of human thioredoxin is active in the rosette inhibition assay. This suggests that thioredoxin could be a redox sensor whose function in the early pregnancy factor system is mediated through the C-terminal portion of the protein, notably cysteine 74. The use of an active catalytic site to modify the protein for another, noncatalytic role is an elegant implementation of gene sharing. Another example of such coupling is the use of GAPDH in the Oct-1 coactivator complex OCA-S, which is responsible for histone H2B transcription.[553]

A different, apparently nonenzymatic role for thioredoxin concerns organelle inheritance (see Warren and Wickner[647]). In particular, thioredoxin is involved in inheritance of the vacuole (lysozome) in yeast. In early S phase of division of *Saccharomyces cerevisiae*, the vacuole prepares for being distributed to the resulting progeny cells by projecting a stream of tubules and vesicles, termed the segregation structure, into the bud. A heterodimeric protein called LMA1 (for low-molecular-weight activity) that is required for proper vacuole inheritance has been isolated from the segregation structure. One of the subunits of LMA1 is thioredoxin;[648] the other is a cytosolic inhibitor of vacuolar proteinase B called I^B_2.[649] The active site of thioredoxin is not necessary for vacuolar inheritance because vacuole fusion is not stimulated when both NADPH and purified thioredoxin are added to an *in vitro* fusion reaction system dependent on LMA1 and, more significantly, the *vac* mutant phenotype (double deletion of both thioredoxin genes) is rescued by thioredoxin in which the active cysteines have been replaced by serine. It has also been shown that the role of LMA1 in vacuolar inheritance does not rely on the inhibition of proteinase B. Thus, both thioredoxin and I^B_2 subunits of LMA1 are double-duty proteins in yeast. Xu and colleagues[649] concluded that LMA1 is a novel vacuole-trafficking factor in yeast.

A fascinating role for a new thioredoxin family member has been discovered for eyespot assembly in the unicellular green alga *Chlamydomonas reinhardtii* (see Roberts et al.[650]). The eyespot of this unicellular alga is present equatorially with respect to two flagellae and comprises "rhodopsin-like photoreceptors" with signal transduction components in the plasma membrane associated with an underlying, orange-colored array of lipid granules filled with carotenoids. The pigmented granules, situated between the inner envelope and thylakoid membranes of the chloroplast, both reflect light back to the photoreceptors and prevent light that traverses the cell from reaching the photoreceptors. Phototransduction by low intensity light leads to an ion flux that results in positive phototaxis (swimming toward the light source) by making the flagellum closer to the eyespot beat more strongly. Phototransduction by high intensity light results in negative phototaxis (swimming away from the light source) because it has the opposite effect on flagellar beating. When *Chlamydomonas* divides, the eyespot disappears and reforms in the sibling alga. A mutant alga, called *eye2–1,* shows 100-fold reduction in phototaxis, orients imprecisely with respect to the direction of light, and has no pigmented chloroplasts associated with the eyespot area. The mutant allele was cloned and encodes a member of the thioredoxin protein family. It can rescue *eye2–1*. Moreover, mutation of the putative active site showed that its function in eyespot development does not depend on catalysis.

Another interesting function of thioredoxin that does not require catalytic sulfhydryls but nonetheless is associated with redox abilities involves protein refolding. *Escherichia coli* thioredoxin binds weakly to and promotes functional refolding of various denatured proteins.[651] These include urea-denatured citrate synthase and α-glucosidase, and guanidine hydrochloride-denatured bacterial galactose receptor (Mg1B). Chaperone function does not require active redox sites: Citrate synthase renaturation occurs equally with reduced or oxidized thioredoxin, and Mg1B lacks cysteine residues. Redox state can affect the chaperone effects associated with thioredoxin, however. The addition of thioredoxin/thioredoxin reductase/NADPH causes synergistic refolding of citrate synthase in the presence of DnaK/DnaJ/GrpE/ATP-dependent citrate synthase refolding. This has led to the suggestion "that the thioredoxin system functions like a redox-powered chaperone machine."[651]

Protein disulfide isomerase (PDI) is another multifunctional protein in the thioredoxin family (see Figure 6.4).[652–654] PDI accumulates in the endoplasmic reticulum of cells where it catalyzes the oxidation and reduction of disulfide bridges. Like thioredoxin, PDI also facilitates protein folding without any redox role. The ability of PDI to enhance refolding of GAPDH[655] and rhodanese[656] are two compelling examples illustrating that the chaperone function of PDI can act independently of its catalytic properties. First, neither protein contains disulfide bridges and second, the active redox site of PDI can be destroyed by alkylation without loss of its chaperone ability.[657]

The accumulation of PDI in the endoplasmic reticulum and its propensity to associate with a multitude of proteins has also led to its being a regulatory component of several heterodimeric enzyme complexes (see Noiva,[653] Ferrari and Soling,[658] and Turano et al.[658a]). Two examples include prolyl-4-hydroxylase[659,660] and the microsomal triglyceride transfer protein[661] (see Berriot-Varoqueaux et al.[662]). PDI is the β subunit in the $\alpha_2\beta_2$ tetrameric prolyl-4-hydroxylase that catalyzes the formation of 4-hydroxyproline in procollagen pro-α-chain prolyl hydroxylation in the rough endoplasmic reticulum, and PDI is the β subunit in the $\alpha\beta$ dimeric microsomal triglyceride transfer protein that accelerates the transport of triglyceride, cholesteryl ester, and phospholipids across membranes. PDI activity decreases substantially when it acts as the β subunit of these enzymes, and it appears that the interaction of the subunits, which is critical for enzyme activity, is independent of the active-site cysteines. It is possible, however, that structural and catalytic properties of PDI combine in prolyl-4-hydroxylase, because thioltransferases have dehydroascorbate reductase activity[663] that could generate one of the cofactors for prolyl-4-hydroxylase.[658]

These multiple roles of thioredoxin proteins, both dependent and independent of their catalytic redox function, are just samplings of the different ways in which these versatile proteins have been exploited. There are more family members and their appearance in different locations strongly suggests additional functions, such as participation in RNA splicing, cell adhesion, and calcium-binding processes.[653,654,658] Many of these cases need additional exploration at the biochemical and gene levels to establish which pertain to gene sharing as defined in this book—having one protein encoded by a single gene performing two or more distinct functions utilizing different properties of the protein—and which may represent specializations of sibling proteins after gene duplication. The numerous uses of thioredoxin and its family members represent fertile ground for examples of gene sharing.

SERUM ALBUMIN: TRANSPORT PROTEIN, ENZYMATIC VASODILATOR, AND DETOXIFIER

Serum albumins are abundant structural proteins secreted from the liver and are commonly known for transporting fatty acids, binding toxic metabolites and contributing to osmotic balance. Serum albumins also have a number of catalytic activities.[664] Using its hydrophobic core as an active site, serum albumin catalyzes the oxidation of nitric oxide and formation of S-nitrothiols.[665] Because S-nitrothiols are relatively stable, this reaction preserves the bioactive, short-lived nitric oxide. Albumin-mediated nitrosylation has a vasodilatory effect that depends on the concentration of low-molecular-weight thiols circulating in the blood stream. Serum albumin also binds and breaks down arachidonic acid metabolites in a complex two-step reaction.[666] A third enzymatic ability of serum albumin involves the detoxification of carbamate insecticides by carboxylesterase activity.[667] Sogorb and colleagues[667] note not only the toxicological and ecotoxicological implications of this enzymatic activity but also the liver's role as the most important organ for detoxifying carbamates via hydrolysis by albumins. In addition, they also point to the medical importance of carbamate detoxification with respect to neurodegenerative diseases.

The Kemp elimination reaction is another enzymatic potential of serum albumins.[668,669] This reaction involves a base-catalyzed removal of a proton from carbon. It occurs in the hydrophobic pocket of albumin, where a lysine side chain is believed to act as a catalytic base. Interestingly, the serum albumin active site is structurally similar to that in antibody 38C2 against a reactive diketone hapten. The structural sites for the Kemp elimination in antibody 38C2 and albumin were, of course, derived convergently during evolution. Unlike the reactions involving S-nitrosylation, arachidonic metabolism, or carbamate detoxification, the Kemp elimination reaction of serum albumin lacks any known physiological significance. The reaction is efficiently performed *in vitro* by albumins of all species examined, but it proceeds with "accidental specificity." That antibodies, structural proteins, or enzymes have inherent and overlapping promiscuous activities adds to the redundant qualities of proteins; it is also consistent with the idea that each protein has many biochemical abilities—which is the heart of the gene-sharing concept.

The Kemp elimination reaction is sensitive to medium effects.[668,670] In contrast to effects from bulk solvents, medium effects result in specific, localized enzyme-like effects and create localized microenvironments that can accelerate reactions tremendously (10^5 fold). It would seem that medium effects have the potential to activate a variety of inherent biochemical activities of proteins placed under different conditions, thereby providing a mechanism for gene sharing and multifunctionality. Some of the accelerated reactions would no doubt be useful and would eventually be selected as part of the protein's functional repertoire. Hollfelder and colleagues[670] (p. 5866) point out the evolutionary implications of medium effects and state as follows: "Supported by the accidental identification of active sites on the surfaces of noncatalytic proteins and the promiscuous activities found in many enzymes, our findings suggest that the interfaces of protein surfaces and their hydrophobic cores provide a microenvironment that is intrinsically active and may serve as a basis for further evolutionary improvements to give proficient and selective enzymes."

Serum albumin may have additional structural properties that have been exploited to biological advantage in the extracellular stroma of the cornea, where it is a major molecular component (~13 percent of the water-soluble protein).[474] One possibility is that the high albumin concentration contributes to the optical properties of the cornea. Perhaps the serum albumin minimizes the refractive index discontinuities of the transparent cornea, as do crystallins in the lens,[454] or acts as a UV light filter, as has been proposed for the abundant corneal ALDH3A1[462] and other abundant water-soluble proteins.[469] Other possibilities for serum albumin functions in the cornea include protection against oxidation[671] or against self-association of partially degraded proteins by a chaperone role.[672] On the other hand, serum albumin may have the more conventional role of a fatty acid carrier in the cornea. Indeed, extracellular serum albumin in the aqueous and vitreous humors of the eye carry fatty acids into the lens cells that are used for lipid synthesis.[673,674] Whatever its natural function in the cornea, albumin may be exploited as a drug carrier to treat corneal disorders; such exploitation would give this protein still another function, though an engineered one.[474]

GELSOLIN: ROLES IN CYTOSKELETAL STRUCTURE, GENE EXPRESSION, CELL DEATH, AND SIGNAL TRANSDUCTION

Gelsolin is a conserved protein best known for its role in regulating cytoskeletal structure[480–482,675,676] (Figure 6.5). Gelsolin severs actin filaments, caps filament ends, and nucleates actin assembly. By regulating the actin cytoskelelton, gelsolin influences cell shape, cell motility, and, possibly, wound healing. Gelsolin, however, has additional functions (see Archer et al.[677]). It has been implicated in the apoptotic pathway by being a substrate for caspase 3[678,679] and by binding and closing the voltage-dependent anion channel of mitochondria in a calcium-dependent manner.[680] Another function of gelsolin involves binding the androgen receptor in the cytoplasm, translocation of the complex into the nucleus, and transcriptional activation of genes: Thus gelsolin plays a role in gene expression.[681] Gelsolin's function in gene expression suggests that it has a role in the known progression of tumors from an androgen-dependent to an androgen-independent state. The switch in androgen sensitivity is associated with an increase in gelsolin expression.

Two other actin-binding proteins involved with cytoskeletal structure, supervillin (a member of the gelsolin protein family)[682] and filamin,[683] also act as transactivating coregulators by associating with the androgen receptor. Filamin, like gelsolin, has been implicated with movement of the androgen receptor from cytoplasm to the nucleus. Filamin binding releases the androgen receptor from heat shock protein 90. Once filamin binds the androgen receptor, it may assume a chaperone role of protecting the androgen receptor as it is trafficked across the cytoplasm. This transient chaperone act itself is an elegant example of gene sharing.

Gelsolin is a member of an extensive actin-binding protein family, many of which act as transcriptional coactivators.[677] Thus, transcriptional regulation seems to be a significant function for this protein family.

Analyses of mice lacking the *gelsolin* gene have shown that gelsolin participates in signal transduction. Although the gelsolin-deficient mice display only minor phenotypes of hemostatic, inflammatory, and fibroblast responses,[684]

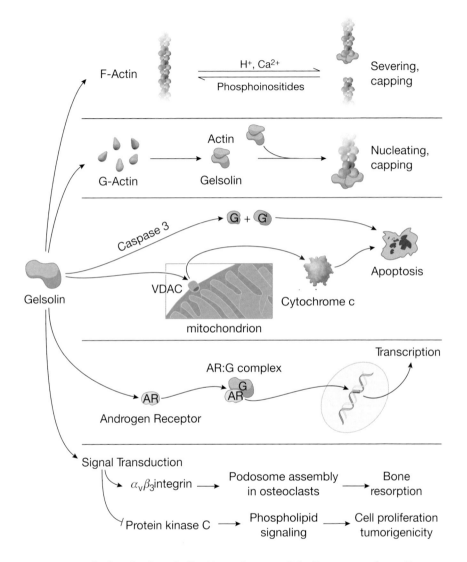

Figure 6.5. Multiple roles for gelsolin. Top to bottom: Gelsolin severs and caps F-actin filaments, nucleates F-actin filament formation from G-actin, affects apoptosis by being a substrate for caspase 3 and by affecting mitochondrial release of cytochrome c, binds androgen receptor, translocates to the nucleus and promotes gene expression, and activates and represses signal transduction pathways.

dermal fibroblasts lacking gelsolin have abnormal filopia (membrane ruffles) in response to serum or epidermal growth factor and reduced motility.[685] Because these growth agents signal through the protein rac (a member of the rho family of GTPases), gelsolin has been identified as essential for rac-mediated actin dynamics. A gelsolin role for regulation of rac signaling was also noted in reduced $\alpha_2\beta_1$ integrin-mediated collagen phagocytosis and calcium mobilization by gelsolin-free fibroblasts.[686] In this case, the signaling role of gelsolin may be connected to its ability to remodel actin filaments and permit collagen-induced calcium entry; the calcium in turn activates rac, which enhances collagen binding to $\alpha_2\beta_1$ integrin. A third signaling role for gelsolin is in osteoclasts: Gelsolin deficiency blocks podosome assembly and $\alpha_v\beta_3$ integrin-stimulated signaling related to motility in osteoclasts.[687] Podosomes are osteoclast-specific cell adhesions in motile cells that are essential for cytoskeletal organization and bone resorption. Mice lacking gelsolin have thicker, stronger bones than normal mice. Finally, gelsolin can suppress tumorigenicity by inhibiting protein kinase C, an activator of cell growth through the phospholipids signal transduction pathway, in a lung cancer cell line (PC10).[688] This growth-suppressing role may make gelsolin an important medical target.

In zebrafish, C/L-gelsolin may participate in the BMP dorsal ventral signaling pathway during development (see Chapter 5).[485] One possibility is that zebrafish C/L-gelsolin affects signaling via exocytosis in some fashion in view of its close homology with adseverin.[1138–1141]

Gelsolin exists extracellularly as well as intracellularly. A secreted form of gelsolin, derived by alternative RNA splicing, is present in the plasma and clears actin fibrils that could hinder blood flow from dying cells.[689,690] This function is clearly an extracellular role related to the ability of an isoform of gelsolin to regulate actin assembly within the cell. A third gelsolin isoform, gelsolin 3, arises by alternative splicing in rats.[691] Gelsolin 3 contains a unique N-terminal amino acid sequence and is expressed primarily in oligodendrocytes in the central nervous system, where its function is not known.

CYTOCHROME C: ROLES IN ELECTRON TRANSPORT, CELL DEATH, AND LIGHT FILTRATION

Cytochrome c (cyt c) is another fascinating multifunctional protein[692] with a role in respiration and apoptosis (Figure 6.6). It is a nuclear encoded protein synthesized in the cytoplasm as an unfolded apoprotein (apo-cyt c) that is directed into the mitochondria by an internal localization signal. A reduced heme group is attached covalently to apo-cyt c by heme lyase in the intermembrane space of mitochondria, converting it to globular holo-cyt c, which acts as an electron carrier during respiration. This respiratory function is dependent on its redox activity—namely, transferring an electron from cytochrome c reductase to cytochrome oxidase. Cyt c is a highly conserved protein

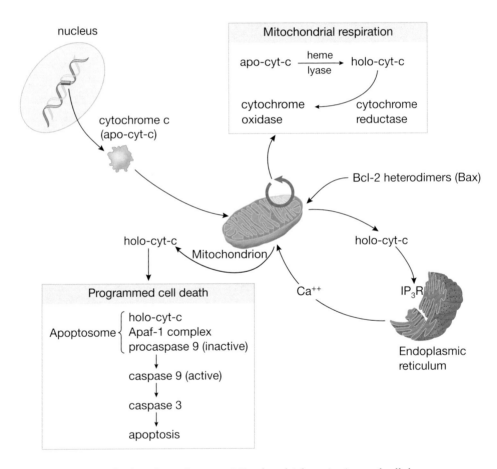

Figure 6.6. Dual roles of cytochrome c: Mitochondrial respiration and cellular apoptosis.

throughout evolution—a characteristic that is consistent with its universal role in electron transport and respiration.

In addition to its respiratory role in the mitochondria, cyt c plays a pivotal role in programmed cell death (apoptosis).[518,692] This is a complex process initiated by release of cyt c, along with other proteins, from the mitochondria.[693] In brief, various heterodimers (some pro-apoptotic, others antiapoptotic) from the Bcl-2 family of proteins regulate the release of holo-cyt c from the mitochondria. Only holo-cyt c with its attached heme group is competent to carry out its apoptotic signaling function. The precise mechanisms of release are not established; but the VDAC channel, pores formed by Bcl-2 proteins, and interactions with phospholipid metabolism in the mitochondria have all been implicated. Once cyt c is released, it can bind specifically with inositol 1,4,5-triphosphate receptors (IP$_3$R) on the endoplasmic reticulum in proximity to the mitochondria, causing the release of calcium.[694] The flood of calcium enters the adjacent mitochondria and this provokes further release of cyt c. The cytoplasmic cyt c then interacts with apoptosis protease-activating factor 1 (Apaf-1); this initial complex grows by addition of more Apaf-1 proteins and procaspase-9 to form the apoptosome. (Caspases are cysteine proteases.) Procaspase-9 is converted to the active caspase 9, which initiates a caspase cascade. As a result, the activated caspase 3, an executioner enzyme that cleaves many intracellular targets at aspartate residues, promotes apoptotic programmed cell death.[695]

The apoptosis role of cyt c is conserved throughout evolution except, unexpectedly, in yeast.[696] The dual roles of cyt c in mitochondrial respiration and in cytoplasmic apoptosis are physically distinct not only in cellular location but also in biochemical mechanism. Cytochrome c proteins in which copper and zinc have been substituted for iron (Fe) have no redox activity and thus cannot function as electron transporters; nevertheless, these modified proteins retain 50 percent of their ability to induce apoptotic changes in a cell-free system.[696] Moreover, different regions of cyt c are employed for interacting with proteins to carry out its electron transport and apoptosis functions.[697] Electron transport is carried out by interactions using the "front face" of cyt c, while apoptosis is carried out by interactions using the front and an opposite surface of cyt c.

Taken together, cytochrome c is both a giver of life in its role in respiration and an executioner in its role in apoptosis. These two contrasting functions use different properties of the same protein and exemplify the power of gene sharing.

Cyt c appears to have also acquired a unique optical role in some fish. Vertebrate retinas have a mitochondria-rich region in the retina called the ellipsoid. The ellipsoid is located in the outer part of the inner segment of the cone cells containing the opsins responsible for color vision.[189] In some birds and reptiles the ellipsoid contains oil droplets that protect the photopigments (opsins) from being damaged by shorter wavelengths of light. In some fish, however, the ellipsoid region has a more structured appearance reminiscent of mitochondria and a spectral absorbance similar to reduced cyt c.[698,699] In view of this difference with the oil droplet–containing ellipsoids of bird and reptile retinas, the cyt c–containing retinal regions are called ellipsosomes. While it is not known whether ellipsosomes contain functional mitochondria, they are located within the path of light traveling to the outer segments of the photoreceptor cells and have a filtering effect like that of the colored oil droplets of birds and reptiles. They would both cut down the intensity of blue light and increase the contrast in the blue-violet region of the spectrum. Thus, optical filtration can be listed with electron transport and apoptosis as still another function of cyt c.

TAKE-HOME MESSAGE

Gene sharing is pervasive, and multifunctional proteins are more the rule than the exception. Perhaps all proteins perform many different functions by employing as many different mechanisms. The functional boundaries for each protein are as open as the structural boundaries of its gene, which include the dispersed regulatory regions. The versatility of proteins challenges us to discover the full range of a protein's function rather than attempt to fit its expected function from past experiences and present knowledge into interpretations of observed phenomena. This is especially important in evolution where orthologous proteins differ in cellular environment, gene expression, and amino acid sequence. The bewildering array of functions performed by each individual protein within and between species testifies to the endlessly rich pragmatism and imagination of Nature.

7

Gene Sharing During Gene Expression

The previous chapters give examples of individual proteins with multiple functions. This chapter portrays how a particular cellular process—gene expression—calls upon existing components to create the complex, functional web that is used to transmit information from DNA to protein. I chose gene expression for this purpose in part due to its central role in biology. In addition, there are other reasons that make gene expression of interest with respect to gene sharing. The combinatorial use of existing proteins, rather than the addition of new proteins, has been recognized as a major contributor to the increasing complexity of species during evolution.[130] Moreover, the protein-coding genes, whatever their number, are organized in a nonrandom fashion within the genome,[700,701,1147] and the chromosomes are nonrandomly positioned in the nucleus as well, as are subnuclear compartments.[702] While it is not yet known exactly how these structural constraints optimize gene expression, they point to the importance of organization as a governing principle for function. This chapter provides a snapshot view of gene expression as a complex organization of proteins that also perform other functions within the cell.

COMPLEXITY OF TRANSCRIPTION

The sheer complexity of transcription increases the likelihood of multifunctional proteins being required. Many proteins (hundreds) must interact at the gene level for proper transcription to occur (see Carlson[703] and Naar et al.[704]). Lemon and Tjian[705] entitled a review of eukaryotic gene transcription an "orchestrated response: a symphony of transcription factors for gene control." The core eukaryotic RNA polymerase II (Pol II) required for basal transcription of protein-encoding genes associates with conserved proteins called general transcription factors (GTFs), which include the accessory proteins TFII

141

A, B, D, E, F, and H. One of these, the TATA-binding protein (TBP), is sufficient for basal gene transcription in a cell-free system. However, binding of Pol II requires the TFIID complex, which includes TBP and TBP-associated proteins called TAFs. Divergent TFIID complexes with different combinations of proteins contribute to specific interactions with enhancers and promoters. TAFs also stabilize the transcription complex, which includes the multiprotein complex of Pol II and the GTFs, by extending contacts with DNA.

Proper gene expression patterns involve multisubunit coactivator and corepressor complexes.[130,704,706] Coactivating complexes comprise a diverse group of proteins that link sequence-specific DNA binding activators to the basal transcriptional machinery and remodel chromatin for transcription. These complexes include metabolic enzymes (for example, histone acetyltransferases activating and deacetylases repressing transcription). Chromatin remodeling is ATP-dependent, which further implicates cellular metabolism in transcriptional processes.

NUCLEAR RECEPTORS

Nuclear receptors (steroid and thyroid receptors) are integral members of modular coactivating complexes.[706] The yeast Mediator was the first such multisubunit complex identified.[707,708] Metazoans have numerous Mediator-like complexes that bridge distal activators (or repressors) and the core promoter initiating transcription. While Pol II and the GTFs are highly conserved throughout evolution, the members of the yeast Mediator and metazoan coactivator complexes are more diverse. Despite their different compositions, yeast, mouse, and human Mediator-like complexes show a similar two-domain structure (head and combined middle tail), suggesting common structural requirements.[709,710] Moreover, recent studies indicate that many coactivators have true counterparts in yeast, fungi, and higher metazoans.[711] We can only wonder to what extent a gene-sharing strategy resulted in different proteins adopting similar functions and structural constraints in order to maximize interplay between gene expression and responsiveness to a variety of environmental conditions among species.

Scientific reports suggest that the use of nuclear receptors in coactivator complexes may have evolved by a gene-sharing strategy. First, it is striking that nuclear receptors vary greatly in number in different species.[712] Nematodes have at least 270 genes that are predicted to encode nuclear receptors, while fruit flies have 21 and humans 50—a variation consistent with the idea that these versatile proteins fulfill different needs in different species. One example is TR3, an orphan (no ligand known) thyroid hormone receptor belonging to the nuclear receptor family of transcription factors.[713] In addition to regulating the transcription of target genes in the nucleus, TR3 translocates from the nucleus to the mitochondria, where it releases cytochrome c to initiate apoptosis. The DNA-binding domain of TR3 is not necessary for cytochrome c release from the mitochondria. Thus, the transcriptional and apoptotic functions of TR3 are physically separated in the protein. Another putative example of gene sharing for a nuclear receptor involves a progesterone receptor that mediates oocyte maturation in *Xenopus* in the absence of a transcriptional role.[714] This cloned receptor functions as a progesterone-regulated transcription factor in heterologous kidney cells (COS cells); but it is strictly cytoplasmic in the *Xenopus* oocyte, where it accelerates oocyte maturation. Still another intriguing case involves an estrogen receptor.[715] Estrogen receptor α (ERα) binds in the presence of the estrogen ligand to the p85α regulatory subunit of phosphatidylinositol-3-OH kinase to provide vascular protective effects via activation of protein kinase B/Akt and endothelial nitric oxide synthase. This function of ERα is independent of transcription and consistent with gene sharing.

METABOLIC ENZYMES AND GENE EXPRESSION

The interplay between cellular metabolism and gene expression is a rapidly growing issue. Metabolic functions are connected through networks of transcriptional regulators[99,716] and metabolic enzymes are directly associated with transcription.[717] The roles of many of these metabolic enzymes in gene expression are not fully understood and may not depend entirely on catalytic activity. Certainly there is ample evidence that chromatin remodeling and transcription factor modification associated with gene expression are governed by active metabolic enzymes (i.e., histone acetyl and deacetyl transferases, methyltransferases, kinases, phosphatases, and ubiquitination enzymes) (see Shi[717]). However, questions have been raised concerning the full functional significance of some enzymes linked with transcription.[718] An example is the methylation of histone 3 at lysine 4 (H3meK4) by the enzyme Set1 in yeast, which may provide a genetic "memory" of recent transcription rather than activate gene expression. In addition, MLL, a human homologue of Set1, activates selected *Hox* genes by methylation; but MLLAF9, a fusion protein generated from this gene in leukemia cells, still activates *Hox* genes but does not methylate the protein H3K4. While this separation of methylation from gene activation does not eliminate the H3meK4 requirement for gene activation under normal circumstances, it suggests that the transcriptional role of this methylase may extend beyond its enzymatic activity *per se*.

A number of metabolic enzymes in plants and animals associated with gene expression appear multifunctional, although their mechanisms of action in transcription remain unknown. As a group these potential transcription regulators have been called metabolism-related transcription factors (MTFs).[717] Many MTFs bind metabolic coenzymes (FAD, NAD, NADP, HEME, acetyl-CoA, ATP, S-adenosyl methionine, and others) that fluctuate in cellular concentration and are indicators of metabolic conditions and stress, such as redox state. It has been proposed that the association of MTFs provides a link between metabolism, reflected by coenzyme levels, and gene expression, allowing cells to respond rapidly at the gene level to satisfy changing cellular requirements.[554,717,1146] The OCA-S coactivator complex described in Chapter 6 is an example. The OCA-S complex regulating histone H2B gene expression contains seven components, four of which are metabolic enzymes (GAPDH, LDH, uracil-DNA glycosylase, and nucleoside diphosphate kinase).[553,717] Thymine-DNA glycosylase and purine-binding transcription factor (Puf), relatives of uracil-DNA glycosylase and nucleoside diphosphate kinase, respectively, also have transcriptional roles in other systems. In addition, the CtBP corepressor complex, also discussed in Chapter 6, contains MTFs, including GAPDH activity. The combined presence of GAPDH and LDH in the OCA-S coactivator complex raises the possibility that NADH produced by the former is oxidized to NAD$^+$ by the latter, as occurs in the glycolytic pathway.[717] Thus, the physical associations of these cytoplasmic enzymes with active nuclear genes does not by itself define multiple biochemical roles, although it would seem that their intimate integration into different pathways of cellular metabolism and gene expression employ different combinations of properties of these proteins.

The multifunctionality of hexokinase has led to its being called a "jack of all trades."[719] This metabolic enzyme switches between many cellular compartments, including cell and endoplasmic reticulum membranes, mitochondria and plastids (in plants), actin filaments, the cytosol, and, importantly for this discussion, regulatory DNA complexes in the nucleus. Hexokinase isoforms dimerize with each other and interact with many other proteins. Hexokinase acts as a catalyst for ATP-dependent phosphorylation of glucose, as a glucose sensor, and as a signal transduction protein. Yeast hexokinase 2 has an internal nuclear localization signal that is necessary for the enzyme to repress signaling to the SUC2 gene.[720] Yeast SUC2 encodes invertase, an enzyme that hydrolyzes sucrose and raffinose and is required for utilization of alternative carbon sources when glucose is limiting. Consequently, glucose suppresses expression of the SUC2 gene. It is in this complex pathway that yeast hexokinase 2 participates in protein complexes that bind the SUC2 promoter and directly regulate the SUC2 gene, a transcriptional function that seems beyond its enzymatic activity as a hexokinase.

Hexokinase (HXK1) function was studied in the *glucose insensitive 2 (gin2)* mutant of the plant *Arabidopsis.*[721] Plants use hexokinase as a glucose sensor to control development and growth by coordinating signaling networks connected to nutrients, light, and hormones (auxin and cytokinin). A gene encoding a catalytically inactive enzyme rescued glucose signaling in the complex *gin2* phenotype. The rescue included multiple events: repression of two promoters of photosynthesis genes, auxin-mediated cell proliferation, and root formation. Moore and colleagues[721] concluded (p. 336) that "*Arabidopsis* HXK1 has a unique function in a broad spectrum of glucose responses including gene expression, cell proliferation, root and inflorescence growth, leaf expansion and senescence, and reproduction" and that these functions "can be mediated at least partially by catalytically inactive mutants, demonstrating the dual functions of HXK1 in glucose signaling and metabolism."

Hexokinase is not alone among cytosolic metabolic enzymes that appear to have noncatalytic roles in gene expression. Arg5,6 is a yeast mitochondrial enzyme that may directly control gene expression at the DNA level.[722] Arg5,6 is cleaved to produce Arg5 (N-acetyl-gamma-glutamyl-phosphate reductase) and Arg6 (acetylglutamate kinase), both enzymes participating in the pathway for ornithine (a precursor to arginine) biosynthesis. Arg5,6 binds preferentially a number of mitochondrial and nuclear genes experimentally *in vitro* as well as in the cell. Transcript levels of the genes that bind Arg5,6 are reduced in yeast carrying an Arg5,6 deletion, which is consistent with a direct role in gene expression for this enzyme. Because enzyme binding occurs throughout the gene, the authors speculate that Arg5,6 function might involve RNA processing of the primary transcript rather than initiation of transcription. How this might work is not clear because Arg5,6 binding to RNA has not been demonstrated.

Another putative example of metabolic machinery being redirected to transcriptional process involves the proteosome.[723] Ubiquitylated proteins are degraded by the 26S proteosomal complex, which is made up of a proteolytic 20S component and a nonproteolytic 19S component. The 19S complex has a base of six ATPases (called APIS) and a multisubunit lid. In addition to being an integral part of the proteosome, the ATPase base associates with promoters of active genes and is required for efficient transcription elongation by RNA polymerase II.[724–726] The 19S complex may be recruited to the promoter by ubiquitylation of promoter activators, and this may subsequently signal their destruction.[727] In addition to the apparent nonproteolytic roles of components of the proteosome in active promoters, the full 26S proteosome associates with RNA polymerase II in the coding and 3′ regions of active genes; moreover, it accumulates in sites of stalled RNA polymerases, where it may resolve difficulties by proteolysis.[728] It is not possible to attribute the proteolytic and transcriptional roles of proteosomal proteins to gene sharing at this time because their biochemical roles in these processes are not known in detail. Future studies may show that the ATPases of the 19S proteosomal complex perform the same reactions with a different biological outcome in the proteosome and on active promoters. Nonetheless, the evolving story illustrates the complex connections between components directing cellular metabolism and gene expression, raising the possibility that individual proteins have modified functions as they perform different biological roles.

Y-BOX PROTEINS

Y-box proteins are a remarkable example of multifunctional proteins that have been incorporated into many roles for gene expression (Figure 7.1). These belong to the conserved nucleic acid–binding cold shock protein family

Figure 7.1. Multiple gene expression functions of the Y-box proteins. A. Diagram of Y-box protein; A/P, alanine/proline rich region; CSD, cold shock domain; B/A repeat, region enriched in basic and aromatic amino acids. B. Y, Y-box protein activated on membrane by physiological stress. C. Y-box protein can activate transcription by first binding DNA or first forming a complex with other transcription factors (TFs). It may also have other roles in modulating DNA structure. D. Y-box protein may influence alternative RNA splicing by binding to an ACE (adenosine/cytosine-rich exon enhancer). E. Y-box protein may affect mRNA translation and stability via various mechanisms described in the text.

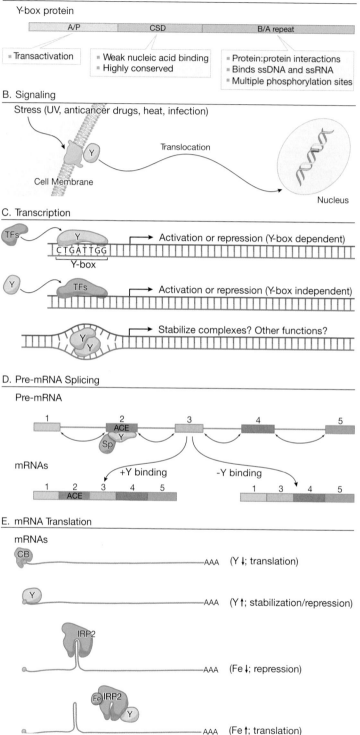

A. Protein

Y-box protein

A/P	CSD	B/A repeat

- Transactivation

- Weak nucleic acid binding
- Highly conserved

- Protein:protein interactions
- Binds ssDNA and ssRNA
- Multiple phosphorylation sites

B. Signaling

Stress (UV, anticancer drugs, heat, infection)

Y

Translocation

Cell Membrane

Nucleus

C. Transcription

TFs

Y

CTGATTGG

Y-box

→ Activation or repression (Y-box dependent)

Y

TFs

→ Activation or repression (Y-box independent)

Y
Y

→ Stabilize complexes? Other functions?

D. Pre-mRNA Splicing

Pre-mRNA

1   2   3   4   5
     ACE
     Y
     Sp

mRNAs

+Y binding   -Y binding

1 2 3 4 5   1 3 4 5
   ACE

E. mRNA Translation

mRNAs

CB ————————————————AAA (Y ↓; translation)

Y ————————————————AAA (Y ↑; stabilization/repression)

IRP2 ————————————————AAA (Fe ↓; repression)

Fe IRP2 Y ————————————————AAA (Fe ↑; translation)

whose members are represented in all bacteria and metazoans except for yeast (see [729–731]). Y-box proteins mediate many biological functions, from cessation of growth in cold-shocked bacteria to a variety of environmental stress responses in eukaryotic cells. The interest in Y-box proteins for the present discussion on gene expression relates to their control of transcription, mRNA transport, and translation. Y-box proteins are also involved in functions not directly related to gene expression, such as DNA repair.

Y-box proteins have a variable alanine/proline rich N-terminus that is probably a *trans*-activation domain, an extremely conserved nucleic acid–binding domain (the cold shock domain), and a C-terminus that interacts with other proteins as well as single-stranded DNA and RNA *in vitro*. The C-terminal domain has multiple phosphorylation sites that may regulate various activities of the protein. Y-box proteins have multiple nucleic acid–binding sites, including the cold shock domain as well as accessory domains enriched in basic and aromatic amino acids called basic/aromatic (B/A) islands. Y-box proteins can activate or repress gene expression directly by binding to a specific sequence motif in promoters (often in growth regulatory genes) or indirectly in a Y-box-independent manner by interactions with other transcription factors. Regulation of transcription by this class of proteins was first shown by the characterization of YB-1 binding to the inverse CCAAT sequence in a major histocompatibility class II promoter.[732] In addition, Y-box proteins bind single-stranded regions of active promoters, where they may stabilize complexes or perform other unknown functions.

Y-box proteins also affect mRNA translation in the cytoplasm. An early example is FRGY2 (also YB-1) in *Xenopus* oocytes.[733,734] FRGY2 accumulates in the oocyte and, in combination with other proteins, represses translation nonspecifically. The dormant masked mRNAs are selectively activated during development. Interestingly, it appears necessary for mRNAs to pass through the oocyte nucleus to be translationally repressed subsequently in the cytoplasm.[735] In addition to nonspecific repression, Y-box proteins can also regulate translation of specific mRNAs. YB-1 binds the 5′ noncoding region of its own mRNA to autoregulate translation.[736] YB-1 also binds to iron-regulatory protein 2 (IRP2) at high concentrations of iron.[737] In this interesting case, both YB-1 and IRP2 repress ferritin mRNA translation individually, but together YB-1 sequesters IRP2 from the mRNA to enhance translation. YB-1 also binds to at least five ribosomal proteins, suggesting possible mechanisms of action for its role in direct translational control. The gene-sharing network during translation becomes even more entangled when we consider that ribosomal proteins themselves have numerous extraribosomal functions![738]

In addition to involvement in transcription and translation, Y-box proteins regulate pre-mRNA splicing. An example is alternative RNA splicing of exon v4 of the human CD44 gene.[739] This exon contains sequences that act as accessory pre-RNA splicing signals known as ACE (for adenosine/cytosine-rich exon enhancer). Splicings signaled by ACEs generally involve a special class of serine-arginine-rich (SR) proteins. However, YB-1 is the major protein binding the ACE in the RNA, including the v4 exon of CD44, and it appears to be responsible for this exon being included in the mature mRNA. Mutagenesis experiments have suggested that the B/A island nucleic acid–binding sites are needed for the splicing function of YB-1. A number of other RNA-binding proteins are known that both regulate splicing and translation, including Sex-lethal, transformer, CUG-binding protein, and stem loop binding protein (see Wilkinson and Shyu[740]).

TRANSCRIPTION FACTORS AS TRANSLATIONAL REGULATORS: BICOID

Y-box proteins are not the only transcription factors that escape the nucleus and regulate mRNA translation. Bicoid (BCD), a *Drosophila* transcription factor that activates segmentation genes directing early development, represses translation of *caudal (cad)* mRNA.[741] BCD binds a sequence within the 3′ untranslated region of the *cad* mRNA. Subsequently the mRNA is believed to bend, allowing interaction between BCD and the translation initiation factor eIF4E, which is attached to the 5′ capped (m7GpppN) end of the mRNA. BCD must be associated with *cad* mRNA to be able to bind eIF4E. Interaction of BCD and eIF4E prevents the latter from interacting with eIF4G, which is required to form an active translation initiation complex. Other transcription factors appear to regulate translation as well. In the mouse, the transcription factor Emx2 forms complexes with eIF4E in the olfactory sensory neuron axons, and the transcription factors Otx2 and Engrailed 2 bind eIF4E in the developing and adult nervous system.[742]

It is also noteworthy to recall here that mRNA translation can be directly regulated by a host of metabolic enzymes. These typically include enzymes that use nucleotide-containing nucleotides (see Chapter 6 for a review).

TRANSLATION FACTORS FOR RNA EXPORT: eIF4

Despite being a cytoplasmic translation initiation factor, eIF4 is present in nuclear bodies in species ranging from yeast to humans and is able to modulate the export of specific mRNAs to the cytoplasm. In cultured myeloid cells, eIF4E interacts with a transcription factor, the proline-rich homeodomain protein PRH, to inhibit mRNA transfer to the cytoplasm.[743] This interaction was demonstrated by overexpressing PRH and noting the consequent specific reduction in cyclin D1 mRNA in the cytoplasm. Computer-assisted analysis has suggested that some 200 other homeodomain proteins may interact with eIF4E independently of their transcriptional roles and contribute to the regulation of mRNA transport and translation in a tissue-specific fashion. It remains unclear how many different functions are carried out by RNA-binding proteins, and further experiments are necessary to determine if and to what extent the biochemical roles of RNA-binding proteins, including translation initiation factors, differ in the nucleus and the cytoplasm. Cellular location is not enough to establish a different function for a protein. For example, translation can occur in the nucleus of mammalian cells and a proof-reading mechanism mediated by RNA polymerase II may exist for nonsense-mediated decay (NMD) of aberrant mRNAs in the nucleus.[744,745]

HOMEOPROTEINS, CHROMOSOMAL PROTEINS AND ACTIN

Structural and secretory data suggest that homeoproteins, which often perform as transcription factors, might function as signaling molecules in both plants and animals.[746] An interesting case for an intercellular transport function of a homeodomain protein exists in plants. *Tomato bushy stunt virus* encodes a cell-to-cell movement protein (P22) that is required for trafficking viral macromolecules between cells through interconnections called plasmodesmata.[747] A novel homeodomain protein is present in tobacco *(Nicotianna tabacuum)* that accumulates during virus infection and that specifically interacts with the cell-to-cell movement protein P22. P22 mutants are associated with movement-defective plants and do not interact with the novel homeodomain protein.

A striking example of an extracellular signaling function for a chromosomal protein is the high mobility group box protein 1 (HMGB1). HMGB1 is an abundant nonhistone structural protein of chromosomes that stabilizes nucleosomes and facilitates gene expression by bending DNA. This chromosomal protein is also an extracellular macrophage-activating factor and functions as a signaling protein and cytokine to induce inflammatory responses.[747a]

A final case I will mention here to support the connection between gene sharing and gene expression is actin, because there is no question about its importance as a cytoskeletal protein affecting cell structure, shape, and motility. β-actin (but not the muscle-specific α-actin) is part of the transcriptional preinitiation complex and is necessary for transcription by RNA polymerase II.[748] Multiple lines of evidence support this claim. These include specific association with the promoters of the interferon-γ-inducible MHC2TA and interferon-α-inducible GIP3 genes, binding to RNA polymerase II, interference of preinitiation complex formation and transcription initiation by β-actin antibodies, and stimulation of transcription by RNA polymerase II. Consequently, Hoffman and colleagues[748] suggest that actin has a fundamental role in the initiation of transcription by RNA polymerase II.

Actin probably has even more roles in gene expression than those coupled to transcription of protein-coding genes by RNA polymerase II (see Hoffman et al.[748] for references). For example, actin has been linked to ribosomal RNA transcription by RNA polymerase I, it associates with small nuclear ribonucleoproteins that are involved in pre-mRNA processing, and it forms complexes with heterogeneous nuclear ribonucleoproteins that contribute to the export of mRNA from nucleus to cytoplasm. Actin is even associated with chromatin remodeling and histone acetyltransferase complexes.

Are the multiple uses of nuclear actin in gene expression examples of gene sharing? The answer depends on mechanisms. The precise functions of actin in any of these gene expression roles are not yet known. One suggested possibility is that actin recruits RNA polymerase II to the preinitiation complex and acts as a bridge between the polymerase and other components of the preinitiation complex.[748] Recruitment and bridging roles depending on protein:protein interactions specific to the nucleus would be gene sharing. A novel, noncytoplasmic role for actin in transcription is also supported by its involvement in *in vitro* transcription, where it has a critical role with only naked DNA and purified transcription factors. On the other hand, actin could stimulate ATPase activity in the chromatin remodeling complexes in view of the ATP-dependence of chromatin remodeling. This nuclear function may be related to the ability of actin to promote myosin hydrolysis of ATP, making the precise mechanism critical to determine whether this is a novel molecular role for actin. It is also possible that actin acts as a molecular motor to drive transcription elongation inasmuch as it has been found in association with myosin IC in the nucleus and myosin antibodies inhibited transcription by RNA polymerase II *in vitro*. If actin plays the role of a molecular motor for transcript elongation through a similar mechanism as it uses in the cytoskeleton, its comparable nuclear biochemistry would not qualify as gene sharing. Rather, it would be an example of a different biological role (phenotype) of a protein used at a different time in a different place (nucleus instead of cytoplasm).

THE DYNAMIC FLUX OF NUCLEAR PROTEINS

Finally, an additional feature that is compatible with the involvement of gene sharing during gene expression is the dynamic nature of the nucleus. Although highly organized and compartmentalized, the discrete nuclear bodies and proteins are in perpetual flux, rapidly intermingling by diffusion.[700,749] Misteli[749] (p. 847) points out that the high mobility of nuclear proteins "favors stochastic, combinatorial use of components and generates a robust system that can respond quickly to external cues." Perhaps the rapid exchange and diffusion of proteins within the nucleus also subjects them to a constantly changing set of interactions and local concentrations and favors the differential use of their properties. Rephrased as a question: Could a high diffusion flux be associated with a corresponding flux of functions? Whatever the final balance of single to multiple uses of proteins, we cannot help but marvel at the extraordinary and pragmatic coordination of functions by versatile proteins engaged in gene expression, cooperating and coevolving to give us life as we know it.

TAKE-HOME MESSAGE

Gene expression, with its dependence on complex assemblages of multifunctional proteins, is a powerful global example of the creative roles of gene sharing during the evolution of what arguably could be considered the central pathway responsible for life. The gene expression pathway illustrates how gene sharing can coordinate different cellular events by reusing the same proteins in different ways in response to biological needs.

8

Gene Sharing As a Dynamic Evolutionary Process: Antifreeze Proteins and Hemoglobins

Phylogenetic studies at the molecular level have established that sequence alterations during evolution are correlated with functional shifts of homologous proteins.[750] The gene-sharing concept postulates that adoption of a new role does not necessarily require the abandonment of the old one and that the transition of functions may involve overlapping periods during which the ancestral and novel functions coexist. I discuss antifreeze proteins (AFPs) in this chapter because they represent examples of proteins (1) that have specialized entirely for a new function and lost the earlier function, (2) that have undergone a functional shift yet have retained the potential to carry out their ancestral function, and (3) that have adopted a new (antifreeze) function without a change in their original function.

I also consider hemoglobins in this chapter because they are perhaps the prototypical example of highly specialized, differentiated proteins. Yet hemoglobins date back to the beginnings of life and have engaged in many functions during their lengthy divergent evolution. Hemoglobins have not only branched to acquire new specialized functions, as unexpected as light shields for vision in parasitic worms, but they have also orchestrated some of their multiple functions to coordinate their roles for oxygen transport. Moreover, AFPs are multiple diverse proteins that arose independently about 15 million years ago, while hemoglobins stem from a single source that had its origins over a billion years ago. These different timescales provide a role for gene sharing in rapid convergent evolution (AFPs) as well as sustained divergent evolution (hemoglobins).

ANTIFREEZE PROTEINS

AFPs were discovered in the blood plasma of Antarctic Notothenioid fish[751,752] (see Cheng,[753] Fletcher et al.,[754] and Ewart and Hew[755]). Initially, AFPs were defined functionally as macromolecules that prevent the freezing of fish due to the growth of environmental ice in the blood. Subsequent studies have shown that AFPs also exist in bacteria, fungi, plants and insects and as well as vertebrates.[756–759] AFPs are of great interest to chemists, physicists, engineers, and industrialists as well as biologists. Practical applications of AFPs are being sought for preserving frozen foods; improving storage of blood, tissues and organs; cryosurgery; protection of crops from frost; and animal husbandry in cold temperatures.[760,761]

Fish and insect AFPs: Fish and insect AFPs are highly diverse proteins that may or may not have repeated motifs and have α-helical, β-helical, or globular structures[753,757,762] (Figure 8.1). Fish have at least five different types of AFPs: AFGP, whose crystal structure has not yet been determined, is a glycoprotein containing tripeptide (Thr-Ala/Pro-Ala) repeats with a disaccharide (galactose-N-acetylgalactosamine) attached to the threonyl residues. Type I contains an alanine-rich α-helix with repeats of an eleven–amino acid sequence; Type II contains a C-type lectin fold of mixed α, β, and loop structures; Type III is a globular protein with short β-strands; and Type IV is predicted to have an α-helix bundle. Insect AFPs are also diverse and differ from those of fish.[763–766]

The fish and insect AFPs function by binding ice crystals and preventing their growth[767] (see Jia and Davies[757]). They are effective noncolligative antifreeze agents, meaning that they can lower the freezing point of solutions hundreds of times further than would be predicted strictly based on the numbers of molecules in solution. Their effect on depressing the freezing point is greater than their effect on lowering the melting point, a gap termed thermal

Figure 8.1. Fish and insect antifreeze proteins (AFPs). AFGP: Fish antifreeze glycoprotein; the structure of this protein is not known yet. The chemical formula (top) is of the sugar moiety that binds the alanine-alanine-threonine repeat structure of the protein. The different AFP structures have been solved except for Type IV AFP, which is predicted on the basis of amino acid sequence. References: for fish AFP structures, Cheng[1114]; for beetle AFP structure, Liou et al.[763]; for moth AFP structure, Graether et al.[764] I am grateful to Dr. Chi-Hing C. Cheng (Department of Animal Biology, University of Illinois, Urbana, Illinois) for materials to make this composite figure.

AFGP

$(-\text{Ala}-\text{Ala}-\text{Thr}-)_n$

Boreogadus saida
(Arctic cod)

Dissostichus mawsoni
(Antarctic toothfish)

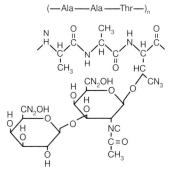

Type I AFP
Pleuronectes americanus
(winter flounder)

Type II AFP
Hermitripterus americanus
(sea raven)

Type III AFP
Licodichthys dearborni
(Antarctic eel pout)

Type IV AFP
Myoxocephalus octodecemspinois
(longhorn sculpin)

Beetle AFP
Tenebrio molitor
(meal worm)

Moth AFP
Choristoneura fumiferana
(spruce budworm)

hysteresis. Larger thermal hysteresis actions translate to more powerful abilities to protect against the growth of ice crystals.

Despite the significant differences in their protein structures, the fish and insect AFPs bind ice using mechanisms based on flat surface complementarity.[768,769] Ice adsorption by AFPs involves hydrogen bonds, van der Waals forces, and hydrophobic interactions.[757,770–772] Experimental modifications of AFP structure have indicated that extension of ice-binding surface area is a mechanism for increasing AFP thermal hysteresis.[773] Distinct AFPs bind different planes of ice, resulting in proteins with unequal antifreeze properties and different isoforms. In general fish AFPs shape ice in hexagonal bipyramids with a single binding face, although more effective dimers with two ice-binding faces[774] and labile hyperactive species[775] also exist. Insect AFPs can have greater thermal hysteresis power than fish AFPs. For example, the insect spruce budworm *(Choristoneura fumiferana)* AFP has multiple isoforms that are ten to thirty times more effective in their antifreeze properties than the fish AFPs.[776,777] The more effective overwintering insect AFPs permit survival at harsher freezing conditions than exist for fish, which inhabit less extreme environments. Both fish and insect AFPs confer freeze avoidance by maintaining body fluids in a liquid state.

Independent derivation of fish AFPs: Even though the precise evolutionary homologues of most AFPs are unknown, comparative studies clearly show that many different proteins have been called upon to provide resistance to freezing conditions.[753,754,778,779] Much more is known about the origin of fish AFPs than of insect AFPs. Type II AFPs and the carbohydrate-recognition domain of calcium-dependent lectins have similar folds, suggesting a possible connection between these proteins; but homology has not been established.[753,780] Type II AFPs have been found in at least three phylogenetically disparate species of fish, which is indicative of parallel evolution of these proteins. Type III AFP shows similarity to the C-terminal region of mammalian sialic acid synthase, further suggesting a link between AFPs and sugars and polysaccharides.[781] Type IV AFP shows sequence similarity to members of the exchangeable apolipoprotein superfamily.[753,762] Indeed, convergent evolution of AFPs appears to be the rule. Even closely related sister species of sculpin from the genus *Myoxocephalus* have entirely different AFPs (Type I and

Type IV), indicating that each was recruited independently from a different gene.[753,762]

The AFGPs provide a compelling case for the independent derivation of fish AFPs[779,782] (Figure 8.2). AFGP was derived from an ancestral *trypsinogen-like* gene in the giant Antarctic Notothenioid toothfish *(Dissostichus mawsoni)* by marked changes in gene structure. This included expansion of a nine-nucleotide threonine-alanine/proline-alanine coding sequence spanning the 3′ end of intron 1 and the beginning of exon 2, and the retention of exons 1 and 6. Thus, noncoding intron sequences of a gene presumably encoding an enzyme were converted to coding sequences for a derived gene encoding an AFP. This major gene remodeling during evolution was accompanied by loss of the putative trypsinogen coding sequences in exons 2–5, resulting in a novel protein lacking enzyme activity. In addition, the translation stop codon at the end of exon 6 moved to the left (5′) region, so most of exon 6 was no longer translated into protein. The modern Antarctic Notothenioid *AFGP* gene comprises multiple protein coding sequences (up to 46) linked in tandem by small spacers encoding cleavable peptides. Consequently, the AFGP protein is synthesized as a polyprotein that is posttranslationally cleaved by a chymotrypsin-like enzyme.

Estimates by sequence analysis suggest that the Antarctic Notothenioid ancestral *trypsinogen-like* gene was converted to the *AFGP* gene 5–14 million years ago. This dating correlates well with the freezing of the Antarctic waters 10–14 million years ago. It also fits with the time at which the Notothenioid fish radiated into five families in Antarctica, allowing them to flourish in these frigid conditions; this radiation is illustrative of the powerful role of natural selection by the environment. The likely assumption that Antarctic Notothenioid AFGP was invented once, followed by gene duplications, has been the basis for surmising an Antarctic evolutionary origin for Notothenioid fishes that live in the temperate waters of New Zealand and South America.[783] At least one New Zealand Notothenioid species retains an active AFGP in its blood after 11 million years, raising the possibility that it might serve another function.

While the Antarctic AFGPs have a single origin, the remarkable story of a Notothenioid-like AFGP in the Arctic cod *(Boreogadus saida)* indicates that very similar structures and analogous functions do not prove common mo-

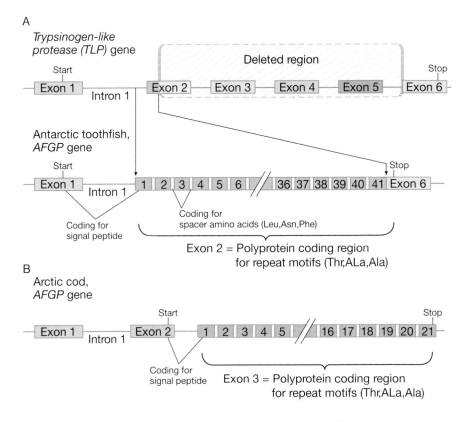

Figure 8.2. Convergent evolution of fish antifreeze proteins. A. The top diagram represents the *trypsinogen-like protease (TLP)* gene that is believed to resemble the ancestral *AFGP* gene for the Antarctic toothfish. Sequences in the 3′ end of intron 1 and the 5′ region of exon 2 expanded, giving rise to the sequences encoding the threonine-alanine-alanine repeat motifs separated by the leucine-asparagine-phenylalanine spacers. Except for exon 6, the remaining exons and introns were lost. B. Arctic cod *AFGP* gene was derived from an entirely different gene although it encodes a polyprotein with a similar repeat motif. The coding sequences for the repeat motifs and the spacer amino acids, as well as the organization of the exons and introns, differ in the Antarctic toothfish and Arctic cod. I am grateful to Dr. Chi-Hing C. Cheng (Department of Animal Biology, University of Illinois, Urbana, Illinois) for materials to make this composite figure. Adapted from Chen et al.[784] and reproduced with permission from the publisher.

lecular ancestry[784] (see Figure 8.2). The AFGPs in Artic cod and Antarctic toothfish are extremely similar, indeed nearly identical, in amino acid composition and multiple threonine-alanine-alanine repeats; nonetheless, they originated from different ancestral genes. The AFGPs from the two families differ in flanking sequences, including a signal peptide, gene organization (introns and exons), codon usage, and spacers between the polyproteins that must be cleaved by different proteases. Convergent functions led to similar structures of entirely different ancestral proteins. The independent derivations of these AFGPs reflect the different timing of evolutionary pressure on the northern and southern species of fish. Freezing conditions developed in the Antarctic more than 10 million years ago, but Arctic glaciation occurred only 2.5 million years ago. These two taxonomic groups of cold water fish diverged at least 40 million years ago, well before the freezing conditions requiring antifreeze proteins; and their AFGPs arose independently when needed instead of being inherited from a common ancestor.

Gene sharing as an evolutionary process—chimeric AFGPs: How is gene sharing relevant to fish AFGPs, which have been derived by gene duplications followed by extensive gene rearrangements, resulting in loss of their original functions? These proteins are clearly specialized for their antifreeze role in protecting fish in their ice-laden habitat, and there is no evidence that they serve more than one function. However, extant AFPs apparently with residual enzymatic activity may represent proteins caught in the act of evolution.[756] The particular gene that was characterized *(Dm7M)* in the giant toothfish has all the earmarks of an expressed gene and encodes a chimeric protein comprising an AFGP polyprotein followed by a trypsinogen-like protease (Figure 8.3). In addition a cDNA was cloned from toothfish pancreatic mRNA that encodes a similar, but not identical, chimeric protein, indicating the presence of more than one such *AFGP/protease* gene in this species. Genomic hybridization studies indicate that these *antifreeze/protease* hybrid genes exist in other Notothenioid species as well. It thus appears that Notothenioid fish express distinct AFGPs and trypsinogens with antifreeze and enzyme function, respectively, as well as chimeric AFGP/trypsinogens that have combined enzyme and antifreeze functions.

If the extant chimeric AFGP/protease proteins perform both antifreeze and protease functions *in vivo,* they represent examples of active gene sharing. In addition, they provide evidence for an evolutionary stage of gene sharing when duplicated *trypsinogen* genes encoded a bifunctional protein before being converted into specialized *AFGP* genes. The presence of modern chimeric *AFGP:trypsinogen* genes argues for the continued usefulness of these bifunctional genes for long periods of time. Finally, the chimeric genes portray gene sharing as an evolutionary process that may result ultimately in complete specialization of one function and loss of the other.

Is it a fluke that an antifreeze function was derived from a pancreatic

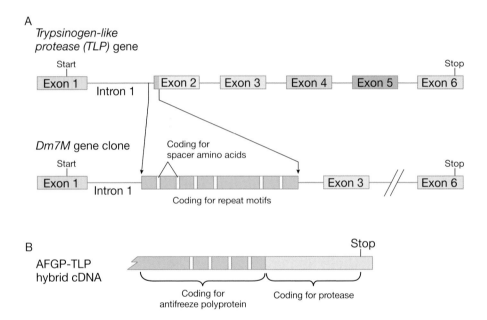

Figure 8.3. Chimeric *AFGP/trypsinogen-like proteinase (TLP)* genes. A. *Dm7M* was cloned from the giant toothfish and contains sequences encoding both an AFGP as well as a trysingongen-like protease. B. A partial chimeric AFGP/TLP cDNA cloned from the toothfish pancreas that encodes the C-terminal region of an AFGP polyprotein as well as a protease. The presence of the *Dm7M* gene and the hybrid cDNA provides evidence for the existence of multifunctional proteins containing both antifreeze and protease activity. I am grateful to Dr. Chi-Hing C. Cheng (Department of Animal Biology, University of Illinois, Urbana, Illinois) for materials to make this composite figure. Adapted from Cheng and Chen[756] and reproduced with permission from the publisher.

trypsinogen gene, or is there a link between these two activities? Chen and colleagues[782] (p. 3815) propose that "conversion of an existing pancreatic enzyme gene into *AFGP* gene and expressing it in the pancreas is both positionally and temporally logical as AFGPs thus could reach the digestive tract, simultaneously with pancreatic enzymes, to prevent the intestinal fluid from freezing while the enzymes perform digestive functions." If this was the basis for converting an ancestral *trypsinogen* gene into the modern *AFGP* gene, it would account for the presence of the presumably bifunctional chimeric genes as gene-sharing intermediates that were useful to maintain.

Seasonal control of fish AFP synthesis: AFP expression is under a variety of control mechanisms, even within a single species. For example, the winter flounder *(Pleuronectes americanus)* produces homologous Type I AFPs: a secreted liver-specific protein and a family of intracellular ubiquitous (liver, skin, scales, fin, and gills) proteins that are differentially under seasonal regulation.[785] The liver AFP is highly expressed in the winter months while the skin (ubiquitous) AFPs are only modestly more abundant in the winter months. Photoperiod operating through pituitary growth hormone release, rather than temperature, appears critical for seasonal control of the winter flounder liver AFP.[754] We might speculate that the differences in expression pattern and seasonal fluctuation reflect a unique, yet-to-be-discovered additional function of the skin AFPs. That would represent gene sharing controlled by differential gene expression.

Plant ice-active proteins as multifunctional enzymes: Plants also have diverse, functionally convergent proteins that protect against freezing (Figure 8.4). The plant proteins are responsible for freeze tolerance rather than freeze avoidance.[759,786,787] Consequently they might be better called ice-active proteins than antifreeze proteins (Cheng, personal communication). Freeze tolerance is achieved by secreted ice nucleators that cause crystallization of water in extracellular spaces in a controlled process that does not damage the plant. This process prevents large chunks of ice from forming by recrystallization.[788,789] Plant ice-active proteins may also change ice morphology to weak spikes, which probably minimizes mechanical stress caused by ice formation. They function at lower concentrations than do fish and insect AFPs because inhibition of ice recrystallization requires 100–150 times less protein than freezing point depression. Controlled ice formation in the extracellular spaces leads to movement of water from the cytoplasm to the growing ice crystals, resulting in a freeze-dehydration that concentrates cyptoplasmic constituents, another obstacle for the plant. Plant ice-active proteins use a flat surface to bind ice, as do the fish and insect AFPs; but they have low thermal hysteresis ability, making them less effective in depressing the freezing point of a solution.

Ice-active proteins of plants are multifunctional, which suggests gene shar-

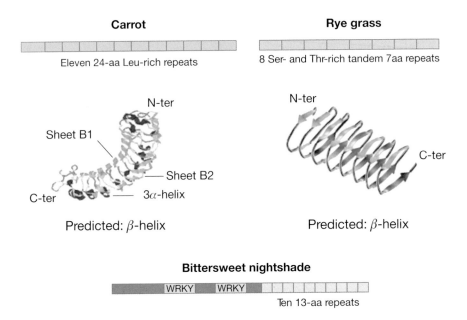

Figure 8.4. Plant ice-active proteins. Computer-assisted modeling was used to predict the structure of the carrot[1115] and rye grass[1116] AFPs. WRKY regions comprise zinc finger domains that bind DNA in a way that is consistent with these proteins being transcription factors. The structure of the bittersweet nightshade AFP has not been predicted yet.[792] I am grateful to Dr. Chi-Hing C. Cheng (Department of Animal Biology, University of Illinois, Urbana, Illinois) for materials to make this composite figure. The image of the predicted structure of the carrot AFP (left) was reproduced with permission from Zhang et al. © The Biochemical Society.[1115] The image of the predicted structure of the rye AFP (right) was reproduced with permission from Kuiper et al.[1114]

ing. A carrot ice-active protein with leucine-rich repeats has sequence similarity to a polygalacturonase inhibitor,[786] although it is not known if this similarity has physiological relevance. Enzymatically active endochitinases, endo-β-1,3-glucanases and thaumatin-like proteins with all the characteristics of ice-active proteins, accumulate in cold-acclimated winter rye *(Secale cereale).*[790,791] Glucanases and chitanases are known as pathogenesis-related proteins because they confer antifungal activity to the plants—another biological function for these enzyme/ice-active proteins. These enzymes apparently evolved subtle structural differences in the cold-resistant winter rye, giving them ice-active activity; the same enzymes induced by pathogens in a cold-sensitive tobacco that is unable to withstand freezing lack ice-active activity.

Pathogenesis-related proteins and freeze-protective proteins merge in still another protein (STHP-64) with thermal hysteresis activity present in winter bittersweet nightshade *(Solanum dulcamara).*[792] STHP-64 has numerous repeat motifs comprising thirteen amino acids in its C-terminus, which is typical of animal AFPs; and it contains two conserved DNA-binding zinc finger motifs that are present in WRKY proteins. WRKY proteins are members of a family of transcription factors that regulate the signal transduction pathway leading to expression of pathogenesis-related proteins. STHP-64 accumulates in winter bittersweet nightshade in the autumn, perhaps due to the upregulation by cold temperatures that is typical of plant ice-active proteins. Unlike the ice-nucleating extracellular plant ice-active proteins, STHP-64 is an intracellular protein that lacks a signal peptide. Possibly it acts as an ice nucleator suppressor, a property of insect AFPs, which prevents ice formation in the cell nucleus where STHP-64 functions as a transcription factor. Further studies are required to verify this intriguing hypothesis, which is consistent with the gene-sharing concept.

Comparisons between AFPs and lens crystallins: Both AFPs and crystallins are diverse, functionally convergent proteins. The different AFPs and crystallins evolved independently in response to the need for survival in a subfreezing environment or for optimal use of the eye for vision, respectively, resulting in many taxon-specific proteins. True and Carroll[32] (p. 64) state that "the diversity of AFPs strongly suggests that AFP protein co-option has been widespread and rapid and, similar to crystallins, provides further evidence that under particular and strong selective conditions, co-option events utilizing a diverse group of gene precursors have frequently occurred during evolution."

In many cases the AFPs and crystallins exploited enzymes for their respective antifreeze and refractive functions. The ancestral *trypsinogen-like* genes were duplicated and profoundly rearranged so the resulting AFGPs appear completely specialized for antifreeze function; however, chimeric *AFGP/protease* genes also exist that encode proteins with antifreeze and enzyme activity. Similarly, *S-crystallin* genes of squid and octopus arose from *glutathione S-transferase* genes that duplicated and rearranged by exon shuffling and mutation to produce lens-specific crystallins specialized for refractive function; here too there is at least one member (S11/Lops4-crystallin) of the refractive S-crystallins that retains a low-level enzymatic activity.[194] It is not known whether the chimeric AFGP/proteases or S11/Losps4-crystallin are presently bifunctional *in vivo* and consequently are undergoing gene sharing, or whether they are performing strictly antifreeze and refractive roles, respectively, and are molecular fossils—situations indicative of earlier evolutionary gene-sharing processes.

By one hypothesis, a pancreatic trypsinogen was selected as a primordial AFGP because of the benefit of protecting digestive functions in Antarctic fish against the dangers of ingested ice in freezing temperatures.[782] Should this be true, it points to an evolutionary stage at which the old and new functions coexisted and cooperated. Whether such dual, cooperative functions exist for the chimeric AFGP/proteases is not known. It is known that cooperation of functions exists for shsp/αB-crystallin of the vertebrate lens. The stress-protective role of this protein suppresses aggregation of partially denatured proteins, which would lead to lens opacification; the high concentration of the protein in the lens affects refractive index and, consequently, image formation on the retina. shsp/αB-crystallin also has additional growth functions required for differentiation and maintenance of lens cells. Thus, both AFPs and crystallins have two or more functions that may have been orchestrated during evolution by the gene-sharing process.

Evolution at the regulation of gene expression is evident in the derivation of AFPs and crystallins. Plant enzymes are induced by cold temperature when they appear to act both as AFPs and enzymes or transcription factors. Crystallins have distinct functions, such as stress proteins and enzymes, depending on concentration and tissue location, both of which are controlled at the level of gene expression (see Chapter 4). Although little is known yet about *AFP* gene regulation, initial studies indicate the presence of specialized transcriptional control mechanisms.[754] For example, an intronic liver-specific enhancer is present in the winter flounder *AFP* gene. This enhancer binds C/EBPα, a known liver-enriched enhancer factor. The homologous skin-type *AFP* gene in the same species has a disrupted C/EBP binding site, consistent with lower-level, ubiquitous expression. Crystallins show many convergent changes in gene regulation that account for high lens expression and refractive function (see Chapter 4). Thus, AFPs, like lens crystallins, show changes in gene regulation coordinated with evolution of new protein function.

HEMOGLOBINS

Perhaps no protein has been studied as intensively as hemoglobin, which is famous for oxygen transport in the circulatory system. Although hemoglobin is an abundant erythrocyte-specific protein in blood, hemoglobin-like proteins pervade living systems and are not restricted to oxygen transport (Figure 8.5).[11,793] Sequence analysis indicates that the *hemoglobin* genes descended from a common ancestral gene. The two major *hemoglobin* gene families (the α- and β-chain clusters) connected directly with oxygen transport in vertebrates are about 50 percent identical in sequence regardless of species, suggesting that they had a common ancestor approximately 450 million years ago. Vertebrate and invertebrate hemoglobins form separate phylogenetic clades that show less amino acid sequence resemblance than structural, three-dimensional homology, especially with respect to the globin-fold that harbors the iron-containing porphyrin heme group. The last common ancestral *hemoglobin* gene before the invertebrate/vertebrate split is estimated to have existed at least 670 million years ago.[793,794]

Leghemoglobin is found in root nodules and is symbiotic with rhizobial bacteria for nitrogen fixation, while other hemoglobins are ubiquitously expressed, nonsymbiotic proteins. Hemoglobins exist in protozoa (e.g., *Tetrahymena*) and alga (e.g., *Chlamydomonas*), and bacteria and yeast have flavohemoglobins with globin folds that appear homologous to vertebrate hemoglobins.[793] Bacterial flavohemoglobins have, as the name implies, a heme-binding domain and often a fused flavin (NADH, FAD) binding domain.[795] The protein of the sliding bacteria *(Vitreoscilla)* lacks a flavin-binding domain although it still groups within the same clade as the flavoproteins. These structural relationships suggest an ancestral hemoglobin gene was present

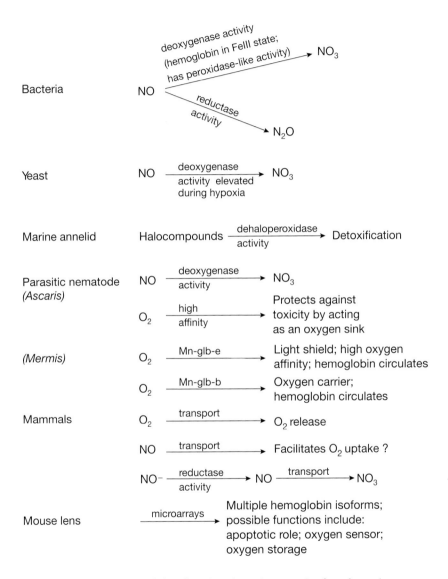

Figure 8.5. Different Hemoglobin functions in various species from bacteria to mammals. The enzyme activities, affinities, or functions of the different hemoglobins are given on the arrow. Substrates are given on the left and products of the reactions and/or functions are given on the right of the arrow.

more than 1,500 million years ago, even before the divergence of plants and animals![793,796] The relevance of all this to the present discussion is that the ancient hemoglobin family of proteins is multifunctional (see Figure 8.5).

Hemoglobins as enzymes: Hemoglobins are associated with enzymatic activities. One major role concerns nitric oxide (NO) metabolism, which has an ancient origin.[38] NO has an important role in signaling, and a potentially damaging effect by nitrosylating proteins. Bacterial flavohemoglobin relieves nitrosative stress via a dioxygenase activity that converts NO to harmless nitrates and an NO reductase activity that generates nitrous oxide (N_2O) under anaerobic conditions.[797–800] At physiological oxygen concentrations, bacterial flavohemoglobin binds NO, not oxygen, and converts it to nitrate.[801] During steady-state turnover of the reaction, the flavohemoglobin is found in the ferric (FeIII) state with peroxidase-like properties. This is consistent with Raman spectroscopic studies demonstrating that the residues lining the heme pocket of *E. coli* flavohemoglobin are designed to activate heme-bound oxygen species, as in peroxidases.[802] In this connection, it is relevant to consider a novel dehaloperoxidase in the marine annelid (polychaete) worm *Amphitrite ornata* that is used to degrade noxious halocompounds secreted by other worms cohabiting the same mudflats.[803] The *A. ornata* dehaloperoxidase has a typical globin fold consistent with an evolutionary relationship with the hemoglobins.

Yeast flavohemoglobin is also a nitrogen-metabolizing protein although, in contrast to its bacterial counterpart, yeast flavohemoglobin is elevated during hypoxia. It metabolizes NO and protects against nitrosylation under both aerobic and anaerobic conditions.[804] Thus, Liu and colleagues[804] (p. 4676) suggested that "the ancient flavohemoglobin (~ billion years ago) originated in the anaerobic mode (or evolved the anaerobic mechanism) to detoxify NO, which was abundant in the earth's primordial atmosphere" and that "as of 1 billion to 1.5 billion years ago (around the time of yeast and plant evolution) the flavohemoglobin was functioning dually as an 'NO reductase-NO oxygenase' to protect against nitrosative stress under anaerobic and aerobic conditions."[804]

The intestinal parasitic nematode *Ascaris lumbricoides* contains an octomeric hemoglobin whose primary function is also in NO metabolism.[805,806] The affinity of *Ascaris* hemoglobin for oxygen is too high to make it suitable as an oxygen carrier. Rather, this hemoglobin has a NADPH-dependent deoxygenase activity that consumes oxygen by converting NO to nitrates. This prevents the damaging effects of oxygen on the anaerobic mitochondrial pathway of this intestinal parasite and, possibly protects against unwanted effects of host NO as well. The symbiotic leghemoglobin of plants also keeps root nodules anaerobic by a different mechanism. Plant leghemoglobin's very high affinity for oxygen acts as an oxygen sink rather than as an enzyme.[806,807] The high oxygen affinity of *Ascaris* hemoglobin and its deoxygenase activity are due to thiol positioning in the heme pocket of the protein. *Ascaris* hemoglobin may be an evolutionary bridge that retains the earlier enzymatic function to control oxygen concentration with NO metabolism.[805]

Moens and colleagues[796] have speculated that ancestral hemoglobin was a redox regulator to protect from nitrosative stress, which probably imposed a serious threat to early life. Today hemoglobin is best known for its non-enzymatic transport of oxygen, but the protein went through a strictly enzymatic stage and currently remains principally, if not exclusively, an enzyme in some species. Thus, "the picture that emerges is one in which redox reactions and peroxidase (ferric heme)-mediated activities govern the interactions of NO with hemoglobins and help rationalize the molecular basis of its evolution. It is evident from studies with NO that the functional distinction between peroxidases, hemoglobins, and cytochromes is increasingly blurred"[801] (p. 10112).

The oxygen transport role of mammalian hemoglobin is accompanied by several enzymatic functions related to its ancestral roles in nitrogen metabolism. Although one of the functions of the tetrameric $\alpha_2\beta_2$-hemoglobin may be to facilitate the uptake of oxygen by transporting NO (which acts as a vasodilator) to oxygen-deficient tissues,[808,809,1148] this hypothesis is controversial.[810] The postulated mechanism is as follows: NO binds to oxyhemoglobin in the lungs and is transferred via an allosteric structural change to cysteine 93 of the β-hemoglobin chain, and the S-nitrosylated hemoglobin is used as a carrier of bioactive NO. This mechanism has been challenged on the basis of numerous arguments.[810–818] The levels of NO diffusion into the erythrocytes are probably too low due to the cell membrane barrier to make this a viable pathway; and experimentally, the S-nitrosylation of mammalian hemoglobin appears to be a negligible event even with the infusion of bolus NO.[813,819] In addition, nitrosylated hemoglobin is very unstable in the reductive environment of the erythrocyte.[814] However, circulating nitrite ions are an important source of NO for vasodilation; the conversion of nitrite ions to NO requires a deoxygenated heme protein, which suggests a novel function of hemoglobin as a deoxygenation-dependent nitrite reductase.[815,820] Moreover, mammalian hemoglobin consumes NO by converting it to nitrates under normoxic conditions,[810,814] a relic of its ancestral properties in a nitrogen-rich environment.

Hemoglobin as a light shield: Hemoglobin functions are not confined to respiration or nitrogen metabolism. Amazingly, hemoglobin has a physical role in the parasitic nematode *Mermis nigrescens*[821] (Figure 8.6). Grasshoppers and other orthopteran insects ingest *M. nigrescens* eggs, which develop within the body cavity until they become adult-sized larvae. The larvae emerge and burrow into the soil where they undergo a final molt. One to two years later, gravid adult females exhibit a positive phototaxis that is used to search for suitable egg-laying sites in the grass. The eggs are ingested by the insects, continuing the cycle.

Dense crystalline oxyhemoglobin accumulates in certain hypodermal cells and forms a cup-shaped pigmented structure in a primitive single ocellus in the anterior tip of *M. nigrescens*.[822-824] This intracellular hemoglobin exists as true crystals that differ from paracrystallin arrays formed by deoxygenated sickle cell hemoglobin.[821] Action spectra have established that the hemoglobin is not used as a visual pigment. The optically dense hemoglobin layer in the eye apparently directs phototaxis by casting shadows onto a photoreceptor when orientation is inappropriate. The anterior tip of the nematode, with continual scanning movements, is preferentially directed toward the light source during phototaxis.[825-828] This ocellar hemoglobin has a high affinity for oxygen, making it a poor candidate for oxygen transport followed by release,[824] but it is well adapted for the high light absorbance required for a shadowing role.[821] Nonetheless, the eye-associated hemoglobin is also expressed outside the eye at lower levels throughout the body, suggesting other functions in other tissues.

There is another *hemoglobin* gene in *M. nigrescens* that is expressed throughout the body but negligibly, if at all, in the ocellus.[821] The eye *hemoglobin* gene is called *Mn-glb-e* and the second gene is called *Mn-glb-b*. The two hemoglobins are 84 percent identical in sequence and their sequence predicts a typical myoglobin folding structure. The highest expression level of Mn-glb-b mRNA occurs during the last molt of the larva (L4). By contrast, Mn-glb-e mRNA expression initiates later, two to three months before visible coloration of the eye and continues at a high level while the pigment accumulates. The high concentration of hemoglobin in the hypodermal cells in the

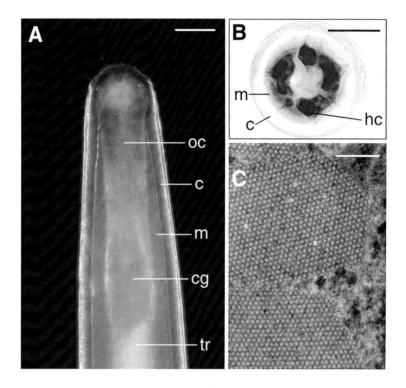

Figure 8.6. Hemoglobin with an optical function in the parasitic nematode *Mermis nigrescens*. A. Whole mount preparation of the anterior tip of the nematode showing the hemoglobin pigment confined to the ocellus. Bar, 100 μm. B. Transverse thick section showing the cylindrical distribution of hemoglobin pigment in expanded hypodermal chords of the ocellus. Bar, 100 μm. C. Electron micrograph taken of a transverse section, stained with uranyl acetate and lead citrate, through the ocellus showing hemoglobin crystals in the cytoplasm of a hypodermal chord cell. Bar, 100 nm. Abbreviations: oc, ocellus; c, cuticle; m, body wall muscle; cg, cerebral ganglion; tr, trophosome (food storage body); hc, hypodermal chord. Taken from Burr and colleagues.[821] Plate is courtesy of A. H. Burr. Reproduced with permission from the publishers.

eye region is reached gradually over approximately eight months. According to Burr and colleagues[821] (p. 4814), "recruitment appears to have occurred simply by changing gene regulation so that hemoglobin accumulates in anterior hypodermal cells to amounts high enough to expand the hypodermal chords and cast a shadow."

Evidence of other hemoglobin functions: Suggestions have been made for other mysterious functions of hemoglobins. For example, microarray analysis shows that genes encoding multiple isoforms of hemoglobin are expressed in the mouse lens (a nonvascularized tissue),[829] as is the gene for the hemoglobin chaperone that stabilizes free hemoglobin A.[830] We do not yet know whether expression of these genes is accompanied by the presence of the proteins. Nonetheless, interesting questions concern the putative functions of the *hemoglobin* genes in the lens. Possible roles include oxygen storage and transport within the avascular lens. Lens hemoglobin may also play an apoptotic role by analogy with its suspected pro-apoptotic function in cancer cell lines via suppression of Bcl-2 and activation of caspases.[831] Hemoglobin could also act as an oxygen sensor, inasmuch as oxygen levels appear to have a significant role in lens fiber differentiation and organelle degradation.[832] Marked reductions in the expression of five mouse *globin* genes and in the *EraF* gene encoding a chaperone involved in hemoglobin folding are correlated with cataract formation in the *Sparc*-deficient mouse, which suggests hemoglobin functions in lens transparency.[833] If hemoglobins act as an oxygen or iron "sink" in the normal lens, their absence could lead to oxidative changes in the crystallins (or other proteins) or in the production of reactive oxygen species; both are associated with cataract formation. Single hemoglobin chains may act as an oxygen sensor in macrophages.[834] Mouse macrophages that are stimulated to produce NO with lipopolysaccharide and interferon-γ also express a minor β-hemoglobin subunit; but the function of this polypeptide is unclear. If we consider all these results, hemoglobin appears multifunctional. Indeed, myoglobin (and other globins) are in many tissues suggesting multifunctionality.[1149,1150]

TAKE-HOME MESSAGE

AFPs evolved independently (convergently) within the last 15 million years by marked changes in gene structure or by using different inherent properties of proteins with entirely different functions (enzymes, transcription factors). By contrast, the last common ancestor of the divergently evolved hemoglobins existed before the invertebrate/vertebrate split, giving these versatile proteins ample time to refine and exploit multiple functions. Together, the convergent, rapid evolution of AFPs and divergent, lengthy evolution of hemoglobins show that gene sharing is a dynamic, pervasive evolutionary process that tinkers with the expression and structure of genes to both generate and orchestrate different functions of their encoded proteins.

9

Gene Duplication and the Evolution
of New Functions

Extensive research has established that gene duplication is intimately connected with the creation of new genes and with the innovation of protein functions during evolution. Early studies noted small intrachromosomal duplications in *Drosophila* and claimed that duplicate genes are responsible for the *Bar* phenotype in flies.[835] Later studies proposed that new genes are created by duplication[836,837]—an hypothesis that was confirmed at the molecular level by showing that myoglobin and the α-, β-, and γ-hemoglobins are products of ancient gene duplications.[2,838]

Ohno[3,839–841] was a strong proponent of the idea that redundant genes are invaluable templates for innovating protein functions. He suggested that there were at least two whole genome duplications (the 2R hypothesis) during vertebrate evolution and that duplicate genes provided the raw genetic source for novel functions. This proposal was extended by Kimura and Ota[4] (p. 2848), who insisted that "gene duplication must always precede the emergence of a gene having a new function." One copy of a duplicated gene would be freed from selective constraint and could accumulate neutral mutations, or even "forbidden" mutations, according to this theory, without immediate consequence.[842] Gene redundancy would allow mutations to accumulate in one member of a duplicated gene without loss of the original function in the other gene. Over time mutations in one of the sibling genes could modify the encoded protein such that it would find a new function.

The molecular era brought new insights into the proposed role of gene duplication for generating new proteins and functions.[5] Mutations are not confined to one of a pair of duplicated genes, mutations are seldom beneficial, duplicated genes can be lost rapidly, and gene regulation is a critical parameter for evolution and protein function.[750,843,1118,1119] Accepting the idea of gene sharing brings another insight: The evolution of new protein functions does not *require* gene duplication.[8,10] This chapter concerns the relationship be-tween gene duplication and gene sharing.

GENE DUPLICATION AND RETENTION OF REDUNDANT GENES

Knowledge of genome sequences leaves no doubt as to the prevalence of du-plications throughout evolution; however, controversies remain as to the ex-tent of whole genome duplications and the mechanisms by which this may have occurred.[5,844,845] Unequivocal evidence for the predicted[846] whole genome duplication in yeast was obtained by comparison of *Saccharomyces cerevisiae* and *Kluyveromyces waltii.*[847] Despite the genomic doubling that occurred in *S. cerevisiae,* this species lost 90 percent of the duplicated genes in small, bal-anced, and complementary regions. At least one copy of each gene of the an-cestral set has been preserved. In 95 percent of 457 remaining gene pairs, one member showed accelerated evolution and many examples of neofunction-alization (the generation of a novel function). This is in line with the idea that gene duplication does indeed free one copy to evolve new functions. Targeted deletions indicated that original functions were carried out by the ancestral gene, and new roles were attributed to derived genes.

The data for fish support the idea of whole genome duplications early in the evolutionary history of vertebrates.[848,1151] Close examination of individual gene clusters, such as the *Hox/ParaHox* genes, indicates that groups of genes have duplicated multiple times during the course of evolution.[849] Other analy-ses question the 2R hypothesis and suggest that tandem duplications of in-dividual genes or chromosomal segments, rather than whole genome du-plications, are the basis of gene redundancy.[850,851] The extent of gene loss, comparisons of distant phyla where homologies are ambiguous, and positive resolutions between similarities and homologies of individual genes compli-cate agreement as to the validity of the 2R hypothesis.

Regardless of mechanism, there is unanimous agreement that genomes are enriched with duplicated genes, particularly vertebrate genomes. The extent of gene duplications in invertebrates is more variable, with *Drosophila* being an example of a species that has relatively few gene duplications compared with yeast and nematodes.[852] In humans at least 5 percent of the genome is es-timated to be segmentally duplicated.[853] This has implications beyond the question of the origin of new functions. For example, genes flanked by dupli-cations may become rearranged, resulting in disease.[853] Moreover, mutations and small-scale chromosomal rearrangements can occur equally on the original or the duplicated gene. Over time and considering demographics of species, multiple gene duplications may contribute to reproductive isolation and speciation.[854,855] It has been estimated in a whole genome–based study that the average rate of gene duplication is 0.01 per gene per million years, although this can vary by tenfold, depending on species.[856] This means that an average gene duplicates every 100 million years. DNA duplications often result in partial gene duplications, further skewing the probability that any one gene duplicate will result in a functional copy.[857] Nonetheless, Lynch and Conery[855] (p. 39) write that "gene duplication is at least as significant as nucleotide substitution as on on-going contributor to genome evolution."

There is also a rapid (in an evolutionary sense) loss of gene duplicates,[856] with variable retention estimated to be between 20 percent and 50 percent.[858] The average half-life of a gene duplicate has been estimated at 4 million years, a number consistent with more than 90 percent of duplicated genes being eliminated within 50 million years.[859] Although many parameters must be considered when creating models of gene loss and retention (population size, mutation rate, fitness time, and ratio of advantageous to deleterious mutations), elimination of gene duplicates has been proposed to be as great as 99 percent.[860]

Despite frequent gene loss, a surprising number of duplicated genes and gene families are maintained.[861,862] The extent to which these might provide the raw material for new functions is coupled to the dynamics of gene loss or retention after duplication. Calculations of the ratio of nonsynonymous (d_N) nucleotide substitutions that cause amino acid replacements to synonymous (d_S) (also called silent) substitutions that do not change the encoded amino acid revealed a brief period of relaxed selection after gene duplication.[10,856,863,864] The period following duplication during which each sibling gene mutates faster than did its parental gene might promote the acquisition of a new function for the encoded proteins before purifying selection (elimination of genes that are either not helpful or are harmful) becomes operative.[69] However, the increased mutations occurring during the relaxed period of selection after duplication would be expected to inactivate one of the two genes relatively quickly (within a few million years), which would restrict the time available for a duplicated gene to innovate a function by a mutation. These dynamics suggest that the evolution of new function plays a minor role in the retention of duplicated genes.[50] Lynch and Conery[855] (p. 35) conclude that "the origin of new function appears to be a very rare fate for a duplicate gene."

Early models suggested that duplicated genes are maintained only if they provide a direct selective advantage.[865] Because most mutations are neutral or detrimental, various ideas have been put forward to explain the retention of gene duplicates that do not rely exclusively on the innovation of a new function. Thomas[866] proposed four functions that selection might act upon to maintain redundancy. The first is the *cumulative* function. In this case the redundant genes allow enough products so as not to be rate-limiting, such as ribosomal RNA. The second is the *fidelity* function. Loss of either duplicate may be slightly detrimental, but loss of both is catastrophic. The third is the *divergent* function. Each duplicate acquires a related but distinct function, while both retain a shared function. Retention of both genes is based on selection for the distinct function, but both are required to maintain the shared function. The fourth is the *emergent* function. Sibling genes perform the same function, but the two together result in a new, synthetic function. Color vision is an example of an emergent property of three distinct pigments expressed in the photoreceptor cells.[867] Another consideration for retention of duplicate genes is overall fitness. For example, in yeast duplicated genes have a higher probability of functional compensation than single-copy genes, and deletion of the more highly expressed duplicate results in a greater probability of a severe fitness effect than deletion of its sibling.[868] These findings suggest that duplicate genes increase robustness and may be favored in evolution.

BIRTH-AND-DEATH OF DUPLICATED GENES

An important aspect of the evolution of gene families and generation of genes with novel functions is the extent to which concerted evolution takes place. In concerted evolution a gene family evolves to some extent as a unit. Concerted evolution occurs by unequal crossover, or intergenic recombination. This would be expected to occur frequently among genes with similar sequences, especially if they are clustered. Concerted evolution was considered the rule in families of closely related sequences (i.e., ribosomal RNA genes) or gene families whose members encode virtually identical proteins within a species (i.e., histone genes). The homogenization of gene sequences by concerted evolution would clearly slow functional diversification among the members of the family. However, exceptions could occur. Imperfect intergenic recombination by unequal crossing-over could result in a modified protein that finds a new function. An interesting case is bovine seminal RNase, a dimeric protein that binds anionic glycolipids. Seminal RNase diverged from a monomeric pancreatic RNase that does not bind lipid by gene duplication approximately 35 million years ago.[869] The early seminal RNase was an unexpressed pseudogene until it was resurrected 5–10 million years ago, apparently by unequal crossing-over during concerted evolution and amino acid substitutions. The active seminal RNase acquired immunosuppressive and cytostatic activities that were apparently not present in the ancestral RNase.[870,871]

Nei and colleagues[872,873] have assessed the extent of concerted evolution by quantifying the proportions of P_N (nonsynonymous mutation affecting amino acid sequence) and P_S (synonymous mutation not affecting amino acid sequence) nucleotide substitutions per nonsynonymous or synonymous site, respectively, of gene families within and between species. A high value for P_S means that different nucleotides specifying the same amino acid are often present in the equivalent codon of member genes. This would not be observed in a gene family that showed appreciable concerted evolution due to the homogenization of gene sequences by homologous recombination. If concerted evolution took place, the DNA sequences of the various members of the gene family would be extremely similar within the same species, but would be expected to differ among species.

Unexpectedly, the results of this test indicated that the major *histocompatibility complex* and *immunoglobulin* gene families underwent little, if any, concerted evolution.[872,873] Instead, birth and death of new genes was hypothesized whereby sibling genes die in high frequency due to dysfunctional mutations.[874] The surviving genes would be subjected to strong purifying selection. In the case of the variable regions of the immunoglobulin genes, the necessary diversity for immune function is thought to be generated by divergent evolution and diversifying selection.[873] Diversifying selection means that some of the amino acid substitutions that have occurred were positively selected (i.e., Darwinian selection) and maintained because of their beneficial effect. Similarly, after examination of 28 species of fungi, plants, and animals, concerted evolution does not appear significant in the *ubiquitin* gene family, despite the tight linkage of these family members and the virtual identity of amino acid sequence of the encoded proteins.[875] Comparable results were obtained with the highly conserved *histone 3*[876] and *histone 4*[877] gene families. Thus, for these families new genes were born by duplication, and some were retained and others died (were deleted or became nonfunctional pseudogenes) by independent processes. Protein homogeneity in these gene families was obtained by purifying selection rather than concerted evolution.

The rates of birth and death differ among families, or even within gene families. For example, the ancient *MADS-box* genes encode developmental DNA-binding transcription factors present in plants, fungi, and animals. In flowering plants (angiosperms), the *MADS-box* genes evolved by a birth-and-death process.[878] However, several lineages developed during evolution, and the type I *MADS-box* genes have undergone more rapid birth-and-death evolution than the type II genes.[879] The heat shock protein superfamily in nematodes is another interesting group showing less distinct patterns. The constitutively expressed members have evolved by a birth-and-death mechanism at different rates in two sibling species of nematodes; by contrast, the heat-inducible members show properties of having undergone both birth-and-death evolution and concerted evolution.[880] These heat-inducible genes are clustered and extremely similar in nucleotide sequence within and between species—a pattern consistent with partial concerted evolution—yet they show higher conservation of amino acid than of DNA sequence, which is consistent with partial birth-and-death evolution.

Taken together, the data indicate that birth-and-death evolution and purifying selection play major roles in preserving members comprising gene families. Concerted evolution does occur, but it seems to have less importance for maintaining gene family members in the genome. Birth-and-death evolution puts each gene on its own, as it were, insulating the group from damaging mutations of individual members. Birth-and-death might also afford greater opportunities than concerted evolution for individual members of gene families to innovate new functions by positive Darwinian selection of certain mutations.

ADAPTIVE EVOLUTION BY POSITIVE SELECTION:
NEW FUNCTIONS AFTER GENE DUPLICATION

There are cases in which positive Darwinian selection accelerates the rate of fixation of advantageous mutations acquired after gene duplication. Positive Darwinian selection differs from purifying selection of duplicated genes, which removes siblings that serve no role or are detrimental. Positive selection also differs from the Dykhuizen-Hartl effect in neutral selection whereby random fixation of neutral mutations are later responsible for causing a functional change in the encoded protein in response to an altered genetic background or environmental change.[881] Positive Darwinian selection can be detected by showing that nonsynonymous nucleotide substitutions (resulting in new amino acids) exceed synonymous nucleotide substitutions (no amino acid change) during the early stages after duplication.[750] Once a new or improved function has been established for the mutated protein, the rate of nonsynonymous substitutions will drop significantly to maintain the beneficial role played by the protein. During neutral evolution (no adaptive benefit for acquiring the mutation), the rate of nonsynonymous substitutions would at most equal synonymous substitutions. Not only does positive Darwinian selection serve to preserve duplicate genes but it also contributes to the fixation of functional modifications or innovations of the encoded protein. With respect to gene sharing, positive Darwinian selection could act to entrench, expand, and/or change the multifunctional status of a gene.

Positive selection of a duplicated gene that has accrued amino acid substitutions is considered adaptive evolution and is usually characterized by the selected protein being strongly associated with a specific environment or being more fit than other comparable proteins with polymorphic amino acids.[750,867,882] Because duplicated genes receive many nucleotide substitutions over time, most of which will be neutral, if not harmful, it is challenging to identify those that are responsible for the modifications that are positively selected. For convincing identification of the critical amino acid(s) changes responsible for the adaptations, we need phylogenetic analysis of the gene coupled with site-directed mutagenesis to reproduce the adaptive changes in the protein, an assay, and ideally a connection with protein structure. Having this information allows reconstruction of the ancient protein, essentially by recreating molecular evolution in the laboratory. A catchy name for such evolutionary reconstructions is "paleomolecular biochemistry."[883] Golding and Dean[867] have reviewed cases in which evolutionary reconstruction has been performed in the laboratory (chymase, RNase A, lactate dehydrogenase (LDH)/malate dehydrogenase (MDH), hemoglobin, and isocitrate dehydrogenase). One of the impressive findings from these genes is the power of one amino acid substitution among hundreds to modify the function of a protein and be the source of positive selection. For example, LDH and MDH differ at approximately 230 out of 320 amino acid sites (not considering deletions and insertions). Nonetheless, changing gluatamine 102 to arginine in LDH of the bacterium *Bacillus stearothermophilus* converts the enzyme to an efficient and specific MDH.[884]

There are a number of proteins from multigene families that have diverged by positive Darwinian selection.[10,885] Ribonucleases (RNases) are an interesting example of proteins encoded by a gene family whose members have been subjected to a variety of selective pressures, including positive selection for adapted functions. Positive Darwinian selection even occurred in seminal RNase after duplication from pancreatic RNase and resurrection from pseudogene status by gene conversion.[869] Another case of positive selection operating on duplicated RNase family members was reported for the divergence of eosinophil cationic protein (ECP) and eosinophil-derived neurotoxin (EDN).[886] EDN tandemly duplicated about 31 million years ago after the divergence of Old World and New World monkeys but before the divergence of hominoids and Old World monkeys. In humans and Old World monkeys, EDN has high RNase activity and apparently functions as an antiviral agent by degrading retroviral RNA. By contrast, ECP has low RNase activity and kills bacterial and parasitic pathogens by making pores in their cell membranes. This antipathogenic activity is independent of RNase activity. New World monkeys have a single *EDN* gene with RNase activity, but no antipathogenic activity. Examination of nonsynonymous and synonymous substitutions has revealed a high rate of the former in the *ECP* gene immediately after duplication—a finding that is consistent with positive selection for a new role as an antipathogen.[886]

The adaptive evolution of a duplicated pancreatic *RNase* gene in a leaf-eating monkey is another instance of positive selection operating on duplicated genes.[111,882,887] Colobines are a subfamily of Old World monkeys that eat leaves rather than fruits and insects; this dietary formula demands that the monkeys digest excessive amounts of bacteria RNA. Non-colobine primates have a single *RNASE1* gene that is expressed in the pancreas and the resulting enzyme is exported to the small intestine to aid digestion. By contrast, the Asian colobine douc langur *(Pygathrix nemaeus)* inherited duplicate genes, *RNASE1* and *RNASE1B*. These genes are products of duplication by unequal crossing-over 4.2 million years ago after separation of colobines from other Old World monkeys. The *RNASE1B* gene is expressed in the intestine of the colobine, where the enzyme is specialized for high activity at the pH of the intestinal environment.

Analysis of the *RNASE1B* coding sequence[888] indicates positive Darwinian evolution for its specialized digestive role.[887] The number of substitutions per nonsynonymous site of the *RNASE1B* gene was significantly greater than that per synonymous and noncoding site. Moreover, the number of radical amino acid substitutions (causing a charge change) per nonsynonymous site outdid (7 out of 9) the number of conservative substitutions (no charge change). Negative charges were strongly favored, indicating nonrandom changes. The resulting RNASE1B is six times more active than RNASE1 at pH 6.3, the pH in the small intestine of the douc langur. RNASE1 has a pH optimum of 7.4. Adaptive changes in pH optimum have been associated with a loss of activity for double-stranded RNA, which is a property of RNASE1 (the physiological significance of this is not known). Thus, digestion was improved in the leaf-eating monkey, allowing it to thrive in its niche, by a gene duplication of *RNASE1* followed by subspecialization of the sibling genes. Because one of the functions (double-stranded RNase activity) of the parental *RNASE1* gene was under relaxed functional constraints, it was expendable in RNASE1B, so its gene mutated sufficiently to specialize for the lower pH of the small intestine. Positive Darwinian selection subsequently favored the probability for fixation of the mutated *RNASE1B* gene. Zhang and colleagues[887] (p. 414) state that their investigation "supports the proposal that gene duplication provides the opportunity for daughter genes to achieve functional specialization" and suggest that "changes in diet and digestive physiology in colobines provided the selective forces for the evolution of a more effective digestive RNase, whereas gene duplication provided the raw genetic material."

In this beautiful example, gene duplication expanded the biological roles of the parental RNase with consequent impact on the biology of the colobine monkey. However, gene duplication did not result (to the best of my knowledge) in the generation of nonenzymatic functions for RNASE1 or RNASE1B and had a greater effect on biological adaptation than on functional innovation of the encoded protein.

SUBFUNCTIONALIZATION AND GENE SHARING

The predominant route to fixation of duplicate genes, according to a proposed model, is not innovation of protein function but rather acquisition of complementary, degenerative mutations in regulatory regions.[50,889] This acquisition leads to a process of subfunctionalization (subdivision of functions) rather than neofunctionalization (development of new function). Subfunctionalization implicates the regulatory modules of gene expression rather than the coding sequences as the primary sites controlling the preservation of duplicate genes. As the name of the duplication-degeneration-complementation (DDC) model indicates, the genes duplicate and then divide the labor of the parental gene by divergent expression patterns. The net result is that the gene expressed in one location complements the function of the other that is no longer expressed in that location. Thus, together the two genes fulfill the function of the original gene, which was expressed in both locations (see [1118,1119] for further discussion).

Subfunctionalization is consistent with mutations being rarely beneficial—in other words, the degenerate mutation in gene regulation is a loss-of-function mutation. The idea that subfunctionalization is more important than neofunctionalization for the retention of duplicate genes is also supported by considerations of adaptations leading to phenotypic diversity in populations. After correlating genome sizes (more duplicated genes) with body masses and population sizes, Lynch and Conery[890] (p. 1403) propose that "much of the increase in gene number in multicellular species may not have been driven by adaptive processes, but rather as a passive response to a genetic environment (reduced population size) more conducive to duplicate-gene preservation by subfunctionalization." In other words, retention of duplicate genes by subfunctionalization fits existing facts of mutagenesis and population genetics and does not require the rapid development of a new function.

The nature of dual functions is important when we consider subfunctionalization with respect to gene sharing. Take, for example, duplication of a hypothetical parental gene encoding a transcription factor that activates target gene A in tissue C and target gene B in tissue D. Subfunctionalization might occur if the expression of one of the sibling genes were lost in tissue C and expression of the other were lost in tissue D: The daughter genes would have subdivided the overall workload of the parental gene, yet each would have remained a transcriptional regulator. This scenario differs from that of a bifunctional parental gene encoding a polypeptide that acts both as a transcriptional regulator and as a membrane protein. The first scenario subdivides biological roles, while the second subdivides molecular functions.

I cover two examples of subfunctionalized expression patterns of duplicated genes encoding transcription factors next. These examples, however, do not necessarily lead to the generation of new biochemical or biophysical functions of the encoded proteins, despite positive Darwinian selection operating on the retained duplicate genes.

Snail/Scratch/Slug genes: The *Snail/Slug* gene family of transcription factors provides a snapshot of extensive gene duplication and subfunctionalization of expression patterns that have taken place in many gene families. *Snail* was first discovered in *Drosophila* and subsequently many family members were identified.[891,892] Phylogenetic analysis has led to the construction of a superfamily comprising two independent families, *Snail* and *Scratch,* which emerged by duplication of an ancestral gene before diversification of the protostomes and deuterostomes. Because no *Scratch* or *Snail* genes have been found in plants or in yeast, they supposedly arose with the metazoans. *Snail* gave rise to *Snail* and *Slug,* and *Scratch* gave rise to *Scatch1* and *Scratch2* as a result of a massive genome duplication at the base of vertebrate evolution.[3,845,893] An additional genome duplication in teleost fish (zebrafish and puffer fish) gave rise to *Snail1* and *Snail2* (also called *snail 1a* and *snail 1b*).[894] *Scratch* and *Snail* gene families proliferated by intrachromosomal duplications. In *Drosophila* there are three linked *Snail* genes (*snail, escargot,* and *worniu*) and three *Scratch* genes (*scratch, scratch-like1,* and *scratch-like2*). Mammals contain two *Snail* family genes (*Snail* and *Slug*) and two Scratch family genes (*Scratch1* and *Scratch2*). A related *Snail* outlier exists in mice (*Smuc*[895]), humans (*Snail3*[896]), and zebrafish/puffer fish (*snail3*[897]). Because chromosomal regions containing these genes show extensive synteny in these species, it was proposed that the genes originated via two rounds of whole genome duplications.[897]

The nonvertebrate chordates *Amphioxus*[898] and *Ciona intestinalis*[899] have only one gene from each family, *Snail* and *Scratch.* Thus, the *Snail/Slug* genes in vertebrates are interesting to study with respect to subfunctionalization after gene duplication because both the orthologous parental gene (*snail*) in invertebrates and nonchordate vertebrates and the daughter genes (*Snail* and *Slug*) in vertebrates are known.

In general Snail/Scratch proteins are conserved tissue-specific transcriptional repressors.[891] *Snail* and *Slug* are involved in many physiological functions, including mesoderm formation, epithelial-mesenchymal transition

(EMT), and tumor progression.[900] *SNAIL-like* expression gives metastatic properties to epithelial cells, adding to the clinical importance of these genes. Snail and Slug proteins are also involved in signaling cascades conferring left-right symmetry, appendage formation, neural differentiation, cell division, and cell survival.

Comparative studies have revealed unexpected changes and reversals of the tissue-specific expression patterns of the sibling *Snail* and *Slug* genes during vertebrate evolution.[900] The protochordates *Amphioxus* and *Ciona intestinalis* express the single parental *Snail/Slug* gene in the mesoderm and at the edges of the neural plate;[898,899] the ancestral preduplicated *Snail/Slug* gene of *Drosophila* is expressed in the precursors of the mesoderm and at gastrulation (see Nieto[891]). In mice and chicken, however, these expression sites are subdivided between the two daughter genes, but differently so in each species.[900] *Snail* is expressed in the embryonic mouse primitive streak and premigratory neural crest cells, while *Slug* shows this expression pattern in the chicken embryo. Mutants of *Snail* or *Slug* show defects corresponding to their expression patterns in mice and chicken. *In situ* hybridization studies have revealed further complexity in the subdivided expression patterns of these two genes. *Snail* is expressed in the cranial premigratory neural crest cells of zebrafish embryos (*snail2*), while *Slug* is expressed in these locations in lizard and turtle embryos. *Snail* is expressed in the tail mesenchyme of zebrafish (*snail1*), lizard, turtle, and mouse embryos, while *Slug* is expressed in these tissues in the cartilaginous dogfish (a species predating bony fish and duplication of *Snail* into *Snail1* and *Snail2*) and chicken. Thus, zebrafish have further subdivided the expression patterns of *Snail* following its duplication into *Snail1* and *Snail2*. The point is this: Duplicated genes subcomparmentalized their expression patterns.

The various expression patterns of *Slug* and *Snail* are consistent with the duplication-degeneration-complementation model of subfunctionalization[889] and highlight differences that may exist between subfunctionalization and gene sharing. In view of the taxon-specific variations in *Snail* and *Slug* expression patterns, it appears as if these genes have maintained functional equivalence throughout evolution. Thus, subfunctionalization of the ancestral *Snail/Slug* expression patterns does not mean that the parental gene was multifunctional before duplication. The differences in expression of *Snail* and *Slug* among species appear to reflect evolutionary dynamism of gene regulation rather than innovation or subdivision of fundamentally different molecular functions of their encoded proteins. Locascio and colleagues[900] suggest that changes in genomic structure or epigenetic modifications may explain these unexpected phylogenetic reversals of the subdivided expression patterns of *Slug* and *Snail.*

Hox genes: *Hox* genes are another example of subfunctionalization of duplicated genes with critical roles in development.[901,902] There are four *Hox* gene clusters in mice and seven in zebrafish due to the genome doubling that has occurred in teleosts.[848,849,1151] The mouse *Hoxa1* and *Hoxb1* genes play divergent, but partially redundant and synergistic, roles in patterning the hindbrain. Among their functions, *Hoxa1* contributes to the appropriate early anterior expression of the *Hoxb1* gene, while *Hoxb1* confers proper identity to the fourth rhombomere. The primary difference between these genes is not in their coding sequences but in their regulated expression. Due to the genomic doubling in zebrafish, zebrafish *hoxa1a* is orthologous to the mouse *Hoxa1* gene, and zebrafish *hoxb1a* and *hoxb1b* are duplicates of the orthologous *Hoxb1* gene in the mouse. Expression studies have indicated that the zebrafish *hoxb1a* and *hoxb1b* duplicate genes have similar developmental roles as the mouse orthologous *Hoxb1* and nonorthologous *Hoxa1* genes—a situation that has been called "function shuffling." In addition to the changes in regulation, there have also been some amino acid sequence alterations of the encoded proteins that account for incomplete biochemical interchangeability in rescue experiments. McClintock and colleagues[902] concluded that initial retention of the duplicates occurred via subfunctionalization of defined *cis*-regulatory elements, which was followed by coding sequence changes that allowed specializations in developmental roles.

The *Snail/Scratch/Slug* and *Hox* gene families represent two of many possible examples of the dynamic function of gene duplication in expanding the roles of genes during development and evolution. However, the genes for these transcription factors continue to encode transcription factors after they have duplicated. The gene duplications enriched source material for the growing complexity of genetic networks, opened opportunities for occupation of new niches in evolution, and contributed to species diversity. This type of subfunctionalization affecting gene expression patterns and affecting primarily phenotype differs from innovation of a new biochemical/biophysical function of the encoded proteins.

RAPID SUBFUNCTIONALIZATION WITH SLOW NEOFUNCTIONALIZATION

Recently He and Zhang[903] provided evidence for rapid subfunctionalization and sustained neofunctionalization of duplicated genes in yeast *(Saccharomyces cerevisiae)* and humans. Analysis of yeast protein:protein interaction data and human expression patterns derived from duplicated and single-copy genes led to a model called subneofunctionalization. The model proposes that subfunctionalization by complementary, degenerate mutations in the regulatory regions of duplicated genes, resulting in their joint levels of expression equal to that of the single parental gene, is generally accompanied by a slow, sustained neofunctionalization. In brief, the subneofunctionalization model was based on the findings that the total number of yeast protein interaction partners and human gene expression sites for duplicates approximate those for randomly picked singletons. He and Zheng[903] (p. 1157) concluded that "enormous numbers of new functions have originated via gene duplication." While it seems incontestable that new protein functions must have arisen as a result of gene duplications, the quantitative relationships between new molecular functions and new biological roles as a consequence of gene duplications are not resolved by this statistical data.

GENE SHARING IS INDEPENDENT OF GENE DUPLICATION

Gene sharing exists when a polypeptide engages in more than one biochemical and/or biophysical function. Multiple molecular functions may be generated, enhanced, subdivided, or reduced before or after gene duplication by mutation and selective pressures depending on the specific circumstances.

The idea that single genes encode multifunctional proteins that can subspecialize after gene duplication is rooted in the early scientific literature. Jensen[13] speculated that primitive enzymes in ancient cells had broad specificity and undeveloped regulation, resulting in substrate ambiguity. He postulated that this led to small amounts of erroneous chemical products that could be exploited for new pathways by subspecialization of daughter genes. Jensen[13] (p. 423) claimed that "gene duplication provided the opportunity for increased gene content and increased specialization of the diverging enzymes, the substrate specialization being further reinforced by the development of regulatory mechanisms." Jensen and Byng[14] also noted that ancestral enzymes may catalyze several reactions that could be subdivided by different proteins after gene duplication. Orgel[17] was another advocate for protein innovation without gene duplication. In a brief report, Orgel[17] (p. 773) wrote the following prophetic words: "In the conventional picture, gene duplication precedes the appearance of a protein with a new function. Here we suggest that an alternative sequence in which the appearance of a new function in a pre-existing protein precedes gene duplication should be considered." However, Orgel added that a new function appearing in a preexisting, single gene would be less likely to survive in haploid prokaryotes than in diploid eukaryotes because multiple gene copies are necessary for neofunctionalization.

After consideration of lens crystallins as well as other multigene families (cf. the hemoglobins and the major histocompatibility complex), Hughes[10] concluded that gene duplication is not necessary for the development of a new protein function. He proposed that single genes typically undergo a period of gene sharing (multifunctionality) *preceding* duplication, allowing daughter genes to subspecialize for a prior function of a bifunctional parental gene. Hughes argued that the time interval before neofunctionalization or subfunctionalization is detrimental to both of the sibling genes due to the accumulation of mutations, few of which are beneficial. Duplication of a bifunctional gene allows each daughter gene to assume one of the two functions of their parent immediately following duplication, eliminating the need for a nonfunctional period before the emergence of a novel role. A multifunctional parental gene would in theory provide an immediate selective basis for maintaining daughter genes.

Similar reasoning has been applied to the evolution of new enzyme activities as a result of gene duplication. O'Brien and Herschlag[664] list enzymes with multiple activities, a situation called catalytic promiscuity. They speculate (p. R100) that "a low level of activity could decrease or eliminate periods of random drift, thereby greatly increasing the probability that the duplicated gene for an enzyme be fixed in the genome and optimized via Darwinian evolution to catalyze a new reaction." In other words, O'Brien and Herschlag[664]

(p. R91) consider that "alternative activities could have played an important role in the diversification of enzymes by providing a duplicated gene a head start towards being captured by adaptive evolution." James and Tawfik[669] (p. 2600) extend this idea and state that "duplicated genes can serve as a starting point for the evolution of a new function only if the protein they encode happens to exhibit some activity towards this new function."

Changes in gene regulation provide one of several possible mechanisms for innovating protein function without loss of the original function.[9,904] The idea that protein function can be modified by changes in gene regulation partially overlaps with the model of complementary, degenerative regulatory mutations that has been proposed for subfunctionalization of duplicated genes.[889] However, gene sharing differs from the subfunctionalization model in (1) eliminating the requirement for gene duplication and (2) allowing the acquisition of new positive gene regulatory capabilities before (or after) duplication. It is also important to emphasize that subfunctionalization is associated with loss of function as well as, in some cases, modification and/or development of new function; gene sharing, by contrast, involves the addition of functions, although switches in function can also occur, as long as the final protein retains more than one molecular function. Subfunctionalization requires gene duplication; gene sharing operates independently of gene duplication.

Finally, protein:protein interaction networks provide indirect support for the idea that gene sharing and functional innovation are independent of gene duplication.[905] Networks link proteins (as well as other molecules or functions) into nonlinear matrices that depend on their structural or functional interactions (see Chapter 10 for a discussion of networks). As a rule biological networks are hierarchical: some nodes (proteins in this case) have many links, while others have few. A highly connected protein (multiple links with other proteins) implies that it interacts with many other proteins and has numerous roles. Networks are useful not only for graphing protein interactions and functions but also for predicting them. Gene duplication contributes to the growth of protein interaction networks during evolution, but only to a minor extent.[906] Most of the growth occurs by new connections with old proteins. While networks do not define the precise functions of the proteins, highly connected proteins are indicative of gene sharing occurring both before and after gene duplication.

LENS CRYSTALLINS: GENE SHARING AT DIFFERENT STAGES OF DUPLICATION

The diverse lens crystallins provide numerous examples of gene sharing at various stages of gene duplication[904] (Figure 9.1). I discuss several examples here; these should be considered in connection with Chapter 4.

Single-copy genes: ε- and τ-crystallins: Lactate dehydrogenase B_4/ε-crystallin[21] and α-enolase/τ-crystallin[369] are encoded in single-copy genes: These proteins

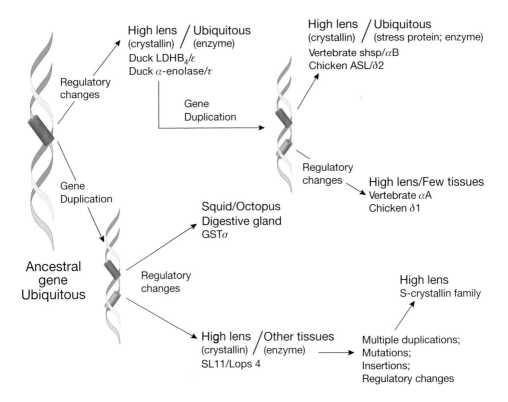

Figure 9.1. Gene duplication in crystallin evolution. Examples are chosen to demonstrate the recruitment of crystallin genes after no duplications (LDHB₄/ε-crystallin and α-enolase/τ-crystallin), one duplication (shsp/α-crystallins and argininosuccinate lyase/δ-crystallins), and multiple duplications (glutathione S-transferase-derived/S-crystallins). Adapted from Piatigorsky.[904]

are ubiquitously expressed at low levels for generating enzymatic functions and highly expressed in the lens of certain species that use these enzymes as crystallins. Thus, these enzyme-crystallins are examples of gene sharing that occurred in the absence of gene duplication. This does not mean, of course, that one or more mutations resulting in an amino acid change did not precede or accompany selection of the enzyme as a lens crystallin. Such a mutational change may have changed the enzyme, either predisposing it to or promoting a crystallin function without eliminating its enzymatic function. Both $LDHB_4$ and α-enolase are members of gene families that were generated before these individual members developed high lens expression and crystallin function.

Lactate dehydrogenase A is another member of the lactate dehydrogenase family that was independently selected to be a lens crystallin (v-crystallin) in the platypus.[907] While it is likely that v-crystallin was recruited for its structural role in the lens by a gene-sharing process, it is not known yet whether v-crystallin has enzymatic activity or whether it is expressed outside of the lens.

Duplicated genes: α- and δ-crystallins: The shsps/α-crystallins involve gene sharing associated with gene duplication and subspecialization. The ancestral gene (probably also a member of the shsp gene family) duplicated at least 500 million years ago.[291] The daughter genes have since migrated to different chromosomes. In humans, the αA-crystallin gene is on chromosome 21[908] and the αB-crystallin gene is on chromosome 11.[909,910] As discussed in Chapter 4, αA-crystallin has subspecialized for very high expression in lenses of vertebrates, where it acts as a refractive protein and a molecular chaperone protecting lens proteins from age-related aggregation. Presumably as a consequence of this subspecialization, the αA-crystallin gene diminished stress-inducibility,[1133] although it is expressed at a low level in various tissues (retina, thymus, spleen) throughout the body. By contrast, the αB-crystallin gene is expressed ubiquitously and is a stress-inducible, authentic shsp; it is also highly expressed in the lens (but less than αA-crystallin), where it has numerous functions (see Chapter 4). Although this scenario is consistent with subfunctionalization of daughter genes by changes in gene regulation, and probably is associated with unknown changes in amino acid sequence, αB-crystallin did not lose its shsp role when it added a crystallin role. Moreover, aA-crystallin continued to be constitutively expressed to some extent in nonlens tissues, where its physiological role is not known. The subfunctionalized *α-crystallin* genes, except for heat inducibility, remained bifunctional. It is also important to note that *both* the *αA-crystallin* and *shsp/αB-crystallin* genes are preferentially and highly expressed in the lens despite their nonlens roles, suggesting that the ancestral gene might have been bifunctional before it duplicated. It is beyond our reach to know at present whether the ancestral gene was used as a lens crystallin before it duplicated. Because duplication was so very long ago, it will be challenging to identify the *αA/αB-crystallin* parental gene (if it still exists) in an ancient species to prove that it had more than one function.

Argininosuccinate lyase/δ-crystallins are another set of crystallins—in this case enzyme-crystallins—that underwent gene duplication and subspecialization. Although the enzyme argininosuccinate lyase is present in all species, its use as δ-crystallin is limited to birds and reptiles.[20,36,911]

There are two, extremely similar, tandemly linked *δ-crystallin* genes in chickens[912,913] and ducks.[914] In chickens, δ-crystallin is expressed thousands of times more highly in the lens than in other tissues.[421] Chicken δ1-crystallin is enzymatically inactive and expressed 50–100 times more highly in the embryonic lens than is the enzymatically active argininosuccinate lyase/δ2-crystallin.[20,420,915–917] Unexpectedly, despite being inactive, the chicken *δ1-crystallin* gene is still expressed at a low level in other tissues, as is the *argininosuccinate lyase/δ2-crystallin* gene.[917,918] δ1-Crystallin has a refractive role in the lens but an unknown, nonenzymatic embryonic function outside of the lens. One possibility is that the δ1-crystallin polypeptide modulates argininosuccinate activity by interacting with the active δ2-crystallin polypeptide as they form tetramers.[919] It is even possible that leakage of *δ1-crystallin* gene expression in nonlens tissues is a relic of past history and serves no useful purpose today, although this is not a satisfying conclusion in view of it having a tissue-preferred expression pattern.

The catalytically inactive δ1-crystallin and active δ2-crystallin are equally expressed in the embryonic duck lens.[370,911] Thus, in contrast to the chicken lens, the duck lens is virtually blazing with argininosuccinate lyase activity.[20,920,921] Low levels of both *δ-crystallin* genes are expressed in the embryonic heart and brain of the duck, with the *argininosuccinate lyase/δ2-crysallin* gene being favored, as in the chicken.

The evidence suggests that the ancestral *argininosuccinate lyase/δ-crystallin* gene became bifunctional before it duplicated. The strongest support for this conclusion is that both δ-*crystallin* genes are highly expressed in the duck lens. This is similar to the situation for the α-crystallins: Both genes are highly expressed in vertebrate lenses. In addition, although the chicken *argininosuccinate lyase/δ2-crystallin* gene is not highly expressed in the lens, it is still easily detectable in the lens and contains a lens-preferred enhancer in its third intron as does the δ*1-crystallin* gene.[419,421] It seems improbable that both δ-crystallin genes would have acquired similar DNA elements for high lens expression independently. The simplest hypothesis is that the ancestral *argininosuccinate lyase* gene began specializing for high lens expression and crystallin function before it duplicated, and that high lens expression of the δ*2-crystallin* gene was reduced transcriptionally or posttranscriptionally after duplication in chickens but not in ducks.

The pattern of argininosuccinate lyase activity among δ-crystallins of various bird and reptile lenses is consistent with the idea that the enzyme became bifunctional before its gene duplicated, and that expression of one of the daughter genes became repressed in the lens only in certain species. In contrast to the chicken and pigeon,[391,922] the African ostrich *(Struthio camelus)*, a member of the flightless Ratitae and considered a living fossil of primitive birds, has an enzymatically active lens δ-crystallin,[390] as does the goose,[923] swan, and ancient caiman[391] (see also Hughes[750]). The sequence and structural similarity of the two δ-crystallin proteins indicate that loss of enzyme activity was probably due to loss of substrate-binding ability.[924]

Phylogenetic analysis also indicates that the *argininosuccinate lyase* gene duplicated before the divergence of duck (Anseriforms) and chickens (Galliformes) but after divergence of birds and mammals.[750] The sequence similarity of the two linked genes in the chicken and duck is believed to reflect homogenization by concerted evolution approximately 24 million years ago in ducks and 60 million years ago in chickens.

Finally, additional strands of data support the idea that the *argininosuccinate lyase* gene was bifunctional before it duplicated. An early hybridization study showed that chicken δ1-crystallin cDNA hybridizes more effectively to genomic DNA of birds (chicken, quail, turkey, and duck) than of reptiles (python, gecko, caiman).[925] As expected, there is little hybridization to DNA of fish (herring, salmon), amphibians (frog, newt) and mammals (mouse, calf). While sequence divergence between the chicken probe and reptilian or other vertebrate *argininosuccinate* genes probably accounts for much of the difference in hybridization intensity, it is possible that the birds have two *δ-crystallin* genes, while certain reptiles have one *argininosuccinate lyase* gene. Fragments of protein data also suggest that some reptiles have a single *argininosuccinate lyase* gene that encodes both the enzyme and lens crystallin. In general at least two *δ*-crystallin polypeptides are resolved by electrophoresis on sodium dodecyl sulfate-polyacrylamide gels,[370,919] a technique that relies principally on size to separate proteins. In contrast, only one electrophoretically unresolved *δ*-crystallin is present under these conditions in the ancient caiman lens.[925] Moreover, peptide analysis of *δ*-crystallin from the tuatara, another ancient reptile from the sphenodonts, points to the presence of a single argininosuccinate lyase/*δ*2-crystallin-like polypeptide.[264] These "straws in the wind" need confirmation and further experimentation, but they suggest that some ancient reptiles may have used a single bifunctional argininosuccinate lyase as an enzyme and lens crystallin.

Multiple-copy genes: S-crystallins: The complex family of *S-crystallins* (homologues of glutathione S-transferase) of cephalopods (squid and octopus) has specialized for lens expression and crystallin function (see Tomarev and Piatigorsky[194]). There is no report of glutathione S-transferase or its derivatives being used as a lens crystallin in other species. *S-crystallins* form a large gene family that may exceed twenty members. Because the multiple *S-crystallin* genes are more similar within than between species of cephalopods, many duplications presumably occurred after the 200–300 million years of divergence between squid and octopus. These genes have drifted apart by sustaining numerous sequence mutations as well as an insertion of a variable-length peptide encoded by an exon placed between exons 3 and 5 by unequal crossing-over.[274,399] The sequence and length variability of the central peptide suggest that it was inserted a number of times independently during evolution, each time inactivating glutathione S-transferase activity.

The *S-crystallin* genes are expressed strictly in the lens. There is, however, an extremely active glutathione S-transferase encoded in a separate gene, *glutathione S-transferase σ (GSTσ)*.[400,925a] GSTσ is expressed highly in the digestive gland but at very low levels in other tissues (mantle, testis, ovary, gills, digestive, gland, heart) and, surprisingly, less in the lens than in any other tissue.

In addition to the collection of inactive S-crystallins that contain an inserted peptide, one S-crystallin lacks an inserted peptide (SL11-crystallin in the octopus *Octopus sloani pacificus,* and its orthologue Lops4-crystallin in the squid *Loligo opalescens*). The *SL11/Lops4-crystallin* gene structure resembles that of *GSTσ.* SL11/Lops4 has enzyme activity, although much less than *GSTσ.*[401] It has been proposed that a *GST* ancestral gene duplicated to produce *GSTσ* and *SLL/Lops4,* and that the daughter *SL11/Lops4-crystallin* gene subsequently proliferated to give rise to the multiple S-crystallins. Because *GSTσ* is so poorly expressed in the lens (if at all), and *SL11/Lops4* is lens-specific, it is unclear whether the former was originally bifunctional and lost lens expression or whether high lens-specific expression evolved with the *SL11/Lops4-crysallin* gene after duplication of the ancestral gene. We might favor the idea that lens expression of the existing *GSTσ* gene was lost secondarily after duplication of the ancestral gene because of the potentially important detoxification and stress-protecting functions that an active GST could have provided the evolving lens. That scenario would give a transient gene-sharing role to the authentic *GST* gene before the complete subspecialization of the multiple S-crystallins for a structural role in the lens. Whatever the pathway, SL11/Lops4 probably serves an enzymatic and structural function in the cephalopod lens today.

TAKE-HOME MESSAGE

Gene duplication often results in the acquisition of new functions (neofunctionalization) for old proteins, but it is not necessary for neofunctionalization to take place. Indeed, gene duplication may not even be the major mechanism for functional innovation of proteins. Gene sharing almost certainly serves as one of the mechanisms for retaining duplicate genes, along with subfunctionalization and Darwinian selection. The acquisition of new functions by gene sharing operates independently of gene duplication and may occur in single-copy genes, duplicate genes, or individual members of gene families. The development of gene sharing does not require changes in the coding sequences (although these may occur) and can be initiated by changes in gene regulation or in the microenvironment of the protein in question. Gene sharing thus provides an important source of new protein functions that operates independently of gene duplication.

10

Gene Sharing and Systems Biology: Implications and Speculations

Earlier chapters of this book defined (Chapter 1), provided examples (Chapters 4–7), and considered the dynamism (Chapters 8, 9) of gene sharing as an evolutionary process. Historically genes have been subject to many definitions and in general these have not been unequivocally resolved by the molecular era (Chapter 3). In particular, ambiguities remain with respect to defining the structure and function of a "gene." The previous chapters have focused mostly on the molecular biology of gene sharing by examining individual proteins, such as metabolic enzymes and various specific proteins commonly identified with one of their physiological roles (for example, crystallins/refraction, hemoglobins/oxygen transport and transcription factors/gene expression). However, gene sharing has many implications at the level of systems biology, where its effects can also be discerned or implied.

This chapter considers the role of gene sharing in the higher echelons of biology and evolution, including networks, evolvability, gene regulation, molecular clocks, gene redundancy, and horizontal gene transfer. I have also taken the liberty to speculate that infrequent gene-sharing events may contribute to cellular noise, which in turn may benefit diversity and evolution.

NETWORKS

Genes and proteins carry out their biological roles in complex cellular (or sometimes extracellular) environments. Thus the challenge is to interpret genomic and proteomic data within cellular contexts rather than as bits and pieces. Although limited examples of gene sharing are presented in this book, many if not most proteins appear multifunctional. Networks seek to diagram and explain these mazes of protein interactions.

Networks, a unified science of "systems biology," are an exciting development of the molecular era consistent with the idea that gene sharing and pro-

195

tein multifunctionality are extensive.[905,1152] Networks integrate different organizational levels of information stored in genes, proteins, and metabolites to create robust, interacting functional units, called modules, that account for cellular behavior.[926–930,1153] Network analysis concerns simultaneous changes in the expression of many genes and can predict new functional associations that would be impossible to connect by conventional biochemical and molecular methods.[931,932] Yeast two-hybrid assays, mass spectrometry, and phage display, among other technical procedures, reveal thousands of potentially relevant protein:protein interconnections in bacteria and eukaryotic cells.[101,105,933–936] Networks transform molecular biology to "modular" biology[927] and link the rules of biology to many different disciplines, including statistical physics, engineering, the Internet, ecosystems, and social interactions.[937,938,1154] Protein interactions, transcriptional regulatory interactions, signaling and developmental interactions, and metabolic interactions have all been woven into networks of biological significance. Biological networks employ a new language and logic taken from engineers and computer scientists that promises a novel understanding of health and disease. It is interesting to consider networks as a higher-order representation of gene sharing or multifunctional proteins. Indeed, at the conceptual heart of these biological networks is the idea that individual proteins play many roles.

Biological networks are not designed; they obey certain evolutionary laws, such as preferential attachment. Networks also retain the concept of tinkering, as do evolutionary processes.[104,905] The conservative nature of evolution is seen in recurring circuit "motifs," which are patterns of interconnections that aggregate into clusters of overlapping functions.[96,937,939] An important feature of protein interaction networks, known as interactomes,[101,940,941] is that many proteins combine with few partners and a few proteins combine with many partners.[936,942,1155,1156] Networks create a global topological organization within which biological functions are performed by modules and integrated by the interactions of nodes. A few highly connected proteins have central roles mediating the interactions among many less connected proteins.[943] These hierarchical networks provide advantages for living systems, such as increasing stability and robustness, which are defined as resistance to perturbations.[106,927,936,944,945] An example of the complex interaction pattern of bacterial proteins is seen in Figure 10.1.

Attempts have been made to correlate protein interactions with the rate of

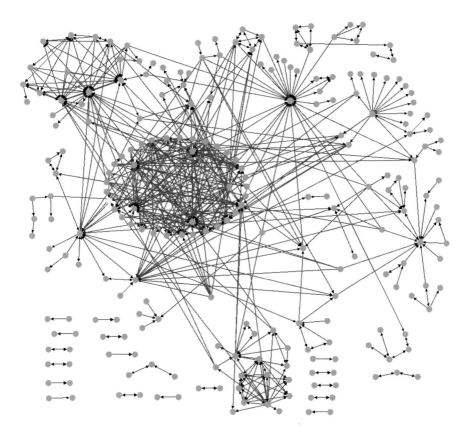

Figure 10.1. Network of validated interacting bacterial protein complexes of *Escherichia coli.* The interaction diagram illustrates directional associations between interacting proteins. In other words, if protein X is affinity purified and copurifies with protein Y, a line is drawn between the two with the arrow pointing from X to Y. Reciprocal interactions are indicated by double-headed arrows. The diagram shows 473 edges (uni- or bidirectional interactions) between 237 proteins (nodes). I am grateful to Dr. Gareth Butland (Banting and Best Department of Medical Research, University of Toronto, Canada) for this figure. Adapted from Butland et al.[936] and reproduced with permission from the publisher.

evolution (sequence divergence). Highly connected proteins were thought to evolve more slowly than those with fewer connections.[946,947] On the other hand, additional studies indicated that this was due to bias toward counting more interactions for slowly evolving abundant proteins and that highly interactive proteins may evolve more rapidly.[948,949] Old proteins that are developing novel functions by way of new protein:protein interactions may be less constrained by natural selection—a possibility that is consistent with the observation that networks are surprisingly dynamic. An estimate from yeast data indicates the addition of more than a hundred protein interactions every million years, some involving previously unconnected proteins.[906] Proteins seem to test new alliances, some of which must surely result in new roles, with the more highly connected proteins gaining interactions preferentially over the less connected proteins.[906,950,951] According to one analysis,[941] multiple interactions occur between recent and more ancient proteins, and Li and colleagues[941] (p. 541) state that "new cellular functions rely on a combination of evolutionarily new and ancient elements, consonant with the classic proposal of evolution as a tinkerer that modifies and adds to pre-existing structures to create new ones." Another study, however, suggests significant partitioning between ancient and younger proteins.[952] Whatever the exact correlation between rates of interactions of proteins that evolved at different times, the data indicate dynamic interactions between many proteins, a conclusion that is consistent with an active role for gene sharing during evolution.

As promising as networks are, difficulties of integrating genomics data must be resolved before they can be interpreted with confidence with respect to functional significance (see Kemmeren and Holstege[953]). The voluminous data used to construct networks are of uneven quality. This is especially true for protein:protein interactions data. There is often disappointing overlap between distinct data sets of protein:protein interactions, and many "false positives" using different techniques cloud the picture. Functions must be interpreted cautiously, because they are often conceived by correlations, which may be misleading. Linking potential associations of proteins without considering cellular locations, intracellular concentrations, and stability is also an error-prone approach, as indicated in a mathematical model of protein-protein interactions in cells.[954] Protein locations are critical to facilitate interactions as well as to prevent detrimental cross-talk. Elegant studies in yeast have shown that protein colocalization is strongly correlated with transcriptional co-expression and, presumably, with biological function.[100,955,956] Greater interaction seems to occur between proteins that co-localize within small compartments (such as microtubules or the actin cytoskeleton) than within larger spaces (such as the cytoplasm), although interactions between subcellular compartments are also apparent. Thus, networks indicate extensive functional interactions, but delineation is still at a formative stage in its application to biology.

EVOLVABILITY

The numerous interactions comprising networks suggest alternative roles for individual genes and proteins, and network growth implies that evolution is associated with greater multifunctionality of genes and proteins. The idea that individual genes and proteins diversify functionally as they specialize for each task during evolution is consonant with the importance of tinkering with the use of existing structures[61] and with the notion that diversity is driven by changes in gene regulation and protein utilization more than by the addition of new genes.[130]

Although controversial, evolvability is another consideration of gene sharing that is consistent with the network property of increasing cross-talk during evolution. Evolvability encompasses the idea that the capacity to evolve—also called evolutionary adaptability—is a selectable trait that facilitates the generation of phenotypic diversity.[945,957,1157] A highly evolvable species is one that can adapt readily to a changing environment through genetic variability. Such a species can tolerate mutations of various sorts such as nucleotide substitutions and DNA additions, deletions, and rearrangements. Those who object to the notion of evolvability claim that selection can only function on present conditions: It does not anticipate coming events. Selection for the *propensity* to evolve suggests selection for the ability to deal with future events. Nonetheless, arguments consistent with gene sharing support the possibility that evolvability is a selectable trait. Computer simulations also indicate that environmental change, especially dramatic change, selects for evolvability by providing selective pressure for genetic change.[958] Earl and Deem[958] (p. 11536) state that "not only has life evolved, but life has evolved to evolve." What does this mean with respect to gene sharing and protein multifunctionality?

Kirschner and Gerhart[957] argue that evolvability minimizes the interdependence of cellular components, reduces constraints on change, and allows the accumulation of nonlethal variation. Evolvability would thus contribute to survival by facilitating the conservation of specialized components while maintaining sufficient flexibility to resist changing environmental challenges and to occupy new niches as they become available. In other words, evolutionary survival depends on genes and proteins being exquisitely adapted for their specialized functions yet able to sustain change without losing their critical roles. This definition fits the criteria of gene sharing. According to Kirschner and Gerhart[957] (p. 8420), "the properties of versatile protein elements, weak linkage, compartmentation, redundancy, and exploratory behavior"—all traits that are consistent with gene sharing—"confer evolvability on the organism by reducing constraints on change and allowing the accumulation on nonlethal variation." Evolvability has resulted in the ability of genes to accumulate mutations and become polymorphic and of proteins to interact with a variety of targets, engage in alternative functions, and be redundant. The success of life on earth indicates that the conflict between specialization and diversity with redundancy has been resolved by mechanisms that have been selected during evolution.

The evolution of adaptability can be seen at the DNA level by the perpetuation of mechanisms that vary the rate of genetic change that may benefit the organism (see Radman et al.[959]). Bacteria have so-called mutator and hyper-rec mutations that link loss of some specific function with increased rates of mutations and/or DNA recombination. On a population level, mutator mutations can result in adaptive alleles that take over the wild type population, probably due to linkage of the mutations with another gene that bestows an advantage under the existing conditions. In other words, selection is based on the linked advantageous gene rather than on the mutator gene itself increasing the mutation rate[945] (see p. 285). In vertebrates, immunoglobulin genes are famous for their ability to generate antibody diversity by having "hotspots" for mutation and DNA rearrangement. Thus rates and patterns of mutations are not always uncontrolled processes dependent entirely on stochastic error.

Many enzymes have promiscuous catalytic activities and broad substrate specificities.[664,960–962] We might imagine that promiscuous enzymes enhance evolvability by having alternative catalytic potentials that could be called upon to varying extents depending on need. That promiscuous activities exist at all for enzymes is incongruent with tight functional constraints and is consistent with the concept of gene sharing. The evolution of enzyme function is believed to involve ancestors with broad substrate specificity.[963] Relevant results were obtained on the process of evolution of non-trypsin-like primary specificities in the trypsin superfamily of serine proteases.[964] Phylogenetic studies on a group of serine proteases involved in cell-mediated immunity (α-and β-chymase, and cathepsin G protease clusters) show that the predicted common ancestor to these enzymes has broad substrate specificity and tolerates mutational changes in the pocket containing the specificity-conferring amino acids. Wouters and colleagues[964] suggest a *despecialization* (my italics) step underlying evolution of new primary specificities in this protease superfamily. The unexpected aspect of this finding is that a new enzymatic function was derived via an intermediate with a structural and functional relaxation of the trypsin-like primary specificity of its predecessor, and that derivation is consistent with a transient step that appears to reverse evolution. Thus, widening and narrowing of promiscuous functions—or changing extents of gene sharing—appear integral to gene and protein evolution. Gene sharing adds to adaptability and evolvability.

Structural plasticity and marginal stability are inherent traits of proteins that broaden specificity and increase functionality.[965] Interestingly, alternative RNA splicing may contribute to intrinsic disorder and functional diversity of proteins.[1158] Selection for marginally stable proteins increases sequence plasticity because they would be less constrained and more tolerant of mutations—more robust—than rigidly stable proteins.[966] Computer-generated models indicate that evolution of robust sequences is a derived property that results without evolutionary pressure for robustness.[1159] Marginally stable proteins with a minimum of constrained amino acids can support more heterogeneous and extensive sequences than highly constrained proteins and should be able to fold flexibly. Disordered regions of polypeptides can be well conserved and are often functional or are important for assembly of functional complexes.[967] In modeling experiments relaxing the stability requirement for proteins resulted in more efficient evolution of function (but see[1160]).[968] This finding agrees with results of experiments that combined mutational analysis of folding kinetics with computer simulations:[969] Protein folding undergoes a rate-limiting transition-state ensemble whose topology is dictated by few key residues in the polypeptide backbone. In summary, genes encoding marginally stable proteins can absorb nucleotide changes due to the reduced constraints for precise amino acid sequence—at least for stretches of amino acid sequence—and many amino acid changes can be tolerated without significant changes in protein structure. Marginally stable proteins, then, favor evolvability, can respond quickly to new stresses, and are consistent with the prevalence of gene sharing and multifunctionality.

Protein promiscuity due to conformational diversity has been called the "new view" of proteins, whereby one sequence adopts multiple structures and functions[15] (Figure 10.2). James and Tawfik[15] suggest that conformational and functional diversity of proteins are evolvability traits that facilitate the evolution of new activities. Conformational diversity has been demonstrated in SPE7, a multispecific IgE antibody against a 2,4 dinitrophenol hapten that binds a number of unrelated ligands.[970,971] This is not due to "hydrophobic stickiness" but to specific interactions of the antibody, with each cross-reactant forming different hydrogen bonds: Specificity and promiscuity are not mutually exclusive properties. James and Tawfik[972] (p. 2191) state that "promiscuous activities need not be nonspecific as such, but rather each protein may have its own unique pattern of promiscuous activities, and each activity, standing alone, is highly specific." This should apply to enzyme specificities as well because these depend on selective binding of transition states for the reactions. Gene sharing is compatible with the idea that promiscuity does not obviate specificity.

A method called dynamic-ensemble refinement, which combines nuclear magnetic resonance relaxation phenomena with molecular dynamics simulations, indicates that proteins do indeed have considerable conformational heterogeneity throughout their structure.[973] Application of the technique to human ubiquitin indicates that each conformation results from tightly packed interior atoms, but overall the ensemble has a "liquid-like character." Lindorff-Larsen and colleagues[973] (p. 131) conclude that "the native state must be considered as a heterogeneous ensemble of conformations that interconvert on the picosecond to nanosecond timescale, as well as populating more expanded conformations arising from rare but large fluctuations on much longer timescales, such as those revealed by hydrogen exchange experiments." This fluid view of proteins, in contrast to a rigid crystallographic

portrait, supports the structural flexibility that is consistent with multifunctionality. Modeling experiments utilizing multiple sequence alignments of a
small protein interaction module (the WW domain) have shown that information on the coevolution of amino acid residues is sufficient to specify se-

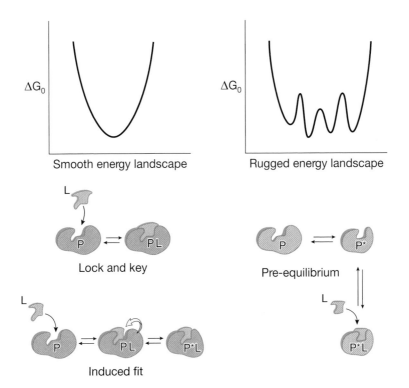

Figure 10.2. Two models for polypeptide interactions. The left diagram represents the
old view: The polypeptide has one primary structure with a smooth energy landscape. When it docks with another protein, it either fits exactly (lock and key) or it
undergoes conformational changes to adopt a close fit with its partner. The right diagram represents the new view: The polypeptide exists in a dynamic state of many
conformations of similar free energy in equilibrium with one another. Each one of
the many conformations can interact with a different partner. I am grateful to Dr.
Dan S. Tawfik, Department of Biological Chemistry, The Weizmann Institute of Science, Rehovot, Israel, for this figure. Adapted from James and Tawfik[15] and reproduced with permission from the publisher.

quences that fold into native structures.[974,975] Based on their analysis, Socolich and colleagues[974] (p. 512) envision "proteins as sparsely coupled architectures with redundant strong constraints linking a few sites, and a great deal of near-independent variation at most sites." Artificial WW sequences have been made that function like their natural counterparts, a finding that Russ and colleagues[976] (p. 579) claim is consistent with the notion that "a small quantity of sequence information is sufficient to specify the global energetics of amino acid interactions." These studies portray polypeptides as being robust to sea changes in amino acid sequence without any loss of original structure and function, but as undergoing enough small changes to be able to interact with many different polypeptides. Thus, proteins are simultaneously robust to change in structure and function yet highly variable—both properties that are conducive to gene sharing.

Finally, mutations may often have large effects on extraneous functions of proteins without significantly affecting their primary native function. This was shown in the laboratory by creating genetic diversity in genes for metabolic enzymes that have promiscuous activities and, after transformation in bacteria, assaying the mutant genes for their natural activities and promiscuous activities.[16] The genes that were mutated included carbonic anhydrase II (which has a promiscuous esterase activity), serum paraoxonase (which has promiscuous lactonase activity and hydrolyzes aryl esters and organophosphates without apparent physiological relevance), and a bacterial phosphodiesterase (which has promiscuous lactonase and esterase activities). Most striking was that the promiscuous activities of the mutant proteins changed much more than the primary activities, which were sometimes not affected at all by the induced mutations. The mutations were generally in the residues that form the walls and perimeter of the active sites of the enzymes, especially on surface loops that are part of the substrate-binding pocket. The versatility of promiscuous proteins may be analogous to that of the transient intermediates with wide substrate specificities that form ancestral proteins during the evolution of new functions.[964] The promiscuous enzyme activities appear to possess unusual plasticity that may favor rapid evolution and evolvability. Thus, the ability of mutations to uncouple promiscuous and native enzyme activities indicates that trade-offs in function are not necessary as evolution proceeds: Mutations can enhance one function without compromising another.[977]

SELECTIVE PRESSURE AFFECTING GENE REGULATION

New functions can arise by novel protein:protein interactions resulting from small changes in protein structure due to genetic mutations in the coding sequence. The altered genes are then fixed in the population or weeded out, depending on the selective pressure on function. Gene sharing encompasses the idea that new functions may arise by fortuitous interactions of proteins with other molecules in a crowded, fluctuating cellular environment. Such cases could involve considerable selective pressure on gene expression.

Correlations between the divergence in levels of gene expression and gene sequence have addressed the question of selective pressure on levels of gene expression.[978] Comparisons between humans and chimpanzees have indicated parallel changes in expression levels and encoded amino acid sequences in genes that are active in many tissues. On the other hand, the correlation between changes in levels of gene expression and amino acid sequences is less tight in genes that are active in fewer tissues. These results were interpreted to be consistent with a model of neutral evolution with negative selection. In a model that is based on *cis*-regulatory elements and that ignores interactions between genes, Khaitovich and colleagues[979] (p. 929) state "that the majority of genes expressed in a certain tissue change over evolutionary time as the result of stochastic processes that are limited in their extent by negative selection rather than as the result of positive Darwinian selection." However, evidence for positive selection on gene expression has been obtained for genes on the X-chromosome expressed in testis.[978] In addition, the results of another study correlating gene expression and gene sequence data indicate that both are subject to the constraints of purifying selection and are influenced by positive Darwinian selection.[980] Current data are consistent with levels of gene expression being subject to neutral changes during evolution and with changes in gene expression levels being associated with functional changes under selective pressure. Gene regulation is thus a focal point for evolving protein function, as predicted by the gene-sharing concept.

FUNCTIONAL SWITCHING AND THE NOTION OF FUNCTIONAL "TRESPASSING"

Functional switching occurs frequently among proteins.[981,982] In general, functional switching depends on changes in protein:protein interactions, and the alternate functions of the protein are mutually exclusive. The polypeptides involved in functional switching are controlled by the frequency of encounters, their local concentration, and the local physicochemical environment.

Beckett[982] has reviewed functional switches for four proteins: dimerization cofactor of hepatocyte nuclear factor (DCoH), β-catenin, plasmid-encoded replication initiator (Rep. protein), and bacterial biotin repressor (BirA). Each of these proteins has an enzymatic function as well as a transcriptional regulatory function, depending on its oligomeric state. The proteins take advantage of molecular mimicry and structural plasticity by utilizing the same surface for their different protein:protein interactions and adopting different conformations for their alternate functions. There is no obvious physiological link between the alternate functions of DCoH, β-catenin, and Rep. protein. By contrast, the dual functions of BirA are elegantly connected. The BirA ligation function involves biotin metabolism and its transcription function regulates biotin biosynthesis. Even the intermediate (biotinoyl-5'-AMP) that ligates biotin to the carboxyl protein subunit (BCCP) of acetyl-CoA carboxylase allosterically activates BirA for binding to the operator DNA regulating biotin transcription (see Beckett[982] for references).

If proteins are constantly bumping into one another within a dynamic cellular environment, do they ever engage in spurious functions as a consequence of sporadic interactions and, if so, to what extent does this affect cellular behavior? Such rarely performed functions of proteins would be akin to a low-level cellular background and might represent a "testing ground" for functions that eventually become prevalent and fixed into the normal repertoire of cellular activities. When such a stochastic behavior of a protein displaces (most probably very inefficiently) the function of another protein, it might be considered as "trespassing" into the pathway normally performed by a different protein. The word "trespassing" is biased (and therefore placed in quotation marks) because an occasional engagement in another function could be an advantageous selected feature, itself falling under the domain of gene sharing. However, if we accept the idea that proteins can spuriously engage in so-called foreign functions, then we recognize that they may not be confined by absolute functional boundaries. Because biology is remarkable and imaginative does not mean that it is neat with clear boundaries. A simple example is the activation of a surface receptor for a specific ligand by a weakly interacting, atypical stimulus—a type of imperfect ligand mimic. Indeed, engineered atypical interactions such as the development of artificial receptor mimics have a role in drug development.[983]

A hypothetical case of functional trespassing would be the infrequent use of a structural cytoskeletal protein (actin perhaps) for a signaling role in a metabolic pathway (Figure 10.3). For the sake of argument, let's say that actin can carry out this function fortuitously, although poorly, as a consequence of an inherent property that would not have been selected for its signaling role in this pathway. An analogy is the promiscuous catalytic potential of an enzyme that has no known physiological role (which does not mean that it has no physiological role). The infrequent, relatively inefficient use of actin as a signaling protein would be a stochastic event dependent on the precise condi-

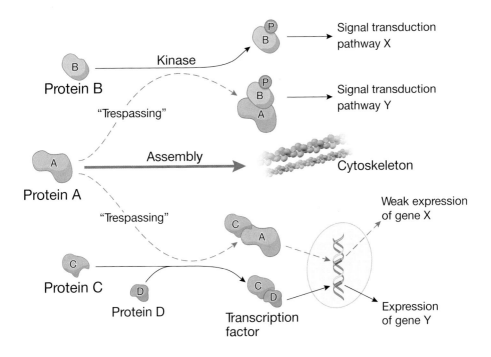

Figure 10.3. Gene sharing and the "functional trespassing" hypothesis. In this hypothetical model, based on the gene-sharing concept, protein A, normally used as a component of the cytoskeleton (center), can occasionally be incorporated into a signal transduction pathway (upper), where it may alter the course of the pathway, or into a transcriptional pathway (lower), where it leads to weak expression of gene Y.

tions within the cell at the moment. This stochastic functional switching of a protein differs from the dual use of a protein that contains two controlled functions, such as β-catenin being both a cytoskeletal anchor and a cotranscriptional activator.[984,985] The different functions of β-catenin are selected for specific roles that take place in a coordinated fashion and in response to intended stimuli. By contrast, the hypothetical functional trespassing of actin is an accidental display of inherent potential.

In short, I am proposing that the gene-sharing concept implies that a protein that normally performs one role may, occasionally, contribute inefficiently to another function. These stochastic trespasses would be difficult to recognize, inasmuch as they are biochemical "accidents." They might account for some of the difficulties that investigators have with so-called contaminants when attempting to purify proteins from cell extracts. Some (probably many) protein:protein interactions in cell extracts are fortuitous; others may have a low-level physiological significance. Functional trespassing would also provide a mechanism for explaining some features of variability that exist for all biological behaviors. The idea that proteins cross lines of specialized functions due to structural similarities and affinities with other proteins, and fluctuations in concentrations resulting from variations in gene expression, is analogous to the idea of molecular mimicry, a suspected (but not proved) cause for autoimmune diseases.[986]

FUNCTIONAL NOISE

The idea that proteins may cross functional barriers suggests that gene sharing may contribute to functional noise by cell-to-cell fluctuations in the expression of specific genes and proteins. The inherently stochastic, quantal expression of genes and proteins is a major factor in the heterogeneity within clonal cell populations in response to environmental signals.[986a,1161–1163] The important and complex effects of noise on developmental pathways, cell cycle control, and optimal levels of metabolites have been recognized.[987,988] Quantitative effects of noise in gene expression in prokaryotes[989,990] and eukaryotes[991,992] are a more recent achievement.[993] Although noise in gene expression is suppressed by various mechanisms, including negative feedback,[993–995] the complexity of transcription and translation provides many steps that together would be expected to lead to significant stochastic fluctuations in the intracellular levels of individual proteins.[996,997]

Gene expression noise is not dependent on the regulatory pathway or absolute rate of expression and is gene-specific.[992] Raser and O'Shea have proposed a model based on expression of yeast PHO5, a gene activated during phosphate deprivation.[992] The model hypothesizes that noise generation of the PH05 promoter depends on the rate of a slow upstream promoter transition involving removal of nucleosomes to make transcription factors accessible. Various mutations altering the noise in gene expression suggest that variability in mRNA levels results from a balance between promoter activation and transcriptional efficiency. Yeast mutants defective in a nuclear inositol polyphosphate kinase impaired PH05 promoter chromatin remodeling—a finding that is consistent with a transcriptional role for this protein.[998] A role for inositol polyphosphates in ATP-dependent chromatin-remodeling complexes regulating gene expression has been found by others as well.[999] In view of the genetic component in noise, Raser and O'Shea have speculated that noise level in gene expression is an evolvable trait.[992] This interesting possibility is in line with the suggestion that evolvability itself is a selectable trait.[958] Earl and Deem have proposed that evolvability may be a factor in selection for somatic hypermutation in the immune system and the evolution of drug resistance and transpositional events in bacteria.[958] By analogy, gene expression noise may be selected for its contribution to coping with change because the variability in protein concentrations may increase opportunities for gene sharing.

Fluctuations in gene expression are not just nuisance background; they contribute to phenotypic variation.[988] Noise has been implicated for random occurrences of lysis or lysogeny of bacteriophage λ, loss of synchrony of circadian clocks, and decreased precision of cell signals (see Ozbudak et al.[989]). The quantitative level of gene expression is a critical determinant of cellular behavior and results in distinct phenotypes. One example is cells with haplo-insufficient tumor-suppressor gene neurofibromatosis type 1 that show increased variation in dendrite formation in cultured melanocytes; this has been attributed to increased noise in cellular regulation.[1000] According to O'Shea[1001] (p. 14314), noise "might be beneficial for cells to diversify phenotypically. The noise that was generated to produce different amounts of proteins in different cells might allow a fraction of the cells to be poised to survive in an unpredictable environment." This statement is consonant with the proposition of Earl and Deem that selection for evolvability is beneficial for

survival in harsh environments.[958] The proposition here is that noise fluctuations in gene expression contribute to putative spurious functional trespasses
of proteins.

GENETIC DIFFERENCES IN LEVELS OF GENE EXPRESSION

Quantitative studies have revealed statistically significant differences in gene
expression within and between species in *Drosophila* species,[1002] yeast,[1003–1005]
fish *(Fundulus)*,[1006] mice, primates, and humans.[1007,1008] In addition, allelic
variation in human gene expression has been reported.[1009,1010] Screening of
normal tissues and cell lines indicates that genes often show unequally expressed alleles.[988] Polymorphic variations in gene expression, if not caused by
epigenetic effects, are of great interest because they are inheritable, which
means that changes in protein function due to differences in the intensity of
gene expression can be passed on in evolution.

Rifkin and colleagues[1002] have compared developmental differences in gene
expression during the start of metamorphosis in *Drosophila simulans, Drosophila yakuba,* and four strains of *Drosophila melanogaster.* Twenty-seven
percent of all the genes tested (12,866) differed in their developmental expression between at least two strains or species. The genes encoding transcription
factors and signal transducers were more stably expressed than the downstream genes encoding enzymes and structural factors. Most of the expression
differences mapped to *trans*-acting loci that were broadly dispersed across
classes of genes with different functions. Although it is difficult at the present
time to assess the meaning of these vast data sets in terms of the biochemical
functions of individual proteins, they provide an inkling of the complex
intra- and interspecific differences in gene expression that exist. The variable
expression of orthologous genes in different species and strains of flies is consistent with the possibility that similar proteins exposed to different slightly
different cellular environments perform different subsets of functions—in
other words, engage in gene sharing.

In humans, quantitative variations in gene expression have been reported
in lymphoblastoid cells among unrelated individuals, among siblings, and
even between monozygotic twins.[1008] More closely related individuals show
less variation in gene expression than unrelated individuals. Such genetic
control of variability allows gene expression phenotypes to be identified at the
DNA level. The most variably expressed genes (top 5 percent) are scattered

throughout the genome. Among the 100 most variably expressed genes were those for cytoskeleton, protein modification, and transport proteins; among the 100 least variably expressed genes were those for signal transduction and cell death/proliferation proteins. Microarrays coupled with genome-wide linkage analysis for quantitative traits have been used to identify the chromosomal loci that control the baseline expressions of different genes.[1011,1012] DNA regulatory sites have been identified for approximately 1000 expression phenotypes in 14 families. The natural variation of gene expression is controlled by both *cis*-acting (DNA sequences near the gene) and *trans*-acting (DNA sequences on different chromosomes) loci. Some loci control few genes, while others (hot spots) control the expression of many genes. According to Morley and colleagues[1012] (p. 747), "normal variation in gene expression is likely to account for a substantial part of human variation . . . and will therefore contribute to differences that are important for understanding essential aspects of human biology, including networks of interacting genetic effects, evolution, and susceptibility to complex diseases." In that conclusion, we can hear echoes of gene sharing as a consequence of the variability in gene expression.

THE MOLECULAR CLOCK

Early comparisons of amino acid substitutions in hemoglobins[1013] and cytochrome c[1014] indicated that the rate of evolution of an individual protein was similar in different lineages. Therefore the rates of amino acid and nucleotide substitutions may reflect molecular clocks, assuming that the mutations occur randomly and accumulate linearly with time. The molecular clock hypothesis was greatly strengthened by Kimura's neutral theory of evolution, which predicts that selection does not affect mutation rate.[842,1015–1017]

The idea of a molecular clock provided hope for a quantitative method to establish evolutionary distances between lineages; yet it has been controversial from the outset.[843,1017–1019] It does not fit with rates of morphological and physiological evolution being variable within and between lineages.[1020] Another objection was that the efficiency of DNA repair may differ among lineages, which would complicate attempts to make a direct connection between the rates of sequence changes and data on the number of sequence differences apparent today. In addition, sequence differences between orthologous proteins may reflect preferential loci for sequence modifications (hot spots) or changes that have been reversed in some cases and not others. Another com-

plication was that concerted evolution can homogenize sequences of duplicated genes over time. Also, species with shorter generation times may incur
more sequence modifications than those with longer generation times, assuming that errors during DNA replication of germ cells is a major source of
mutations (which is also balanced by DNA repair). Still another difficulty that
has been raised was that species with higher metabolic rates can be expected
to generate more oxygen radicals and these would cause nucleotide changes.
Population size, which is subject to change during evolution, also affects the
rate of fixation of nucleotide substitutions; random drift is higher in small
populations, where even slightly deleterious mutations can become fixed.

Some studies support these objections to a molecular clock but others do
not; moreover, the incomplete data remain problematic.[843,1164] Despite the ambiguity, the potential usefulness of a molecular clock has not been abandoned
due to the growing abundance of sequence data, which can be analyzed quantitatively.[1017,1021] After reviewing the "mirages" of the molecular clock, Ayala[1022]
(p. 71) concluded that "molecular evolution is dependent on the fickle process of natural selection. But it is a time-dependent process, so that accumulation of empirical data often yields an approximate clock, as a consequence of
the expected convergence of large numbers."

Because natural selection acts at the level of phenotype, allowing neutral
changes to take place at the sequence level, researchers have hypothesized that
separating rates of change in synonymous nucleotide substitutions (those
that do not result in amino acid replacements) from nonsynonymous substitutions (those that cause amino acid replacements) would approximate a true
molecular clock that ticks at the gene level. To test this theory, they compared
nucleotide differences in synonymous and nonsynonymous codons within
and between populations.[1023] This approach has proved very useful but it
is not error-free. Synonymous mutations can be altered if selection pressure changes on the protein, thereby obscuring the number of changes that
have actually occurred. In addition, changes may occur rapidly and saturate
the gene in question, so differences noted between species (especially more
evolutionarily distant species) reflect multiple changes. It is also unlikely that
all synonymous nucleotide changes are free of selective pressures even if selection on synonymous changes is less stringent than on nonsynonymous
changes.[1024] Together, these complexities confound a simple relationship between nucleotide changes and evolutionary time.

Although the evolution of gene expression appears to be subject to selective pressures,[978,980] there does appear to be a considerable amount of neutrality and clock-like behavior in the divergence of gene expression level during evolution.[978,979,1025–1027] These properties are relevant to the gene-sharing concept, which postulates that changes in gene expression levels can affect the function of the encoded proteins. That variations in gene expression may be at least partially neutral and linearly progressive creates a possible link between the molecular clock and gene sharing due to the latter's connection to protein concentrations. A gene expression molecular clock would be reflected in dispersed DNA sequences controlling variations in gene expression rather than in localized protein-coding sequences; thus, it would be difficult to examine, with our present state of knowledge, at the level of gene structure.

Finally, gene sharing may affect idiosyncrasies of the molecular clock. The molecular clock requires that nucleotide substitutions occur randomly and stochastically.[1022] This neutrality predicts that the substitutions display a Poisson distribution. In statistical terms, the mean, M, should equal the variance, V, so that $V/M = 1$, a property called the index of dispersion. In most cases, the clock is overdispersed, which means that V/M is greater than unity; this is an argument against the regularity of the molecular clock.[1027a] Difficulties in interpreting the molecular clock have not resolved overdispersion. More confounding facts remain as well. It is accepted that the molecular clock for each protein or gene of a species runs on a different timescale—the clocks "tick" at different rates, as it were. Some proteins change very little during evolution (histones, cytochrome c), while others are modified much more rapidly (fibrinopeptides) depending on the selective pressures that apply specifically to those genes and proteins.[57] But evolution of glycerol-3-phosphate dehydrogenase (GPDH) and Cu,Zn superoxide dismutase (SOD) highlights a new dimension.[1022,1028,1029] The clocks tick in opposite directions for these two enzymes: The evolutionary rate of the *Gpdh* gene increases while that of the *Sod* gene decreases in *Drosophila* when comparisons are extended among species.

GPDH is a conserved NAD-dependent cytoplasmic enzyme that participates in the glycerophosphate cycle to provide energy for flight in thoracic muscles of *Drosophila*; it has homologues throughout plants and animals. SOD disarms cells from the dangers of free oxygen radicals and is present throughout eukaryotes and in some bacteria. Sequence data indicate that the

Gpdh gene has evolved slowly (1.1 × 10⁻¹⁰ amino acid replacements per site per year) when comparisons are made within *Drosophila* species in the last 55 million years. This evolutionary rate increased (4.5 × 10⁻¹⁰ amino acid replacements per site per year) when *Gpdh* sequence comparisons are made between mammals within the last 70 million years, between Dipteran families within the last 100 million years, between animal phyla within the last 650 million years, or between multicellular kingdoms (animals, plants, fungi) within the last billion years.[1029] Moreover, sequence comparisons indicate that *Gpdh* evolution within flies has been erratic (slow in *Drosophila,* much faster in the related genus *Chymomyza* and even faster in the medfly *Ceratitis*). A different result is obtained for SOD evolution when similar comparisons are made for the *Sod* gene. The data indicate rapid evolution when *Sod* gene sequence comparisons are made between *Drosophila* species (16.2 × 10⁻¹⁰ replacements per site per year), between Dipteran families (15.9 × 10⁻¹⁰ replacements per site per year), or between mammals (17.2 × 10⁻¹⁰ replacements per site per year); by contrast, the sequence data indicate much slower evolution when comparisons are made between animal phyla (5.3 × 10⁻¹⁰ replacements per site per year) or between animals, plants, and fungi (3.3 × 10⁻¹⁰ replacements per site per year). Stated differently, 10 amino acid replacements occurred in the *Sod* gene during the last 75 million years, although only 21 replacements occurred in the previous 600 million years.[1030]

Because these differences in clock direction for *Gpdh* and *Sod* were identified in the same species, we must discount the general obstacles to a molecular clock I have already outlined (DNA repair, generation time, metabolic rates) as causes for the discrepancies. Ayala[1029] considered differences in the molecular clock between *Gpdh* and *Sod* "disquieting." One attempt to account for the erratic behaviors of *Gpdh* and *Sod* invoked the covarion (*co*ncomitantly *vari*able cod*ons*) hypothesis, which postulates that only a limited number of amino acid sites can be substituted at a given time in any one lineage, although the composition of the set of invariable sites changes in time and between lineages as the protein evolves.[1029–1031] Covarion considerations have been helpful for understanding the *Sod* clock, which slows when the species compared are more remote;[1031] but they have been less helpful for explaining the changing rate for the *Gpdh* clock, which increases as the evolutionary distances between the compared species grows longer.[1029]

From the perspective of gene sharing, one of the obvious parameters that could account for erratic behavior of the molecular clock is changes in protein function during evolution. For example, changes in the molecular clock would be expected during the positive and negative (purifying) selection of lysozyme c, when it was initially recruited (increase in mutational rate) and later adapted (decreases in mutational rate) for digestion in ruminants;[1032] lysozyme of primates also underwent episodes of adaptive evolution.[1033] Because gene sharing leads to a mosaic of functions that may differ in species, it may well impact molecular clocks. The difficulty of knowing which gene-sharing functions appeared suddenly, and the length of time they lasted, detracts from the practical use of gene sharing for normalizing a molecular clock. On the other hand, deviant behavior of a molecular clock might provide a flag to investigators: The protein in question may have acquired a new function via a gene-sharing strategy.

GENE KNOCKOUT EXPERIMENTS

Gene deletion by homologous recombination is a powerful method for identifying and/or proving protein function. In some instances gene knockout produces unambiguous phenotypes that directly link gene to function. An example is deletion of the highly lens-preferred *αA-crystallin* gene in the mouse, resulting in cataract.[294] Feeling satisfied that mice carrying a gene deletion acquire the expected phenotype, such as *αA-crystallin*-deficient mice being afflicted with cataract, can lead to false confidence that the complex role of the encoded protein is understood. If gene sharing is at work, the deleted protein may have other "hidden" functions. For example, in the case of *α*A-crystallin, the gene knockout revealed the transparency role of the protein in the lens but not its antiapoptotic role (see Chapter 4; Figure 4.5).

Not uncommonly, deleting a gene from a mouse by homologous recombination has no apparent phenotype. This lack of resulting phenotype is generally attributed to functional redundancy; in other words, the function of the deleted protein is rescued by another member of the same gene family (a paralogous gene). The myogenic basic helix-loop-helix genes (*MRF4*, *Myf5*, *myogenin*, and *MyoD*) exemplify hierarchical and overlapping functions of a family of genes. These genes regulate skeletal muscle development and may substitute for one another under experimental situations.[1034,1035] Deletion of the gene encoding prolactin-like protein A (PLP-A) is another case of functional redundancy by a paralogous gene.[1036,1037] *PLP-A* is one of 26 genes of the prolactin family in the mouse. It is normally expressed in the natural killer cells of the uterus that are associated with the uteroplacental vasculature. PLP-A is an atypical prolactin in that it does not function via the prolactin receptor (perhaps a clue that PLP-A might have unknown functions). Mice lacking the *PLP-A* gene appear normal; however, in contrast to wild type mice, they are unable to adapt when challenged with hypobaric hypoxia (11 percent oxygen) during gestation. It is unlikely that PLP-A was selected for reproductive survival under low-oxygen conditions, yet it can perform that physiological function under experimental hypoxic conditions, thereby illustrating the difficulty in revealing the many possible functions of a gene and protein.

Gene redundancy is not a completely satisfying explanation for the absence of or weak phenotypes after gene deletion. Many duplicated genes have diverged in function and expression profile. This divergence makes it difficult to explain how the function of gene *A* can be compensated by its duplicate, gene *B*, whose regulation and function have changed over time. Also striking, many single-copy genes can be deleted in yeast and other species without apparently affecting the traits under study.[868,1038] Of course, it is possible that other, unstudied traits would not have escaped unscathed from deletion of a duplicated or single-copy gene. Wagner[945] (pp. 239–242) argues that the lion's share of resilience of an organism to a deleted gene is due to a phenomenon called "distributed robustness" rather than to gene redundancy. Due to the robustness of networks, one part can compensate for another failed part although the specific functions of the two parts are not identical. Networks can reroute metabolic fluxes and pathways through parts that are unaffected by a gene deletion. Thus, robustness—the ability to withstand perturbations—emerges from cooperation and cross-talk of proteins with different activities. The problem of identifying protein function by gene deletion is not trivial.

In theory, although a gene has a vital function, the encoded protein could also have other functions that go unnoticed in gene knockout experiments because the mice die. An example is deletion of the *transketolase* gene, which is necessary for nucleic acid synthesis; mice lacking the *transketolase* gene suffer an early embryonic death.[473] However, mice containing a single copy of the *transketolase* gene have unexpected phenotypes. These include reduction of adipose tissue despite an enormous amount of enzyme in the remaining fat cells and female fertility problems. The mystery of the abundance of transketolase in the cornea also remains unsolved in the *transketolase* hemizygous mice (see Chapter 5). In brief, gene sharing should not be overlooked in gene knockout experiments, whether the affected offspring have the expected phenotype, appear normal, or are nonviable.

GENE DELETION OF *β-catenin*

It is instructive to examine mice lacking the well-studied *β-catenin* gene as an example of the difficulty in interpreting the roles of multifunctional genes.[984,985,1039] *β*-catenin is located in adherens junctions of the cell, where it binds cadherins (cell surface adhesion proteins) through its central armadillo repeat sequence (the segment polarity gene, *armadillo*, is the *Drosophila* homolog of *β-catenin*). The amino-terminal end of *β*-catenin binds *α*-catenin to bridge the cytoplasmic domain of the cadherins to the actin cytoskeleton. This anchoring function of *β*-catenin mediates cellular adhesion, and reduction in cadherin-catenin complexes due to expression of a truncated *β*-catenin correlates with the loss of cell adhesion.[1040] *β*-catenin complexes with the colon tumor suppressor protein (adenomatous polyposis coli), a scaffold protein (axin), and glycogen synthase kinase-3*β* (GSK-3*β*). GSK-3*β* phosphorylates *β*-catenin at its amino terminus, marking it for ubiquitination and subsequent proteosomal degradation. The cytoplasmic cell:cell adhesion role of the unstable *β*-catenin is lost when a family of surface receptors, called Frizzled, respond to secreted glycoproteins to activate the Wnt pathway (named for *wingless* in *Drosophila;* see Cadigan and Nusse[1039]). The canonical Wnt pathway stabilizes *β*-catenin in the cytoplasm (there are also noncanonical Wnt pathways that work differently; see Veeman et al.[1041]). The canonical Wnt pathway operates by inhibiting phosphorylation by GSK-3*β* and fostering interaction with another protein, called 14–3–3*ζ*, which promotes phosphorylation by the survival kinase Akt.[1042] The stabilized *β*-catenin moves to the nucleus, where its central armadillo repeats interact with transcription factors of the lymphoid enhancer factor (LEF)/T-cell factor (TCF). Complexed in the nucleus, *β*-catenin activates a number of target genes by virtue of its carboxy-terminal sequence. Thus, *β*-catenin is a prime example of gene sharing, being both a member of a submembranous cytoskeletal structure mediating cell adhesion and a nuclear transcription factor activating gene expression.

What happens if the *β-catenin* gene is deleted in mice? The initial report showed that β-catenin-deficient mice die at gastrulation due to detachment of the ectodermal cell layer.[1043] This result is consistent with the role of β-catenin in cell adhesion. A subsequent report indicated that β-catenin-deficient embryos show an earlier defect in anterior-posterior axis formation, resulting in the absence of mesoderm and head structures.[1044] Chimeric embryos (embryos containing a mixture of β-catenin-deficient and normal cells) were used to confirm the importance of β-catenin for ectodermal cell layer viability. Interestingly, intercellular adhesion was maintained by functional redundancy in the early β-catenin-deficient embryos, presumably by increased levels of another protein, plakoglobin (also known as γ-catenin), a member of the armadillo family preferentially associated with desmosomes. In contrast to its cell adhesion function, a role for β-catenin signaling in axis formation and mesoderm production is consistent with the induction of a secondary embryonic axis by β-catenin overexpression in *Xenopus*.[1045] The central armadillo region alone was shown to be sufficient to promote axis duplication, a result that eliminates a cell adhesion role in this function.

The cell adhesion and signaling functions of β-catenin were teased apart through tissue-specific gene deletion experiments. β-catenin deletion in endothelial cells of mouse embryos led to cell death in the uterus due to defects in vascular morphogenesis; specifically, vascular patterning was affected in selected regions (head, vitelline, umbilical vessels, and placenta). Moreover, intercellular junctions, α-catenin levels, and actin cytoskeleton structure were altered in cultured endothelial cells from the β-catenin-deficient embryonic mice.[1046] This endothelial function of β-catenin appeared to be independent of its Wnt signaling/transcriptional role. On the other hand, conditional ablation of the β-catenin gene in the ventral endoderm of the mouse hindlimb disrupted limb formation by interfering with the Wnt signaling pathway.[1047,1048] Targeted removal of the β-catenin gene from the lung also implicated the Wnt pathway for the formation of the distal (but not the proximal) airways.[1049]

Analysis of the targeted β-catenin gene knockout mice provided additional surprises and distinguished between cell adhesion and Wnt signaling functions. Deleting β-catenin from thymocytes and T cells showed that this protein is essential for normal splenic T cell development.[1050] However, despite much evidence that maturation of blood cells requires Wnt signaling, deleting the β-catenin gene in bone marrow progenitor cells did not affect hematopoiesis or lymphopoeisis.[1051] It is not known whether differentiation of the bone marrow cells in the absence of β-catenin is due to functional redundancy. More interestingly, the targeted removal of *β-catenin* from the hippocampal pyramidal neurons revealed that β-catenin is necessary for localizing synaptic vesicles at the presynaptic active zone.[1052] This β-catenin function is not brought about by a cell adhesion or a Wnt signaling mechanism, but by β-catenin acting as a scaffolding protein and linking clusters of cadherins to P/DZ domain-containing proteins by using its own PDZ binding motif. Thus, β-catenin functions are not limited to cell adhesion or Wnt signaling/transcription—a finding that is consistent with β-catenin being able to interact with many different proteins and mediate multiple functions (see Bamji et al.[1052] for further discussion and references).

Taken together, *β-catenin* knockout experiments illustrate the issues associated with identifying the function of a gene by removing it: The gene's encoded protein plays different roles in different cells and at different times of development. In other words, the expression of the gene plays a critical role in defining its function. We will need to keep this principle in mind during the design of future gene knockout experiments as only $\sim$10 percent of the $\sim$25,000 mouse genes have been deleted.[1053]

HORIZONTAL GENE TRANSFER

The vertical transmission of genes (from parent to child) forms the basis of evolutionary lineages. Genes, however, are transferred horizontally among species as well. Horizontal gene transfer (also called lateral gene transfer) has even been considered a stepping stone for the origins of species.[1054] Horizontal gene transfer is accepted in the evolution of prokaryotic genomes[1055–1059] and apparently occurred preferentially for metabolic genes.[1060] Horizontal gene transfer is believed to have been especially important when life emerged as primitive loosely connected cells 3 to 4 billion years ago.[1061,1062] Although molecular phylogenetic analyses suggest that horizontal gene transfer extends to higher species, including humans,[125] reservations exist as to its extent and whether it includes transfers from bacteria to vertebrates.[1057,1063,1064] Nonetheless, the present evidence suggests the occurrence of horizontal gene transfer throughout evolution, including bacteria to eukaryotes.[1065,1066]

Expected consequences of and reasons for fixation of horizontally transmitted genes include the transmission of new metabolic capacities to the recipient[1066] and the formation of new genes by structural rearrangements, especially because mobile DNA elements and bacteriophage may be involved.[1067,1068] Antibiotic resistance is an example in which horizontal gene transfer confers a metabolic ability to the host that has a profound effect on survival as well as major clinical implications.[1069] A potential example of metabolic innovation resulting from horizontal gene transfer is the evolution of cell-cell signaling in animals. It has been suggested that bacterial enzymes that were horizontally transferred to animal cells have extended biochemical pathways to yield numerous diffusible messengers (catecholamines, indoles, histamine, acetylcholine, and nitric oxide) in signaling reactions.[1070] Another possible biological benefit of horizontal gene transfer is the conversion of parasitism to symbiosis, which is mutually advantageous to parasite and host.[1067]

The origins of cellular mitochondria and chloroplasts represent well-known cases of horizontal gene transfer. The former was derived from ingested α-proteobacteria[1071] and the latter from ingested cyanobacteria.[1072–1074] Over time, many of the endosymbiont genes were transferred to the host cell nucleus; and after synthesis in the cytoplasm, the expressed proteins were retargeted to the mitochondria and chloroplasts. The result was complete integration of mitochondria and chloroplasts into cellular structure and function. There has also been widespread horizontal transfer of mitochondrial genes among plants.[1075,1076] In addition, horizontally transferred genes by secondary engulfment of alga have contributed significantly to the evolution of eukaryotic algae.[1077,1078] The host nucleus acquires hundreds of genes transferred from endosymbionts; many of the proteins encoded by the nuclear genes derived from the endosymbionts are directed back to the plastids.[1079] Thus, according to Doolittle[1080] (p. 308), "their eating habits, not their sex habits, are most relevant to the understanding of 'lateral transfer' into the genomes of unicellular eukaryotes that survive through the ingestion of bacteria." It is fascinating to consider that the ancient origins (~2 billion years ago) of mitochondria are associated with a rise in atmospheric oxygen, presumably due to activity of photosynthetic cyanobacteria in the oceans; thus, the original usefulness of mitochondria was perhaps more for oxygen detoxification than energy production.[1071]

The hydrogenosome is a less well known organelle whose origin is tied to mitochondria.[1081] Hydrogenosomes are double-membrane structures that generate hydrogen while making ATP. They are found in hypoxic eukaryotic microbes, such as protozoa. It has recently been reported that hydrogeno-somes of an anaerobic ciliate *(Nyctotherus ovalis),* which lives in the hindgut of cockroaches, have a small genome that encodes components of the mito-chondrial electron transport chain.[1082] These structures are probably derived secondarily from mitochondria and thus have an evolutionary history involving lateral gene transfer.

Phylogenetic studies have yielded numerous examples consistent with horizontal transfer of individual genes from bacteria or other microorganisms to plants and animals.[1083] A famous situation with agricultural implications is the induction of crown gall formation in plants by the horizontal transfer of DNA (T-DNA) from several ubiquitous soil bacteria, most notably *Agro-bacterium tumefaciens.*[1084,1085] Another example attributed to horizontal DNA transfer from bacteria to plants is glycerol transporters, which were probably acquired from bacterial aquaporins approximately 1200 million years ago.[1086] This adaptive recruitment of a bacterial aquaporin gene was important for the further evolution of plants, whose common ancestor apparently lacked this gene. Interestingly the plant glycerol transporters underwent similar convergent changes throughout their parallel evolution. Cellulose synthase is another protein whose gene may have been transferred horizontally from bacteria, fungi, or slime molds to metazoans.[1087] Urochordates (ascidians) are the only metazoans that are capable of making cellulose, which is a product of their epidermis and strengthens an outer coat. Alone the fact that no other animal makes cellulose is suggestive (but not proof) that urochordates acquired the necessary cellulose synthase from other sources. The cDNA-deduced cellulose synthase of the ascidian *Ciona sayignyi* resembles the enzyme from a wide variety of bacteria, a slime mold *(Dictyostelium discoideum),* and a few fungi. Matthysse and colleagues[1087] thought it likely that urochordates acquired this enzyme from a microbial source very early in their evolutionary history. Other interesting examples credited with horizontal gene transfer are glutamate dehydrogenases,[1088] enolase,[1089] and glyceraldehye-3-phosphate dehydrogenase.[1090]

These cases represent examples where presumably the original biochemical function was transmitted horizontally. The relevant question with respect to gene sharing is whether a different function results when a protein is placed in a foreign cellular environment as a consequence of horizontal gene transfer. Interpretation of function before and after gene transfer is necessarily limited by the uncertainty of the original function and by the consequences of mutations acquired after the gene was transmitted a long time ago. Nonetheless, there are intriguing cases suggesting functional shifts after horizontal gene transfer. One concerns polγ, a mitochondrial DNA polymerase comprising a catalytic subunit (polγA, which is the orthologue of bacterial DNA polymerase I) and an accessory subunit (polγB, which increases processivity of the polymerase reaction).[1091] Mitochondrial polγB appears to have been derived by horizontal transfer of the *glycyl-tRNA synthetase* gene of bacteria of the *Thermus-Deinococcus* group. Wolf and Koonin[1091] (p. 433) conclude that "the case of polγB is rare in that there seems to be strong evidence of horizontal gene transfer from a specific bacterial lineage to a specific eukaryotic lineage, with subsequent adaptation of the acquired gene for a completely new function." In general, amino acid-tRNA synthetases have been horizontally transmitted to a considerable extent because their activity is limited to a small set of tRNAs and particular amino acids, making them universal agents able to function in different environments.[1061,1092,1093]

Apoptosis, or programmed cell death, is another possible example of functional shifts of bacterial proteins after horizontal transfer to eukaryotes.[1094] The borrowing of ancient bacterial proteins for apoptosis is consistent with the fact that mitochondria, which were themselves derived by transfer of α-proteobacteria,[1071] have a major role in apoptosis and release critical proteins—cytochrome c being the classic case—that drive apoptosis. Phylogenetic studies indicate that many of the core proteins of the apoptotic pathway, such as caspases, are homologous to bacterial proteins that might have interacted originally in the bacteria with each other as parts of signaling pathways.[1094] What is not known, if this scenario is true, is when the functional shifts took place in evolution and whether periods of overlapping functions occurred. Overall, these data support the possibility that the derivation of apoptosis, a major eukaryotic innovation that had a critical role in the evolution of multicellularity, involved gene-sharing processes via horizontal gene transfer. Koonin and colleagues conclude[1094] (p. 402) that "much of the glory of eukaryotic ascension to the ultimate complexity of higher plants and animals might owe to a lucky choice of bacteria with complicated differentiation processes as the primary promitochondrial, and perhaps subsequent symbionts." And perhaps, we might add, to the adaptability of promiscuous proteins that can perform entirely different functions when placed in different cellular environments, thanks to gene sharing.

TAKE-HOME MESSAGE

Gene sharing has far-reaching consequences at the level of systems biology. It contributes to the complexity of biochemical interactions that can be analyzed as networks and it may affect the episodic and variable molecular clock during evolution. The prevalence of gene sharing may obscure the interpretation of gene knockout experiments, with the likely result being underestimation of functions for proteins. The origins of complex protein interactions and development of new functions in gene sharing may reside in part in the variability of gene expression and cellular noise.

The propensity for gene sharing is advanced by selection for marginal stability rather than rigidity of proteins, resulting in conformational diversity of individual polypeptides. Marginal stability increases the ability of proteins to interact widely and favors the accumulation of (apparently) neutral mutations, conferring genetic variability, traits that may contribute to what arguably may be considered the highest level of evolution—evolvability, or the ability to evolve. Thus the gene-sharing concept is applicable at different levels of organization: near the bottom of biological reductionism by virtue of its reliance on molecular interactions, at mid-level by virtue of extensive and dynamic networks of protein:protein interactions, and near the top of systems biology by virtue of its indirect role in possibly promoting the capacity to evolve and adapt.

The crux is as follows: Enlarging the view of gene sharing from molecular biology to systems biology reinforces the paradox that the functional diversity of individual polypeptides increases as they specialize and as species gain complexity. The challenge remains to distinguish between molecular and phenotypic functions in each case, which is a difficult gap to bridge.

11

Recapitulations: Ambiguities and Possibilities

Gene sharing concerns many aspects of biology: It is about the structure, expression, and nature of genes, about multiple functions of single polypeptides, and about existing metabolic pathways and evolutionary processes. Gene sharing contributes to sculpting new images from old forms; it is about molecular function but affects phenotypes resulting from complex processes, cellular behavior, and species diversity. The domain of gene sharing is the vast space between molecular events and systems biology. Mutations both in amino acid sequences and regulatory motifs affect gene sharing. Gene sharing can be analyzed by reductionism—the molecular details of gene expression and protein:protein interactions—or holistically as one of many processes that comprise the complexities of development, evolution, and systems biology.

An area as encompassing as gene sharing evokes ambiguities, allows association of ideas, and suggests future possibilities. In this last chapter, I amplify on ambiguities and implications that I believe are of central importance to the concept of gene sharing, and I end the book by imagining silhouettes on the horizon as targets for the future.

AMBIGUITY OF CAUSE AND EFFECT

Chapter 1 considers the multiple uses of a chair, and how an alien might fail to list "seat" among the functions that he/she observes for the chair. The point was meant to introduce the concept of gene sharing by the use of an analogy conveying the idea that structures—chairs or proteins—are used in many pragmatic ways. Using a chair analogy for gene sharing, however, is misleading: A chair is designed to sit on and other uses are serendipitous, which means this analogy is better suited for exaptation[49] than gene sharing. The concept of gene sharing obviates the question of original intent. A more apt analogy might be an individual who blends professions such as law and medicine, for example, with each subdiscipline leading to increased knowledge about the other. If you were to meet this person, you would not be able to tell whether he was a physician who became attracted to the legal profession and switched to law specializing in medical cases, or he was a lawyer who learned a great deal about medicine. But this is not a perfect analogy either. A lawyer with expertise in medicine is practicing law, not medicine, and thus is engaged in one career, not two. Perhaps a better analogy would be a writer by night supporting herself as a waitress by day. In that case, however, we may ask, was she a writer who needed money or a waitress who was driven to writing by a chance encounter with an artist? Without complete knowledge, the causes for the many functions of an individual would remain ambiguous.

The causes leading to multiple functions of genes and proteins are subject to even greater ambiguity. In evolution, we can only guess (albeit sometimes with considerable insight) the complex functional history of any protein. The impossibility of recreating the totality of past events is eloquently stated by Stach[1094a] (p. 14) in his biography of Franz Kafka, the master of ambiguity: "Even the most precise imagination, armed with knowledge...remains in the dark. No mind, not even the most powerful, can conquer the frustration of not knowing, the progressive fading of historical memory, the fact that what is past is past. The best we can do is produce evidence, sharpen the contours, and increase the dimensions of the image. The best we can say is, It may have, could have, must have been this way."

NATURAL SELECTION AND RANDOM DRIFT

Darwinian evolution,[44] the subsequent classical Mendelian population genetics,[1095–1097] and the synthetic theory of evolution[1098,1099] have considered the natural selection of randomly generated variations (mutations) as the cornerstone of change and adaptation.[26,69,453,843,1100] By contrast, the neutral theory of evolution[842,1015,1101] treats (as does the synthetic theory) undirected mutation as the ultimate source of variability but proposes that a random process of genetic drift is largely responsible for fixing the mutations within the populations. The neutral theory does not eliminate natural selection, which is ever present, but acknowledges chance effects of drift for fixing the mutation within the population. Thus the neutral theory of evolution allows for an internal (mutation) driving force for the acquisition of new molecular traits during speciation instead of a strict reliance on the environment for adaptive changes. In small populations, even presumably advantageous mutations can be lost and deleterious alleles (individual variants of the same gene) can become fixed by drift, although with different probabilities, damping the effect we might expect from selection (see Li[843] and Weinreich et al.[1165] for further discussion).

The proportional roles of natural selection and neutral drift for the fixation of traits have been studied extensively. Convincing conclusions have come from relating quantitative data on synonymous and nonsynonymous mutations in codons with the fixation of known polypeptide functions (see Chapter 9). Gene sharing increases the layers of difficulty in establishing adaptive selection or neutral drift as the basis for some of the multiple functions of a given protein for several reasons. First, the changes in gene regulation that affect polypeptide function are influenced by environmental factors, genes encoding transcription factors and various small RNAs, and regulatory motifs. These are currently impossible to analyze quantitatively in the manner that codon mutations have been analyzed (see Chapter 10). Second, gene sharing contributes to the uncertainty concerning the number of molecular functions carried out by individual polypeptides, which are subject to a plethora of fluctuations and different intracellular microenvironments and environmental conditions. This complexity increases the ambiguity regarding the number and relative strengths of selective pressures that exist for each protein. In addition, the role of gene sharing in "testing" for new functions is ambiguous. Because polypeptide interactions are so extensive, which interactions are fixed for advantageous functions and which are fortuitous? The ambiguity between fortuitously occurring and fixed interactions is amplified when we consider stochastic events, which affect all cellular processes[988] and contribute to gene sharing.

GENE SHARING AND ROBUSTNESS: WHEN IS A MUTATION NEUTRAL?

Gene sharing raises ambiguity about the neutrality of mutations (see Wagner[945]). In general, biological systems are robust to mutations. Many nucleotide changes do not cause known changes in protein function or fitness of the organism. Proteins (and RNAs[1157]) can sustain amino acid changes without significant alteration in three-dimensional structure or function. This invaluable robustness of biology, necessary for survival, results in many mutations appearing neutral. However, each mutation affecting amino acid sequence or gene regulation potentially changes the possibilities for gene sharing and functional innovation. An amino acid change that is considered neutral because it does not modify a known function may still confer a small localized change in structure or charge in the polypeptide, which then initiates a new interaction that may promote a different function. The converse is also true. The opposite of robustness is fragility, or greater sensitivity to perturbations, a likely consequence of mutation. In contrast to a robust system, a fragile system will vary in structure or function in response to a mutation. Because neutrality cannot be assessed properly without knowledge of all the functions of a polypeptide (both robust and fragile), extensive gene sharing increases the uncertainty, already high due to robustness, that any mutation is truly neutral.

Gene sharing is affected by robustness to environmental perturbations as well as to mutational perturbations.[945] Molecular chaperones, for example, mask deleterious mutations in proteins. Heat shock proteins are a class of such chaperones that are induced under cellular stress and can protect cells by preventing proteins from aggregating or losing function when they are weakly affected by mutations or subjected to denaturing conditions such as heat or radiation. The presence of heat shock protein 90 (Hsp 90) in flies[1102] or plants[1103] allows proteins to accumulate mutation-driven changes in amino acid sequence without damage to their function. Thus, the mutations appear neutral because they are not associated with a change in the function of the protein. If, however, the Hsp 90 protein itself incurs a temperature-sensitive mutation (a mutation that is only manifest with consequent loss of protein function when the temperature reaches a threshold level), a plethora of abnormal phenotypes suppressed by the normal Hsp 90 suddenly appear as the temperature is raised.[1102] Hsp 90 exemplifies how robustness to environmental insult can mask a growing potential to engage in gene sharing.

INCONSISTENCY WITH DESIGN

The ambiguity in protein structural and functional history leaves a crack of space for believers in intelligent design.[1104] Gene sharing, of course, cannot provide proof for or against original intent or the existence of a "designer," which in my opinion is outside the realm of science; but gene sharing is hard to reconcile with design.[1105] Use of the same polypeptide for diverse molecular functions is contrary to the idea that it was designed for an original purpose or elaborated for any single function.

As stated in the take-home message of Chapter 1, simultaneous specialization and diversification of protein functions, the paradigm of gene sharing, is intuitively paradoxical. This paradox may be considered as an example of the "unnatural nature of science."[1106] As Wolpert (p. 17) points out, "common sense provides no more than some of the raw material required for scientific thinking." The "obvious" at first blush is often incorrect. The notion that specialization of a protein or of a tissue arises from design—that each painting is a single image necessarily produced by an artist—may be intuitively satisfying, but it does not apply to polypeptide functions. The multifunctional nature of proteins as a consequence of gene sharing tells us that specialization does not limit the many different activities of a polypeptide.

The argument that structural components, such as bricks, can be used to construct a variety of edifices differs from the idea of gene sharing, which refers to functional innovation, not reutilization of the same roles with different outcomes. Gene sharing is about creative tinkering with polypeptides in the sense that Jacob[61] (p. 43) meant when speaking about organisms: "Diversification and specialization of organisms thus appear to result not so much from the appearance of new components as from a different use for the same components." Proteins, like organisms, are not designed for a specific purpose. The use of a metabolic enzyme present throughout the animal kingdom to bestow refractive properties to the transparent eye lens in selected species suggests pragmatism and flexibility rather than design and purpose. For recent discussion on science versus intelligent design and the reasons for the persistence of the debate on this matter (which in itself is surprising), see Ruse.[1107]

NAMING IS NOT KNOWING

It is necessary to name identified structures to maintain order and have a reference to the entity. This is as true for proteins as for any other entity. In the case of proteins, names are often based on their first-noted or most commonly observed function. However, the knowledge required to name a protein does not limit the range of diverse molecular functions and biological roles in which the protein participates. A good deal of the fascination for gene sharing may spring from the simple fact that we have succumbed to the erroneous belief that we know a protein's function because we know its name. Thus, by instinct, we resist the idea that a protein carries out a variety of functions unrelated to its name. Gene sharing shows us that the name of a multifunctional protein is not definitive of its functions: An enzyme may be a structural component, or a transcription factor may be membrane protein. Consequently, all genes and their encoded polypeptides represent the ambiguity that naming is not knowing.

THE QUESTION OF TISSUE HOMOLOGY

The question of tissue homology is another contentious issue touched upon by gene sharing. I discuss this issue in Chapter 4 with respect to whether the eye was invented once or multiple times during evolution. Proponents of a single invention (divergent evolution) say that variations in eye morphology and biochemistry were brought about by intercalated steps, with new molecules being inserted during the course of evolution;[210] proponents of multiple derivations (convergent evolution) say that similar components were assembled independently at different times in evolution.[206,240] Functional similarity has always been a consideration for homology, even though similar functions do not establish homology. For example, the wings of insects and of birds are both used for flying, but they are not homologous structures; insect wings did not evolve into bird wings. Wings for flying developed convergently in insects and birds.

Gene sharing suggests that tissue homology may not have to be all-or-none, especially at the molecular level. The same protein can perform different functions in the same species and, conversely, an orthologous protein can perform different functions in different species. The relationship between gene expression and polypeptide function, a central tenet of gene sharing, also adds to the sources of ambiguity with respect to tissue homology. Mutations affecting the complex process of gene expression are more difficult to identify than those affecting protein structure. Because independently derived changes in expression pattern can lead to new functions for proteins, as indicated by the use of entirely different proteins for lens crystallins in different species, we must weigh the role of convergence (separately acquired similarities in function) as a serious contributor to the derivation of structures with common functions in distinct lineages. Gene sharing does not answer the question of monophyletic versus polyphyletic evolution of the eye or indeed of any other structure; but it does support the idea of partial tissue homology, especially at the molecular level. The concept of partial tissue homology concerns the extent to which biological structures with similar or related functions are homologous in different species and raises the question, What is the level of homology between two tissues or organs?[198,199] There may be both common denominators and, due in part to gene sharing, different proteins used for critical roles in tissues that have similar functions, and the challenge is to sort these out and understand their evolutionary derivations.

PHYLOGENETIC TREES: THE COMPLICATION OF FUNCTION

Gene sharing raises the question as to which proteins have engaged in species-specific differences in functions in the past, or continue to do so in modern times. These multiple functions, which are subject to different selective pressures in different species, may have been transient or may have persisted for considerable lengths of time. The prevalence of gene sharing, and the possible changes in gene sharing for a given polypeptide during evolution, must add to the ambiguity concerning rates of sequence change during the long histories of orthologous proteins in species living under different and variable conditions. Thus, the number of amino acid differences between two proteins may not accurately characterize the original phylogenetic relationship or relative rates of evolution between those proteins. Given that we will never know the gene-sharing history of any protein since its inception, such ambiguity may never be completely eradicated.

DEFINING AND COUNTING GENES

Counting genes has been a preoccupation of the molecular era (see Chapter 3). Surprisingly, gene numbers do not increase with species complexity—though learning more about differential gene regulation, alternative RNA splicing, and posttranslational modifications of polypeptides for generating multiple proteins and diverse phenotypes has made this concept easier to accept. The complexity of genomic organization has raised many issues concerning an acceptable definition for a "gene" and even whether it makes sense to think of a structurally invariant "gene."[1121,1122]

The open concept of a "gene" including the multiple DNA regulatory regions that determine expression patterns for a given multifunctional polypeptide is compatible with gene sharing. This concept, however, makes gene boundaries ambiguous because regulatory regions are shared among genes, change with expression pattern, and are scattered throughout the genome. Even *cis*-regulatory motifs are not always situated on the same chromosome as the genes they regulate.[188] The role of gene sharing in tightening the close association of polypeptide function with differential gene expression compounds the difficulty of defining a gene as a structural entity. We may even consider the possibility that one polypeptide executing alternative functions as a consequence of differential gene expression is encoded by different structural genes due to the combinatorial use of various regulatory elements. Such an argument makes counting genes ambiguous until all the regulatory elements are identified and their combinatorial use for differential gene expression delineated. The difficulty of formulating a comprehensive gene concept raises the radical idea of defining the genome not as a collection of discrete genes but rather as a large network of interacting components. According to Burian[85] (p. 173), who attributes Dan Hartl, we might consider changing terminology from "the molecular biology of the gene . . . to the molecular biology of the genetic material."

DEFINITION OF POLYPEPTIDE FUNCTION: THE AMBIGUITY OF MOLECULAR MECHANISM

One of the most troublesome ambiguities concerning gene sharing is the definition of polypeptide "function." The difficulty lies in knowing the mechanism of action. For example, actin, a cytoskeletal protein, is also a transcription factor (see Chapter 7). The question as to whether actin participates in gene sharing is not whether its biological roles in cell motility and gene expression are different, which they certainly are, but whether actin performs the same *molecular* functions in the cytoplasm and in the nucleus. Is actin using similar properties that allow it to affect chromatin structure in the nucleus and cytoskeletal structure in the cytoplasm? It is a reasonable guess that actin is associating with different proteins and acting in a different molecular fashion in the cytoplasm and nucleus. However, we cannot be certain about this characterization until the precise mechanisms of actin's functions are established in both cases. In view of the complexity of biological reactions, differences in molecular mechanisms employed when a polypeptide is engaged in alternate functions must often be inferred. This necessity generates ambiguity regarding the extent of gene sharing—of different molecular function—when a polypeptide performs two or more roles with different biological outcomes.

BETWEEN GENOTYPE AND PHENOTYPE

Gene sharing is affected by genotype—the nucleotide sequences encoding amino acid sequence and directing gene expression—and influences the phenotypes of cellular and organismic functions. The many interactions between polypeptides, their multiple molecular functions in the same and different tissues, and the resulting phenotypes of cellular expression are what make gene sharing such a remarkable and rich manifestation of biology. Gene sharing contributes to the fuzzy layers that exist between the genome and the organism. It adds meaning to the genetic material and gives life via inanimate molecules. Gene sharing links the molecular with the living. It does not, however—and this is important—clarify the nature of the link. Gene sharing is identified and studied at the molecular level, but ambiguity remains concerning its various connections to cellular and organismal phenotypes.

For the sake of argument, we may consider the ambiguity between molecular and higher-level phenotypic functions analogous to the ambiguity between micro and macro phenomena. Some physical phenomena, such as the quantum mechanical duality between wave and particle properties of electrons, are not detectable until examined at the appropriate level of resolution. This concept is not new to physicists and cosmologists. These properties of matter do not disappear, but they may lose immediate relevance at one or another scale of reference. It is interesting to consider that the dynamics of the gene-sharing process is a microscale phenomenon of molecular dimensions that, in some instances, jumps to the macroscale phenomenon of cellular phenotype. When the jump occurs and how it is enacted are among the great mysteries and ambiguities of gene sharing.

GENE SHARING AND THE IMPORTANCE OF RESEARCH ON DIVERSE SPECIES

There are many reasons for using diverse species—from microbes to humans—for basic and medical research. The enormous success of widespread, free explorations of living systems speaks for itself and does not need justification. Bacteria gave us operons, bread mold led to penicillin, and the famous fruit fly, *Drosophila,* provided a model beyond expectations for vertebrate and human genetics. As Carlson[74] (p. 256) writes, "In surveying the history of the gene concept, I have found one point of view, frequently alluded to, that I should like to advocate vigorously. This is the necessity for a 'comparative genetic outlook.' This outlook is based on the premise that research with a single organism or a single technique restricts the possible universality of new concepts." He goes on to state (p. 258) that "the 'comparative outlook' does more than generate models. It permits the geneticist to seek ways out of the genetic impasse that often results from the presence of contending models in one system . . . the exposure to the concepts generated

from experiments with different organisms can give a geneticist a more enriched way of looking at his own material and its implications."

Gene sharing gives us another reason for investigating every nook and
cranny of the living world: alternate uses of genes and proteins. Indeed, the
use of different proteins, including physiological stress proteins and enzymes,
as crystallins was discovered by comparing the same tissue (lens) of many
species. This book provides other examples. Gene sharing tells us to recognize
that each species has the potential to teach us new tricks for old proteins. Free,
open-minded exploration of biological diversity is not only fascinating for its
own sake; it is a core for technical innovation and a bulwark to threatening
diseases.

MEDICAL IMPLICATIONS

Gene sharing has numerous implications for medicine. Drug design must be
tailored to target the appropriate function of a protein in order to be effective
and to avoid potentially deleterious side effects. This is a challenging task.
First we must identify the various molecular functions of a protein, which is
not trivial, and second, we must be able to target the drug to the correct function—a daunting job in bioengineering. Correlating mutations with different protein:protein interactions provides a starting point for identifying the
amino acids, and consequently drug targets, involved in specific functions.
The prevalence of gene sharing increases the likelihood that drug toxicity is
due to interference with multiple functions of a protein rather than (or in addition to) interference with a number of different proteins.

Gene sharing may unify apparently unrelated symptoms or diseases. If the
same protein has multiple molecular functions depending on tissue location
or other conditions, it is likely to cause a plethora of symptoms when it sus-
tains a lesion of some kind. Conversely, the multifunctional nature of proteins
that are engaged in gene sharing complicates the identification of the function that is responsible for a disease once a protein has been implicated by genetic or biochemical methods.

The prevalence of gene sharing also introduces a caveat for medical research. When researchers are developing nonhuman models for human disease, care must be taken to account for the possibility that in some situations
entirely different proteins may perform similar functions. The taxon-specificity of lens or corneal crystallins comes to mind (see Chapters 4 and 5).
Clearly, taxon-specific gene sharing will affect the interpretation of experi-

mental results on animal models with respect to the etiology and eventual treatment of disease.

Gene sharing also highlights the necessity of paying strict attention to gene expression when linking a disease to a particular protein or when attempting gene therapy. Given that location and concentration affect protein function, tissue-specificity and level of gene expression may be important considerations when diagnosing disease or attempting gene therapy.

Gene sharing encompasses the many interactions of individual polypeptides (for example, networks), alternate functions for similar proteins (see Chapters 6 and 7), and similar roles for different proteins (note especially the diverse lens crystallins). Gene sharing adds to the possibilities of manipulating biology to advantage. Theoretically it should be possible in some circumstances to consider repairing a lesion in one protein by tampering with another, or substituting one protein, or a part of one protein, for a different protein that is defective. The robustness of metabolic networks may be underlaid by a corresponding robustness of protein:protein interactions.[945] Gene sharing may thus provide a network of molecular functions resembling dynamic interacting signaling pathways, which have been singled out as important points for drug intervention.[1108] Thus, gene sharing may have a place in the new view of treatment for disease proposed by Fishman and Porter.[1108] They state (p. 493): "Historically, diseases have been categorized by organ pathology. Today we are moving towards pangenomic assays of gene expression, protein levels, and the ways in which proteins become modified after they have been produced from mRNA." We may anticipate that the concept of gene sharing will become increasingly important in this scientific world of multifunctional proteins, interactive pathways, and genomic medicine. Finally, the modular nature of proteins, like the modular nature of all biological organizations,[1109] suggests that it will be possible to engineer proteins to assume new, potentially therapeutic functions without deleterious effects on their normal functions—in essence, to create new roles for endogenous polypeptides to treat and/or prevent diseases.

Overall, the paradox of gene sharing—that proteins specialize and diversify simultaneously—opens countless possibilities and lays as many minefields. The trick is to observe Nature without bias and appreciate its resourcefulness, think imaginatively, dare to take a chance, and expect the unexpected.

Easier said than done!

Glossary

References

Index

Glossary

Alternative RNA splicing   The process used during gene expression by which exon sequences in the primary RNA transcript are arranged in various combinations during the maturation of messenger RNA in the nucleus; this usually involves skipping the incorporation of one or another exon sequence, and may involve translated or untranslated sequences; it also may involve the inclusion of intron sequences into the messenger RNA.

Amino acid   A molecule that is a building block for polypeptides; there are twenty different amino acids that can be used to make a polypeptide.

Apoptosis   Programmed cell death.

Allele   An alternative variant state of a gene.

Base   A nucleotide used in DNA or RNA; see nucleotide.

Catalyst   A substance that increases the rate of a chemical reaction: an enzyme is a catalyst.

Chromosome   Condensed nuclear material (chromatin) that contains the genomic DNA associated with proteins.

Chromatin   Decondensed nuclear material consisting of double-stranded DNA that is packaged by proteins.

Coding region (or sequences)   A stretch of DNA or RNA whose nucleotide sequence dictates the order of amino acids of a protein.

Codon   Three consecutive nucleotides in DNA or messenger RNA that specify a particular amino acid.

Convergent   Changes that have occurred independently in different lineages but have led to a similar result.

Crystallin   A protein comprising at least 5 percent of the water-soluble protein of a transparent eye lens; also may refer to a similarly abundant protein of the cornea.

Cytoplasm   The material within the cell membrane of an intact cell; contrasts with the nucleoplasm, which is the material within the nuclear membrane.

Cytoskeleton   The collection of structural entities in the cytoplasm that contribute to cellular shape and motility; actin, tubulin, and many other proteins comprise the cytoskeleton.

Diploid   Possessing two copies (one from the male and one from the female) of each chromosome characteristic of the species.

237

Divergent   Changes that have occurred in inherited structures during evolution during the phylogenetic lineage of animals.

DNA   Deoxyribonucleic acid; the genetic material composed of purine and pyrimdine nucleotides (bases) linked by phosphodiester bonds.

Enhancer   DNA sequences that potentiate transcription of DNA; these are often responsible for tissue specificity of transcription during gene expression.

Enzyme   A protein that catalyzes a biochemical reaction; RNA can also act as an enzyme in some circumstances.

Eukaryote   An organism whose cells contain a defined nucleus.

Exaptation   A term coined by Gould to refer to the exploitation of various properties of a protein to biological advantage; complex issue, see text.

Exon   A DNA nucleotide sequence that is transferred to mature messenger RNA; exons are contiguous DNA sequences that are separated by introns.

Gene expression   The entire process involving transfer of information from DNA to a polypeptide.

Genetic code   The total set of codons, or triplet sets of nucleotides, that specify amino acids to be incorporated into polypeptides during translation of messenger RNA.

Genome   The entire genetic material, or DNA content, of a cell or organism.

Genotype   Defines a cell or organism by its DNA sequence. The genotype is generally invariant; however, DNA rearrangements (for example in immunoglobulin rearrangement) can occur that would change the genotype of that particular cell.

Haploid   Possessing one copy of each chromosome characteristic of the species.

Heterozygous   A state when two alleles (different versions of the same gene) are both present in a diploid genome.

Homologous   Refers to genes or proteins in different species that appear generally similar because they descended from a common ancestor.

Homologous recombination   Recombination of DNA based upon DNA sequence similarity; homologous recombination is the basis of replacing or deleting genes in gene knockout experiments.

Homozygous   A state in which two identical alleles are both present in a diploid genome.

Horizontal gene transfer   Lateral DNA transfer between cells or species; contrasts with vertical DNA transfer, which refers to DNA transfer to linear descendents via gametes in sexual reproduction or fission.

Hydrogen bond   A weak electrostatic interaction resulting from the sharing of a hydrogen atom by two oxygen or nitrogen atoms, or an oxygen and nitrogen atom; hydrogen bonds are responsible for holding the two DNA strands together in a double helix.

Intervening sequences   See intron.

Intron   Noncoding intervening DNA sequences that separate exons; introns are transcribed into the primary transcript but are not incorporated into mature messenger RNA.

Lateral gene transfer   See horizontal gene transfer.

Messenger RNA   The cytoplasmic RNA containing the coding region that is translated into protein; messenger RNAs also contain untranslated sequences in front of (5′ to) and behind (3′ to) the coding sequence.

Microtubule   A cylindrical organelle in the cytoplasm composed of protein subunits called tubulin; microtubules contribute to cell structure and participate in cell movements, transport processes, and cell division.

microRNA   A small noncoding RNA that regulates translation or stability of specific messenger RNAs.

Mitochondrion   A complex DNA-containing structure in the cytoplasm that is responsible for cellular respiration.

mRNA   Abbreviation for messenger RNA.

Mutant   A gene or organism carrying a gene that has been modified and does not lead to a normal polypeptide or adult trait or traits; a recessive mutation is only manifest when it occurs in two copies in a diploid genome, while a dominant mutation is manifest when it occurs in a single copy.

Mutation   Alteration of DNA structure, due to a change in nucleotides, an addition or deletion of nucleotides, or a rearrangement of nucleotides; a mutation can also refer to an abnormality in the expression of a gene.

Neofunctionalization   Refers to the acquisition of a new gene function after gene duplication.

Nonsynonymous mutation   A nucleotide change in a protein-coding gene that changes the amino acid specified by the codon.

Nucleotide   The chemical building blocks of DNA or RNA; these are purines (adenine, guanine) and pyrimidines [thymidine (DNA only), uridine (RNA only) and cytidine]; nucleotides are abbreviated A, G, T, C, U.

Nucleus   A membrane-bound organelle in the cytoplasm of eukaryotic cells that contains the genetic material; an idiosyncratic usage is "lens nucleus," which refers to the nonnucleated central fiber cell in the vertebrate lens.

Orthologous   Refers to a direct ancestral gene or protein in different species.

Paralogous   Refers to two or more genes in the same species that are so similar in nucleotide sequence that they are assumed to have arisen by gene duplication; paralogous genes could also be considered sibling genes.

Paramutation   A mutation that is inherited epigenetically, that is without a change in DNA sequence.

Pathogenic   Causing disease.

Phenotype   Defines a cell or organism by traits; the same genotype may exhibit different phenotypes under different conditions.

Polypeptide   A linear chain of amino acids linked by peptide bonds.

Primary transcript   The entire RNA that is obtained when a DNA segment is transcribed by RNA polymerase; usually refers to an RNA strand before it is spliced to remove introns and becomes a mature messenger RNA.

Prokaryote   A bacterial cell that lacks a defined nucleus.

Promoter   The DNA sequence responsible for initiating transcription of the adjacent nucleotide sequence.

Protein   A folded polypeptide that is assumed or shown to have a biological function; a protein can be a single polypeptide or consist of two or more interacting polypeptides that act as a unit; the interacting polypeptides, or subunits, may be identical or different in individual proteins.

Pseudogene   A gene copy that is not in its normal chromosomal location and cannot be expressed into a functional protein due to alterations in its regulatory sequences or coding sequences.

Ribosomal RNA   The structural RNAs of ribosomes that are used during the process of translating messenger RNA into a polypeptide.

RNA   Ribonucleic acid; generally a single-stranded sequence of nucleotides linked by phosphodiester bonds and containing a sugar (ribose) backbone.

RNA editing   The process of altering a nucleotide in cytoplasmic messenger RNA; this usually changes the amino acid encoded in the messenger RNA so that it is no longer the same as that specified by the DNA codon in the gene.

Splicing   The process of removing intron sequences from the primary transcript and joining the exon sequences together; the final result of splicing is a mature messenger RNA.

Start codon   The codon initiating polypeptide synthesis on messenger RNA; the start codon always specifies methionine and is AUG in the RNA and ATG in the DNA.

Stop codon   The codon terminating polypeptide synthesis on messenger RNA; the stop codon does not specify an amino acid and can be UAG, UAA, or UGA in the RNA and TAG, TAA, or TGA in the DNA.

Subfunctionalization   Refers to the separation of gene functions after duplication.

Synonymous mutation   A nucleotide change in a protein-coding gene that does not change the amino acid specified by the codon; usually the mutation occurs in the third position of the triplet codon.

Taxon   A group of animals that belong to a particular class or phylum.

Taxon-specific   Occurring only in selected species belonging to a particular class or phylum.

Transfer RNA   RNA that carries individual amino acids to the ribosome:messenger RNA complex during translation; may be abbreviated as tRNA.

Translation   The process of making a polypeptide with an amino acid sequence dictated by the coding sequence of the messenger RNA.

Transcription   The process of making a RNA strand directly from DNA by using the enzyme RNA polymerase.

tRNA synthetase   The enzyme that catalyzes the formation of a peptide bond during messenger RNA translation.

Untranslated sequences   Refers to nucleotide sequences in a gene or messenger RNA that are not translated into a polypeptide; these occur in front of (5′) and behind (3′) the coding sequences of the messenger RNA; introns are also referred to as untranslated gene sequences.

References

1. Horowitz, N. H. Biochemical genetics of *Neurospora*. *Adv. Genet.* **3**, 33–71 (1950).

2. Ingram, V. M. *Hemoglobins in Genetics and Evolution* (Columbia University Press, New York, 1963).

3. Ohno, S. *Evolution by Gene Duplication* (Springer Verlag, New York, 1970).

4. Kimura, M. & Ota, T. On some principles governing molecular evolution. *Proc Natl Acad Sci U S A* **71**, 2848–52 (1974).

5. Meyer, A. & Van de Peer, Y. (eds.) *Genome Evolution. Gene and Genome Duplications and the Origin of Novel Gene Functions* (Kluwer Academic Publishers, Dordrecht, 2003).

6. Hughes, A. L. Gene duplication and the origin of novel proteins. *Proc Natl Acad Sci U S A* **102**, 8791–2 (2005).

7. Hughes, M. K. & Hughes, A. L. Evolution of duplicate genes in a tetraploid animal, *Xenopus laevis*. *Mol Biol Evol* **10**, 1360–9 (1993).

8. Piatigorsky, J. & Wistow, G. The recruitment of crystallins: new functions precede gene duplication. *Science* **252**, 1078–9 (1991).

9. Piatigorsky, J. Lens crystallins. Innovation associated with changes in gene regulation. *J Biol Chem* **267**, 4277–80 (1992).

10. Hughes, A. L. The evolution of functionally novel proteins after gene duplication. *Proc R Soc Lond B Biol Sci* **256**, 119–24 (1994).

11. Goodman, M., Moore, G. W. & Matsuda, G. Darwinian evolution in the genealogy of haemoglobin. *Nature* **253**, 603–8 (1975).

12. Uetz, P. & Finley, R. L., Jr. From protein networks to biological systems. *FEBS Lett* **579**, 1821–7 (2005).

13. Jensen, R. A. Enzyme recruitment in evolution of new function. *Annu Rev Microbiol* **30**, 409–25 (1976).

14. Jensen, R. A. & Byng, G. S. The partitioning of biochemical pathways with isozyme systems. *Isozymes Curr Top Biol Med Res* **5**, 143–74 (1981).

15. James, L. C. & Tawfik, D. S. Conformational diversity and protein evolution—a 60-year-old hypothesis revisited. *Trends Biochem Sci* **28**, 361–8 (2003).

16. Aharoni, A. et al. The "evolvability" of promiscuous protein functions. *Nat Genet* **37**, 73–6 (2005).

241

17. Orgel, L. E. Gene-duplication and the origin of proteins with novel functions. *J Theor Biol* **67**, 773 (1977).

18. Wistow, G. J., Mulders, J. W. & de Jong, W. W. The enzyme lactate dehydrogenase as a structural protein in avian and crocodilian lenses. *Nature* **326**, 622–4 (1987).

19. Wistow, G. & Piatigorsky, J. Recruitment of enzymes as lens structural proteins. *Science* **236**, 1554–6 (1987).

20. Piatigorsky, J. et al. Gene sharing by δ-crystallin and argininosuccinate lyase. *Proc Natl Acad Sci U S A* **85**, 3479–83 (1988).

21. Hendriks, W. et al. Duck lens ε-crystallin and lactate dehydrogenase B4 are identical: a single-copy gene product with two distinct functions. *Proc Natl Acad Sci U S A* **85**, 7114–8 (1988).

22. Wistow, G. J. & Piatigorsky, J. Lens crystallins: the evolution and expression of proteins for a highly specialized tissue. *Annu Rev Biochem* **57**, 479–504 (1988).

23. Piatigorsky, J. & Wistow, G. J. Enzyme/crystallins: gene sharing as an evolutionary strategy. *Cell* **57**, 197–9 (1989).

24. de Jong, W. W., Hendriks, W., Mulders, J. W. & Bloemendal, H. Evolution of eye lens crystallins: the stress connection. *Trends Biochem Sci* **14**, 365–8 (1989).

25. Karniely, S. & Pines, O. Single translation—dual destination: mechanisms of dual protein targeting in eukaryotes. *EMBO Rep* **6**, 420–5 (2005).

26. Simpson, G. G. *The Meaning of Evolution* (Yale University Press, New Haven, 1949).

27. Wilkins, A. S. *The Evolution of Developmental Pathways* (Sinauer Associates, Inc., Sunderland, MA, 2002).

28. Hekerman, P. et al. Pleiotropy of leptin receptor signalling is defined by distinct roles of the intracellular tyrosines. *Febs J* **272**, 109–19 (2005).

29. Kullo, I. J. et al. Pleiotropic genetic effects contribute to the correlation between HDL cholesterol, triglycerides, and LDL particle size in hypertensive sibships. *Am J Hypertens* **18**, 99–103 (2005).

30. Sung, Y. H., Choi, Y. S., Cheong, C. & Lee, H. W. The pleiotropy of telomerase against cell death. *Mol Cells* **19**, 303–9 (2005).

31. Simpson, A. et al. The structure of avian eye lens δ-crystallin reveals a new fold for a superfamily of oligomeric enzymes. *Nat Struct Biol* **1**, 724–34 (1994).

32. True, J. R. & Carroll, S. B. Gene co-option in physiological and morphological evolution. *Annu Rev Cell Dev Biol* **18**, 53–80 (2002).

33. Jeffery, C. J. Moonlighting proteins. *Trends Biochem Sci* **24**, 8–11 (1999).

34. Bray, D. Genomics. Molecular prodigality. *Science* **299**, 1189–90 (2003).

34a. Ptacek, J. et al. Global analysis of protein phosphorylation in yeast. *Nature* **438**, 679–84 (2005).

35. Evans, T. C., Xu, M. Q. & Pradhan, S. Protein splicing elements and plants: from transgene containment to protein purification. *Annu Rev Plant Biol* **56**, 375–92 (2005).

35a. Davidson, E. H. & Erwin, D. H. Gene regulatory networks and the evolution of animal body plans. *Science* **311**, 796–800 (2006).

36. Piatigorsky, J. δ Crystallins and their nucleic acids. Mol Cell Biochem 59, 33–56 (1984).

37. Sophos, N. A. & Vasiliou, V. Aldehyde dehydrogenase gene superfamily: the 2002 update. *Chem Biol Interact* **143–144**, 5–22 (2003).

38. Durner, J., Gow, A. J., Stamler, J. S. & Glazebrook, J. Ancient origins of nitric oxide signaling in biological systems. *Proc Natl Acad Sci U S A* **96**, 14206–7 (1999).

39. Thomas, D. D. et al. Hypoxic inducible factor 1{α}, extracellular signal-regulated kinase, and p53 are regulated by distinct threshold concentrations of nitric oxide. *Proc Natl Acad Sci U S A* **101**, 8894–9 (2004).

40. Reynaert, N. L. et al. From the cover: nitric oxide represses inhibitory {κ}B kinase through S-nitrosylation. *Proc Natl Acad Sci U S A* **101**, 8945–50 (2004).

41. Marshall, H. E., Hess, D. T. & Stamler, J. S. S-nitrosylation: physiological regulation of NF-{κ}B. *Proc Natl Acad Sci U S A* **101**, 8841–2 (2004).

42. Brown, G. C. Nitric oxide and mitochondrial respiration. *Biochim Biophys Acta* **1411**, 351–69 (1999).

43. Brown, G. C. & Borutaite, V. Inhibition of mitochondrial respiratory complex I by nitric oxide, peroxynitrite and S-nitrosothiols. *Biochim Biophys Acta* **1658**, 44–9 (2004).

44. Darwin, C. *On the Origin of Species by Means of Natural Selection, or Preservation of Favored Races in the Struggle for Life* (Murray, London, 1859).

45. Bock, W. J. Preadaptation and multiple evolutionary pathways. *Evolution* **13**, 194–211 (1959).

46. Goldschmidt, R. *The Material Basis of Evolution* (Pageant Books, Paterson, NJ, 1960).

47. Simpson, G. G. *The Major Features of Evolution* (ed. Dunn, L.) (Columbia University Press, New York, 1953).

48. Bock, W. J. The role of adaptive mechanisms in the origin of higher levels of organization. *Syst Zool* **14**, 272–87 (1965).

49. Gould, S. J. *The Structure of Evolutionary Theory* (Harvard University Press, Cambridge, MA, 2002).

50. Stoltzfus, A. On the possibility of constructive neutral evolution. *J Mol Evol* **49**, 169–81 (1999).

51. Gould, S. J. & Vrba, E. S. Exaptation—a missing term in the science of form. *Paleobiology* **8**, 4–15 (1982).

52. Williams, G. C. *Adaptation and Natural Selection. A Critique of Some Current Evolutionary Thought.* (Princeton University Press, Princeton, NJ, 1966).

53. Gould, S. J. & Lewontin, R. C. The spandrels of San Marco and the Panglossian paradigm: a critique of the adaptationist programme. *Proc R Soc Lond B Biol Sci* **205**, 581–98 (1979).

54. Gould, S. J. The exaptive excellence of spandrels as a term and prototype. *Proc Natl Acad Sci U S A* **94**, 10750–5 (1997).

55. Weiss, M. A. et al. Protein structure and the spandrels of San Marco: insulin's receptor-binding surface is buttressed by an invariant leucine essential for its stability. *Biochemistry* **41**, 809–19 (2002).

56. Carroll, S. B., Grenier, J. K. & Weatherbee, S. D. *From DNA to Diversity. Molecular Genetics and the Evolution of Animal Design* (Blackwell Science, Madison, WI, 2001).

57. Wilson, A. C., Carlson, S. S. & White, T. J. Biochemical evolution. *Annu Rev Biochem* **46,** 573–693 (1977).

58. Wilson, A. C., Ochman, H. & Prager, E. M. Molecular time scale for evolution. *Trends in Genetics* **3,** 241–7 (1987).

59. King, M. C. & Wilson, A. C. Evolution at two levels in humans and chimpanzees. *Science* **188,** 107–16 (1975).

60. Prager, E. M. & Wilson, A. C. Slow evolutionary loss of the potential for interspecific hybridization in birds: a manifestation of slow regulatory evolution. *Proc Natl Acad Sci U S A* **72,** 200–4 (1975).

61. Jacob, F. Evolution and tinkering. *Science* **196,** 1161–6 (1977).

62. Jacob, F. *The Possible and the Actual* (Pantheon Books, NY, 1982).

63. Britten, R. J. & Davidson, E. H. Gene regulation for higher cells: a theory. *Science* **165,** 349–57 (1969).

64. Davidson, E. H. & Britten, R. J. Organization, transcription, and regulation in the animal genome. *Q Rev Biol* **48,** 565–613 (1973).

65. Davidson, E. H. *Genomic Regulatory Systems. Development and Evolution* (Academic Press, San Diego, CA, 2001).

66. Britten, R. J. & Davidson, E. H. Repetitive and non-repetitive DNA sequences and a speculation on the origins of evolutionary novelty. *Q Rev Biol* **46,** 111–138 (1971).

67. Galliot, B. & Miller, D. Origin of anterior patterning. How old is our head? *Trends Genet* **16,** 1–5 (2000).

68. Suga, H. et al. Extensive gene duplication in the early evolution of animals before the parazoan-eumetazoan split demonstrated by G proteins and protein tyrosine kinases from sponge and hydra. *J Mol Evol* **48,** 646–53 (1999).

69. Nei, M. *Molecular Evolutionary Genetics* (Columbia University Press, New York, 1987).

70. Conway Morris, S. The Cambrian "explosion": slow-fuse or megatonnage? *Proc Natl Acad Sci U S A* **97,** 4426–9 (2000).

71. Morgan, T. H., Sturtevant, A. H., Muller, H. J. & Bridges, C. B. *The Mechanism of Mendelian Heredity* (Henry Holt and Company, New York, 1915).

72. Morgan, T. H. *The Theory of the Gene* (Hafner Publishing Company, New York, 1926 (reprinted 1964)).

73. Stadler, L. J. The gene. *Science* **120,** 811–9 (1954).

74. Carlson, E. A. *The Gene: A Critical History* (W.B. Saunders, Philadelphia, 1966).

75. Dawkins, R. *The Selfish Gene* (Oxford University Press, Oxford, 1989).

76. Waters, C. K. Genes made molecular. *Philosophy of Science* **61,** 163–85 (1994).

77. Beurton, P. J., Falk, R. & Rheinberger, H.-J. (eds.) *The Concept of the Gene in Development and Evolution: Historical and Epistemological Perspectives.* (Cambridge University Press, Cambridge, 2000).

78. Burian, R. M. On the internal dynamics of Mendelian genetics. *C R Acad Sci III* **323,** 1127–37 (2000).

79. Keller, E. F. *The Century of the Gene* (Harvard University Press, Cambridge, MA, 2000).

80. Keller, E. F. *Making Sense of Life. Explaining Biological Development with Models, Metaphors, and Machines* (Harvard University Press, Cambridge, MA, 2002).

81. Portin, P. Historical development of the concept of the gene. *J Med Philos* 27, 257–86 (2002).

82. Moss, L. *What Genes Can't Do* (The MIT Press, Cambridge, MA, 2003).

83. Burian, R. M. Molecular Epigenesis, Molecular Pleiotropy, and Molecular Gene Definitions. *Hist Philos Life Sci* 26, 59–80 (2004).

84. Amundson, R. *The Changing Role of the Embryo in Evolutionary Thought: Roots of Evo-Devo* (Cambridge University Press, New York, 2005).

85. Burian, R. M. *The Epistemology of Development, Evolution, and Genetics* (Cambridge University Press, New York, 2005).

86. De Vries, H. *Intracellular Panagenesis* (Open Court Library of Living Philosophers, Chicago, 1910).

87. Stern, C. & Sherwood, E. R. (eds). *The Origin of Genetics. A Mendel Source Book* (W.H. Freeman and Company, San Francisco, 1966).

88. Falk, R. The dominance of traits in genetic analysis. *J. Hist. Biol.* 24, 457–84 (1991).

89. Bateson, W. & Saunders, E. R. Experimental studies in the physiology of heredity: Report 1, in *Reports to the Evolution Committee of the Royal Society,* 1–160 (Harrison & Sons, London, 1902).

90. Burian, R. M. "Historical realism," "contextual objectivity," and changing concepts of the gene, in *The Philosophy of Marjorie Grene* (eds. Auxier, R. E. & Hahn, L.), 340–363 (Open Court Library of Living Philosophers, Chicago, 2000).

91. Watson, J. D. & Crick, F. H. Molecular structure of nucleic acids; a structure for deoxyribose nucleic acid. *Nature* 171, 737–8 (1953).

92. Beadle, G. W. & Tatum, E. L. Genetic control of biochemical reactions in *Neurospora. Proc Natl Acad Sci U S A* 27, 499–506 (1941).

93. Srb, A. M. & Horowitz, N. H. The ornithine cycle in *Neurospora* and its genetic control. *J Biol Chem* 154, 129–39 (1944).

94. Jacob, F. & Monod, J. Molecular and biological characterization of messenger RNA. *J Mol Biol* 3, 318–56 (1961).

95. Brenner, S., Jacob, F. & Meselsohn, M. An unstable intermediate carrying information from genes to ribosomes for protein synthesis. *Nature* 190, 576–80 (1961).

96. Shen-Orr, S. S., Milo, R., Mangan, S. & Alon, U. Network motifs in the transcriptional regulation network of *Escherichia coli. Nat Genet* 31, 64–8 (2002).

97. Guet, C. C., Elowitz, M. B., Hsing, W. & Leibler, S. Combinatorial synthesis of genetic networks. *Science* 296, 1466–70 (2002).

98. Almaas, E., Kovacs, B., Vicsek, T., Oltvai, Z. N. & Barabasi, A. L. Global organization of metabolic fluxes in the bacterium *Escherichia coli. Nature* 427, 839–43 (2004).

99. Lee, T. I. et al. Transcriptional regulatory networks in *Saccharomyces cerevisiae. Science* 298, 799–804 (2002).

100. Huh, W. K. et al. Global analysis of protein localization in budding yeast. *Nature* 425, 686–91 (2003).

101. Giot, L. et al. A protein interaction map of *Drosophila melanogaster*. *Science* **302**, 1727–36 (2003).

102. Hinman, V. F., Nguyen, A. T., Cameron, R. A. & Davidson, E. H. Developmental gene regulatory network architecture across 500 million years of echinoderm evolution. *Proc Natl Acad Sci U S A* **100**, 13356–61 (2003).

103. Bray, D. Molecular networks: the top-down view. *Science* **301**, 1864–5 (2003).

104. Alon, U. Biological networks: the tinkerer as an engineer. *Science* **301**, 1866–7 (2003).

105. Yeger-Lotem, E. et al. Network motifs in integrated cellular networks of transcription-regulation and protein-protein interaction. *Proc Natl Acad Sci U S A* **101**, 5934–9 (2004).

106. Li, F., Long, T., Lu, Y., Ouyang, Q. & Tang, C. The yeast cell-cycle network is robustly designed. *Proc Natl Acad Sci U S A* **101**, 4781–6 (2004).

107. Singer, M. & Berg, P. *Genes and Genomes: A Changing Perspective*. (University Science Books, Mill Valley, CA, 1991).

108. Waterston, R. H. et al. Initial sequencing and comparative analysis of the mouse genome. *Nature* **420**, 520–62 (2002).

109. Williams, B. A., Slamovits, C. H., Patron, N. J., Fast, N. M. & Keeling, P. J. A high frequency of overlapping gene expression in compacted eukaryotic genomes. *Proc Natl Acad Sci U S A* **102**, 10936–41 (2005).

110. Cavalier-Smith, T. Nuclear volume control by nucleoskeletal DNA, selection for cell volume and cell growth rate, and the solution of the DNA C-value paradox. *J Cell Sci* **34**, 247–78 (1978).

111. Hughes, A. L. Adaptive evolution after gene duplication. *Trends Genet* **18**, 433–4 (2002).

112. Dawkins, R. *The Ancestor's Tale: A Pilgrimage to the Dawn of Evolution* (Houghton Mifflin Co., Boston, 2004).

113. Harrison, P. M., Kumar, A., Lang, N., Snyder, M. & Gerstein, M. A question of size: the eukaryotic proteome and the problems in defining it. *Nucleic Acids Res* **30**, 1083–90 (2002).

114. Snyder, M. & Gerstein, M. Genomics. Defining genes in the genomics era. *Science* **300**, 258–60 (2003).

115. Southan, C. Has the yo-yo stopped? An assessment of human protein-coding gene number. *Proteomics* **4**, 1712–26 (2004).

116. Rubin, G. M. et al. Comparative genomics of the eukaryotes. *Science* **287**, 2204–15 (2000).

117. Blattner, F. R. et al. The complete genome sequence of *Escherichia coli* K-12. *Science* **277**, 1453–74 (1997).

118. Stover, C. K. et al. Complete genome sequence of *Pseudomonas aeruginosa* PA01, an opportunistic pathogen. *Nature* **406**, 959–64 (2000).

119. Galagan, J. E. et al. The genome sequence of the filamentous fungus *Neurospora crassa*. *Nature* **422**, 859–68 (2003).

120. Goffeau, A. et al. Life with 6000 genes. *Science* **274**, 546, 563–7 (1996).

121. C. elegans Sequencing Consortium. Genome sequence of the nematode *C. elegans: a platform for investigating biology. Science* **282**, 2012–8 (1998).

122. Adams, M. D. et al. The genome sequence of *Drosophila melanogaster. Science* **287**, 2185–95 (2000).

123. Myers, E. W. et al. A whole-genome assembly of *Drosophila. Science* **287**, 2196–204 (2000).

124. Okazaki, Y. et al. Analysis of the mouse transcriptome based on functional annotation of 60,770 full-length cDNAs. *Nature* **420**, 563–73 (2002).

125. Lander, E. S. et al. Initial sequencing and analysis of the human genome. *Nature* **409**, 860–921 (2001).

126. Venter, J. C. et al. The sequence of the human genome. *Science* **291**, 1304–51 (2001).

127. Claverie, J. M. Gene number. What if there are only 30,000 human genes? *Science* **291**, 1255–7 (2001).

128. Imanishi, T. et al. Integrative annotation of 21,037 human genes validated by full-length cDNA clones. *PLoS Biol* **2**, E162 (2004).

129. The ENCODE (ENCyclopedia Of DNA Elements) Project. *Science* **306**, 636–40 (2004).

130. Levine, M. & Tjian, R. Transcription regulation and animal diversity. *Nature* **424**, 147–51 (2003).

131. Epp, C. D. Definition of a gene. *Nature* **389**, 537 (1997).

132. Wain, H. M. et al. Guidelines for human gene nomenclature. *Genomics* **79**, 464–70 (2002).

133. Carninci, P. et al. The transcriptional landscape of the mammalian genome. *Science* **309**, 1559–63 (2005).

134. Katayama, S. et al. Antisense transcription in the mammalian transcriptome. *Science* **309**, 1564–6 (2005).

135. Lu, C. et al. Elucidation of the small RNA component of the transcriptome. *Science* **309**, 1567–9 (2005).

136. Willingham, A. T. et al. A strategy for probing the function of noncoding RNAs finds a repressor of NFAT. *Science* **309**, 1570–3 (2005).

137. Eddy, S. R. Non-coding RNA genes and the modern RNA world. *Nat Rev Genet* **2**, 919–29 (2001).

138. Eddy, S. R. Computational genomics of noncoding RNA genes. *Cell* **109**, 137–40 (2002).

139. Novina, C. D. & Sharp, P. A. The RNAi revolution. *Nature* **430**, 161–4 (2004).

140. Dykxhoorn, D. M., Novina, C. D. & Sharp, P. A. Killing the messenger: short RNAs that silence gene expression. *Nat Rev Mol Cell Biol* **4**, 457–67 (2003).

141. Lee, R. C. & Ambros, V. An extensive class of small RNAs in *Caenorhabditis elegans. Science* **294**, 862–4 (2001).

141a. Macrae, I. J. et al. Structural basis for double-stranded RNA processing by Dicer. *Science* **311**, 195–8 (2006).

142. Ambros, V. The functions of animal microRNAs. *Nature* **431**, 350–5 (2004).

143. Bartel, D. P. MicroRNAs: genomics, biogenesis, mechanism, and function. *Cell* **116**, 281–97 (2004).

144. Bartel, D. P. & Chen, C. Z. Micromanagers of gene expression: the potentially widespread influence of metazoan microRNAs. *Nat Rev Genet* **5**, 396–400 (2004).

144a. Kwon, C., Han, Z., Olson, E. N. & Srivastava, D. MicroRNA1 influences cardiac differentiation in *Drosophila* and regulates Notch signaling. *Proc Natl Acad Sci U S A* **102**, 18986–91 (2005).

144b. Wu, L., Fan, J. & Belasco, J. G. MicroRNAs direct rapid deadenylation of mRNA. *Proc Natl Acad Sci USA* **103**, 4034–9 (2006).

145. Jopling, C. L., Yi, M., Lancaster, A. M., Lemon, S. M. & Sarnow, P. Modulation of hepatitis C virus RNA abundance by a liver-specific microRNA. *Science* **309**, 1577–81 (2005).

146. Pillai, R. S., Artus, C. G. & Filipowicz, W. Tethering of human Ago proteins to mRNA mimics the miRNA-mediated repression of protein synthesis. *RNA* **10**, 1518–25 (2004).

147. Pillai, R. S. et al. Inhibition of translational initiation by Let-7 microRNA in human cells. *Science* **309**, 1573–6 (2005).

148. Mansfield, J. H. et al. MicroRNA-responsive "sensor" transgenes uncover Hox-like and other developmentally regulated patterns of vertebrate microRNA expression. *Nat Genet* **36**, 1079–83 (2004).

149. Liu, C. G. et al. An oligonucleotide microchip for genome-wide microRNA profiling in human and mouse tissues. *Proc Natl Acad Sci U S A* **101**, 9740–4 (2004).

150. Dermitzakis, E. T. et al. Evolutionary discrimination of mammalian conserved non-genic sequences (CNGs). *Science* **302**, 1033–5 (2003).

151. Masse, E., Majdalani, N. & Gottesman, S. Regulatory roles for small RNAs in bacteria. *Curr Opin Microbiol* **6**, 120–4 (2003).

151a. Sanchez-Elsner, T., Gou, D., Kremmer, E. & Sauer, F. Noncoding RNAs of Trithorax response elements recruit *Drosophila* Ash1 to ultrabithorax. *Science* **311**, 1118–23 (2006).

152. Wilderman, P. J. et al. Identification of tandem duplicate regulatory small RNAs in *Pseudomonas aeruginosa* involved in iron homeostasis. *Proc Natl Acad Sci U S A* **101**, 9792–7 (2004).

153. Graveley, B. R. Alternative splicing: increasing diversity in the proteomic world. *Trends Genet* **17**, 100–7 (2001).

154. Schmucker, D. et al. *Drosophila* Dscam is an axon guidance receptor exhibiting extraordinary molecular diversity. *Cell* **101**, 671–84 (2000).

155. Modrek, B. & Lee, C. J. Alternative splicing in the human, mouse and rat genomes is associated with an increased frequency of exon creation and/or loss. *Nat Genet* **34**, 177–80 (2003).

156. Lee, C., Atanelov, L., Modrek, B. & Xing, Y. ASAP: the Alternative Splicing Annotation Project. *Nucleic Acids Res* **31**, 101–5 (2003).

157. Xu, Q., Modrek, B. & Lee, C. Genome-wide detection of tissue-specific alternative splicing in the human transcriptome. *Nucleic Acids Res* **30**, 3754–66 (2002).

158. Mangus, D. A., Evans, M. C. & Jacobson, A. Poly(A)-binding proteins: multifunctional scaffolds for the post-transcriptional control of gene expression. *Genome Biol* **4,** 223 (2003).

159. Parker, R. & Song, H. The enzymes and control of eukaryotic mRNA turnover. *Nat Struct Mol Biol* **11,** 121–7 (2004).

160. Chen, C. Y. & Shyu, A. B. AU-rich elements: characterization and importance in mRNA degradation. *Trends Biochem Sci* **20,** 465–70 (1995).

161. Gingerich, T. J., Feige, J. J. & LaMarre, J. AU-rich elements and the control of gene expression through regulated mRNA stability. *Anim Health Res Rev* **5,** 49–63 (2004).

162. Yeap, B. B., Wilce, J. A. & Leedman, P. J. The androgen receptor mRNA. *Bioessays* **26,** 672–82 (2004).

163. Amrani, N. et al. A faux 3′-UTR promotes aberrant termination and triggers nonsense-mediated mRNA decay. *Nature* **432,** 112–8 (2004).

164. Soukup, J. K. & Soukup, G. A. Riboswitches exert genetic control through metabolite-induced conformational change. *Curr Opin Struct Biol* **14,** 344–9 (2004).

165. Barrick, J. E. et al. New RNA motifs suggest an expanded scope for riboswitches in bacterial genetic control. *Proc Natl Acad Sci U S A* **101,** 6421–6 (2004).

166. Mandal, M. et al. A glycine-dependent riboswitch that uses cooperative binding to control gene expression. *Science* **306,** 275–9 (2004).

167. Sudarsan, N., Barrick, J. E. & Breaker, R. R. Metabolite-binding RNA domains are present in the genes of eukaryotes. *RNA* **9,** 644–7 (2003).

168. Joyce, G. F. The antiquity of RNA-based evolution. *Nature* **418,** 214–21 (2002).

169. Schibler, U. & Sierra, F. Alternative promoters in developmental gene expression. *Annu Rev Genet* **21,** 237–57 (1987).

170. Schmitz, F., Konigstorfer, A. & Sudhof, T. C. RIBEYE, a component of synaptic ribbons: a protein's journey through evolution provides insight into synaptic ribbon function. *Neuron* **28,** 857–72 (2000).

171. Piatigorsky, J. Dual use of the transcriptional repressor (CtBP2)/ribbon synapse (RIBEYE) gene: how prevalent are multifunctional genes? *Trends Neurosci* **24,** 555–7 (2001).

172. Gonzalez, P. et al. Comparative analysis of the ζ-crystallin/quinone reductase gene in guinea pig and mouse. *Mol Biol Evol* **11,** 305–15 (1994).

173. Lee, D. C., Gonzalez, P. & Wistow, G. ζ-crystallin: a lens-specific promoter and the gene recruitment of an enzyme as a crystallin. *J Mol Biol* **236,** 669–78 (1994).

174. Lu, B. & Hanson, M. R. Fully edited and partially edited nad9 transcripts differ in size and both are associated with polysomes in potato mitochondria. *Nucleic Acids Res* **24,** 1369–74 (1996).

175. Lu, B., Wilson, R. K., Phreaner, C. G., Mulligan, R. M. & Hanson, M. R. Protein polymorphism generated by differential RNA editing of a plant mitochondrial rps12 gene. *Mol Cell Biol* **16,** 1543–9 (1996).

176. Benne, R. et al. Major transcript of the frameshifted coxII gene from trypanosome mitochondria contains four nucleotides that are not encoded in the DNA. *Cell* **46,** 819–26 (1986).

177. Hanrahan, C. J., Palladino, M. J., Ganetzky, B. & Reenan, R. A. RNA editing of the Drosophila para Na(+) channel transcript. Evolutionary conservation and developmental regulation. *Genetics* **155**, 1149–60 (2000).

178. Palladino, M. J., Keegan, L. P., O'Connell, M. A. & Reenan, R. A. A-to-I pre-mRNA editing in *Drosophila* is primarily involved in adult nervous system function and integrity. *Cell* **102**, 437–49 (2000).

179. Palladino, M. J., Keegan, L. P., O'Connell, M. A. & Reenan, R. A. dADAR, a *Drosophila* double-stranded RNA-specific adenosine deaminase is highly developmentally regulated and is itself a target for RNA editing. *RNA* **6**, 1004–18 (2000).

180. Hoopengardner, B., Bhalla, T., Staber, C. & Reenan, R. Nervous system targets of RNA editing identified by comparative genomics. *Science* **301**, 832–6 (2003).

181. Kawahara, Y. et al. RNA editing and death of motor neurons. *Nature* **427**, 801 (2004).

182. Strohman, R. C. The coming Kuhnian revolution in biology. *Nature Biotech.* **15**, 194–200 (1997).

183. Gelbart, W. M. Databases in genomic research. *Science* **282**, 659–661 (1998).

184. Griffiths, P. E. & Neumann-Held, E. M. The many faces of the gene. *Bioscience* **49**, 656–662 (1999).

185. Dawkins, R. *The Extended Phenotype* (W. H. Freeman, Oxford, 1982).

186. Sarkar, S. From the *reaktionsnorm* to the adaptive norm: the norm of reaction, 1909–1960. *Biology and Philosophy* **14**, 235–252 (1999).

187. Dobzhansky, T. & Spassky, B. Genetics of natural populations. Xxxiv. Adaptive norm, genetic load and genetic elite in *Drosophila pseudoobscura*. *Genetics* **48**, 1467–85 (1963).

188. Spilianakis, C. G., Lalioti, M. D., Town, T., Lee, G. R. & Flavell, R. A. Interchromosomal associations between alternatively expressed loci. *Nature* **435**, 637–45 (2005).

189. Walls, G. *The Vertebrate Eye* (Cranbook Institute of Science, Bloomfield Hills, MI, 1942).

190. Ali, M. A. (ed.) *Photoreception and Vision in Invertebrates.* (Plenum Press, New York, 1984).

191. Cronin, T. W. Photoreception in marine invertebrates. *American Zoologist* **26**, 403–415 (1986).

192. Van Dover, C. L., Szuts, E. Z., Chamberlain, S. C. & Cann, J. R. A novel eye in "eyeless" shrimp from hydrothermal vents of the Mid-Atlantic Ridge. *Nature* **337**, 458–60 (1989).

193. Land, M. F. & Fernald, R. D. The evolution of eyes. *Annu Rev Neurosci* **15**, 1–29 (1992).

194. Tomarev, S. I. & Piatigorsky, J. Lens crystallins of invertebrates—diversity and recruitment from detoxification enzymes and novel proteins. *Eur J Biochem* **235**, 449–65 (1996).

195. Wolken, J. J. *Light Detectors, Photoreceptors, and Imaging Systems in Nature* (Oxford University Press, New York, 1995).

196. Land, M. F. & Nilsson, D.-E. *Animal Eyes* (Oxford University Press, Oxford, 2002).

197. Fernald, R. D. Evolution of eyes. *Curr Opin Neurobiol* **10**, 444–50 (2000).

198. Fernald, R. D. Eyes: variety, development and evolution. *Brain Behav Evol* **64**, 141–7 (2004).

199. Land, M. F. The optical structures of animal eyes. *Curr Biol* **15**, R319–23 (2005).

200. Nilsson, D. E. Eye ancestry: old genes for new eyes. *Curr Biol* **6**, 39–42 (1996).

201. Foster, K. W. & Smyth, R. D. Light antennas in phototactic algae. *Microbiol Rev* **44**, 572–630 (1980).

202. Francis, D. On the eyespot of the dinoflagellate, *Nematodinium*. *J Exp Biol* **47**, 495–501 (1967).

203. Greuet, C. *Leucopsis cylindrica,* Nov. Gen., Nov. Sp., reridinien *Warnowiidae* lindemann considerations phylogenetiques sur les *Warnowiidae. Protistologica* **4**, 419–422 (1968).

204. Arendt, D. & Wittbrodt, J. Reconstructing the eyes of Urbilateria. *Philos Trans R Soc Lond B Biol Sci* **356**, 1545–63 (2001).

205. Arendt, D. Evolution of eyes and photoreceptor cell types. *Int J Dev Biol* **47**, 563–71 (2003).

206. Fernald, R. D. Evolving eyes. *Int. J. Dev. Biol.* **48**, 701–5 (2004).

207. Oakley, T. H. & Huber, D. R. Differential expression of duplicated opsin genes in two eyetypes of ostracod crustaceans. *J Mol Evol* **59**, 239–49 (2004).

208. Arendt, D., Tessmar-Raible, K., Snyman, H., Dorresteijn, A. W. & Wittbrodt, J. Ciliary photoreceptors with a vertebrate-type opsin in an invertebrate brain. *Science* **306**, 869–71 (2004).

209. Gehring, W. J. Historical perspective on the development and evolution of eyes and photoreceptors. *Int. J. Dev. Biol.* **48**, 707–17 (2004).

210. Gehring, W. J. New perspectives on eye development and the evolution of eyes and photoreceptors. *J Hered* **96**, 171–84 (2005).

211. Packard, A. Cephalopods and fish: the limits of convergence. *Biol. Rev.* **47**, 241–307 (1972).

212. Kumar, J. P. & Moses, K. Eye specification in *Drosophila:* perspectives and implications. *Semin Cell Dev Biol* **12**, 469–74 (2001).

213. Salvini-Plawen, L. V. & Mayr, E. On the evolution of photoreceptors and eyes. *Evol Dev* **10**, 207–263 (1977).

214. Oakley, T. H. & Cunningham, C. W. Molecular phylogenetic evidence for the independent evolutionary origin of an arthropod compound eye. *Proc Natl Acad Sci U S A* **99**, 1426–30 (2002).

215. Nilsson, D. E. & Pelger, S. A pessimistic estimate of the time required for an eye to evolve. *Proc R Soc Lond B Biol Sci* **256**, 53–8 (1994).

216. Dawkins, R. Evolutionary biology. The eye in a twinkling. *Nature* **368**, 690–1 (1994).

217. Nordstrom, K., Wallen, R., Seymour, J. & Nilsson, D. A simple visual system without neurons in jellyfish larvae. *Proc R Soc Lond B Biol Sci* **270**, 2349–54 (2003).

218. Quiring, R., Walldorf, U., Kloter, U. & Gehring, W. J. Homology of the *eyeless* gene

of *Drosophila* to the *Small eye* gene in mice and *Aniridia* in humans. *Science* **265**, 785–9 (1994).

219. Callaerts, P., Halder, G. & Gehring, W. J. *PAX-6* in development and evolution. *Annu Rev Neurosci* **20**, 483–532 (1997).

220. Pichaud, F., Treisman, J. & Desplan, C. Reinventing a common strategy for patterning the eye. *Cell* **105**, 9–12 (2001).

221. Ton, C. C. et al. Positional cloning and characterization of a paired box- and homeobox-containing gene from the aniridia region. *Cell* **67**, 1059–74 (1991).

222. Glaser, T., Walton, D. S. & Maas, R. L. Genomic structure, evolutionary conservation and aniridia mutations in the human *PAX6* gene. *Nat Genet* **2**, 232–9 (1992).

223. Hanson, I. M. PAX6 and congenital eye malformations. *Pediatr Res* **54**, 791–6 (2003).

224. Cvekl, A. & Tamm, E. R. Anterior eye development and ocular mesenchyme: new insights from mouse models and human diseases. *Bioessays* **26**, 374–86 (2004).

225. Halder, G., Callaerts, P. & Gehring, W. J. Induction of ectopic eyes by targeted expression of the *eyeless* gene in *Drosophila*. *Science* **267**, 1788–92 (1995).

226. Tomarev, S. I. et al. Squid Pax-6 and eye development. *Proc Natl Acad Sci U S A* **94**, 2421–6 (1997).

227. Callaerts, P. et al. Isolation and expression of a *Pax-6* gene in the regenerating and intact Planarian *Dugesia*(G)*tigrina*. *Proc Natl Acad Sci U S A* **96**, 558–63 (1999).

228. Loosli, F., Kmita-Cunisse, M. & Gehring, W. J. Isolation of a *Pax-6* homolog from the ribbonworm *Lineus sanguineus*. *Proc Natl Acad Sci U S A* **93**, 2658–63 (1996).

229. Glardon, S., Callaerts, P., Halder, G. & Gehring, W. J. Conservation of Pax-6 in a lower chordate, the ascidian *Phallusia* mammillata. *Development* **124**, 817–25 (1997).

230. Glardon, S., Holland, L. Z., Gehring, W. J. & Holland, N. D. Isolation and developmental expression of the amphioxus *Pax-6* gene (*AmphiPax-6*): insights into eye and photoreceptor evolution. *Development* **125**, 2701–10 (1998).

231. Chow, R. L., Altmann, C. R., Lang, R. A. & Hemmati-Brivanlou, A. Pax6 induces ectopic eyes in a vertebrate. *Development* **126**, 4213–22 (1999).

232. Gehring, W. J. & Ikeo, K. Pax 6: mastering eye morphogenesis and eye evolution. *Trends Genet* **15**, 371–7 (1999).

233. Gehring, W. J. *Master Control Genes in Development and Evolution: The Homeobox Story.* (Yale University Press, New Haven, 1998).

234. Gehring, W. J. The genetic control of eye development and its implications for the evolution of the various eye-types. *Int J Dev Biol* **46**, 65–73 (2002).

235. Treisman, J. E. A conserved blueprint for the eye? *Bioessays* **21**, 843–50 (1999).

236. Ashery-Padan, R. & Gruss, P. Pax6 lights-up the way for eye development. *Curr Opin Cell Biol* **13**, 706–14 (2001).

237. Baker, N. E. Master regulatory genes; telling them what to do. *Bioessays* **23**, 763–6 (2001).

238. van Heyningen, V. & Williamson, K. A. *PAX6* in sensory development. *Hum Mol Genet* **11**, 1161–7 (2002).

239. Simpson, T. I. & Price, D. J. Pax6; a pleiotropic player in development. *Bioessays* **24**, 1041–51 (2002).

240. Conway Morris, S. *Life's Solutions. Inevitable Humans in a Lonely Universe* (Cambridge University Press, Cambridge, 2003).

240a. Kozmik, Z. Pax genes in eye development and evolution. *Curr Opin Genet Develop* **15**, 430–8.

241. Zuckerkandl, E. Molecular pathways to parallel evolution: I. Gene nexuses and their morphological correlates. *J Mol Evol* **39**, 661–78 (1994).

242. Arthur, W. The emerging conceptual framework of evolutionary developmental biology. *Nature* **415**, 757–64 (2002).

243. Relaix, F. & Buckingham, M. From insect eye to vertebrate muscle: redeployment of a regulatory network. *Genes Dev* **13**, 3171–8 (1999).

244. Heanue, T. A. et al. Synergistic regulation of vertebrate muscle development by *Dach2, Eya2,* and *Six1,* homologs of genes required for *Drosophila* eye formation. *Genes Dev* **13**, 3231–43 (1999).

245. Kardon, G., Heanue, T. A. & Tabin, C. J. The *Pax/Six/Eya/Dach* network in development and evolution, in *Modularity in Development and Evolution* (eds. Schlosser, G. & Wagner, G. P.) 59–80 (University of Chicago Press, Chicago, 2004).

246. Zuber, M. E., Gestri, G., Viczian, A. S., Barsacchi, G. & Harris, W. A. Specification of the vertebrate eye by a network of eye field transcription factors. *Development* **130**, 5155–67 (2003).

247. Sivak, J. G. Through the lens clearly: phylogeny and development: the Proctor lecture. *Invest Ophthalmol Vis Sci* **45**, 740–7; 739 (2004).

248. Swamynathan, S. K., Crawford, M. A., Robison, W. G., Jr., Kanungo, J. & Piatigorsky, J. Adaptive differences in the structure and macromolecular compositions of the air and water corneas of the "four-eyed" fish (*Anableps anableps*). *Faseb J* **17**, 1996–2005 (2003).

249. Coates, M. M. Visual ecology and functional morphology of Cubuzoa (Cnidaria). *Int Comp Biol* **43**, 542–8 (2003).

250. Nilsson, D. E., Gislen, L., Coates, M. M., Skogh, C. & Garm, A. Advanced optics in a jellyfish eye. *Nature* **435**, 201–5 (2005).

251. Harding, J. J. & Dilley, K. J. Structural proteins of the mammalian lens: a review with emphasis on changes in development, aging and cataract. *Exp Eye Res* **22**, 1–73 (1976).

252. Bloemendal, H. *Molecular and Cellular Biology of the Eye Lens* (John Wiley & Sons, New York, 1981).

253. Bloemendal, H. Lens proteins. *CRC Crit Rev Biochem* **12**, 1–38 (1982).

254. McDevitt, D. S. & Brahma, S. K. Ontogeny and localization of the crystallins in eye lens development and regeneration, in *Cell Biology of the Eye* (ed. McDevitt, D. S.) 143–191 (Academic Press, New York, 1982).

255. Robinson, M. L. & Lovicu, F. J. in *Development of the Ocular Lens.* (eds. Lovicu, F. J. & Robinson, M. L.) 3–26 (Cambridge University Press, Cambridge, 2004).

256. Benedek, G. B. Theory of transparency of the eye. *Applied Optics* **10**, 459–473 (1971).

257. Delaye, M. & Tardieu, A. Short-range order of crystallin proteins accounts for eye lens transparency. *Nature* **302**, 415–7 (1983).

258. Bettelheim, F. A. & Siew, E. L. Biological basis of lens transparency, in *Cell Biology of the Eye* (ed. McDevitt, D. S.) 243–297 (Academic Press, New York, 1982).

259. Bettelheim, F. A. & Siew, E. L. Effect of change in concentration upon lens turbidity as predicted by the random fluctuation theory. *Biophys J* **41**, 29–33 (1983).

260. Piatigorsky, J., Horwitz, J., Kuwabara, T. & Cutress, C. E. The cellular eye lens and crystallins of cubomedusan jellyfish. *J Comp Physiol [A]* **164**, 577–87 (1989).

261. Piatigorsky, J. Lens differentiation in vertebrates. A review of cellular and molecular features. *Differentiation* **19**, 134–53 (1981).

262. Duncan, M. K., Cvekl, A., Kantorow, M. & Piatigorsky, J. Lens crystallins, in *Development of the Ocular Lens* (eds. Robinson, M. L. & Lovicu, F. J.) (Cambridge University Press, New York, 2004).

263. Zigler, J. S., Jr., Horwitz, J. & Kinoshita, J. H. Studies on the low molecular weight proteins of human lens. *Exp Eye Res* **32**, 21–30 (1981).

264. Wistow, G. *Molecular Biology and Evolution of Crystallins: Gene Recruitment and Multifunctional Proteins in the Eye Lens* (Springer, New York, 1995).

265. Mörner, C. T. Untersuchungen der Protein-substanzen in den lichtbrechenden Medien des Auges. *Hoppe Seylers Z Physiol Chem* **18**, 61–106 (1894).

266. Clayton, R. M. Comparative aspects of lens proteins, in *The Eye* (eds. Davson, H. & Graham, L. T.), 399–494 (Academic Press, New York, 1974).

267. Rabaey, M. Electrophoretic and immunoelectrophoretic studies on the soluble proteins in the developing lens of birds. *Exp Eye Res* **1**, 310–6 (1962).

268. Rabaey, M. Lens proteins during embryonic development of different vertebrates. *Invest Ophthalmol* **44**, 560–78 (1965).

269. Zwaan, J. Electrophoretic studies on the heterogeneity of the chicken lens crystallins. *Exp Eye Res* **7**, 461–72 (1968).

270. Halbert, S. P. & Fitzgerald, P. L. Studies on the immunologic organ specificity of ocular lens. *Am J Ophthalmol* **46**, 187–94; discussion 194–6 (1958).

271. Halbert, S. P. & Manski, W. Organ specificity with special reference to the lens. *Prog Allergy* **7**, 107–86 (1963).

272. Manski, W. & Halbert, S. P. Immunochemical investigation on the phylogeny of lens proteins, in *Protides of the Biological Fluids. Proceedings of the Twelfth Colloquium* (ed. Peeters, H.) 117–134 (Elsevier Publishing Company, Amsterdam, 1964).

273. Siezen, R. J. & Shaw, D. C. Physicochemical characterization of lens proteins of the squid *Nototodarus gouldi* and comparison with vertebrate crystallins. *Biochim Biophys Acta* **704**, 304–20 (1982).

274. Tomarev, S. I. & Zinovieva, R. D. Squid major lens polypeptides are homologous to glutathione S-transferases subunits. *Nature* **336**, 86–8 (1988).

275. Zinovieva, R. D., Tomarev, S. I. & Piatigorsky, J. Aldehyde dehydrogenase-derived omega-crystallins of squid and octopus. Specialization for lens expression. *J Biol Chem* **268**, 11449–55 (1993).

276. Piatigorsky, J. et al. Omega-crystallin of the scallop lens. A dimeric aldehyde dehydrogenase class 1/2 enzyme-crystallin. *J Biol Chem* **275**, 41064–73 (2000).

277. Komori, N., Usukura, J. & Matsumoto, H. Drosocrystallin, a major 52 kDa glycoprotein of the *Drosophila melanogaster* corneal lens. Purification, biochemical characterization, and subcellular localization. *J Cell Sci* **102** (**Pt 2**), 191–201 (1992).

278. Piatigorsky, J., Horwitz, J. & Norman, B. L. J1-crystallins of the cubomedusan jellyfish lens constitute a novel family encoded in at least three intronless genes. *J Biol Chem* **268**, 11894–901 (1993).

279. Piatigorsky, J. et al. J3-crystallin of the jellyfish lens: similarity to saposins. *Proc Natl Acad Sci U S A* **98**, 12362–7 (2001).

280. Clayton, R. M., Campbell, J. C. & Truman, D. E. A re-examination of the organ specificity of lens antigens. *Exp Eye Res* **7**, 11–29 (1968).

281. Clayton, R. M. Problems of differentiation in the vertebrate lens. *Curr Top Dev Biol* **5**, 115–80 (1970).

282. Bower, D. J., Errington, L. H., Cooper, D. N., Morris, S. & Clayton, R. M. Chicken lens δ-crystallin gene expression and methylation in several non-lens tissues. *Nucleic Acids Res* **11**, 2513–27 (1983).

283. Zwaan, J. Lens-specific antigens and cytodifferentiation in the developing lens. *J Cell Physiol* **72**, Suppl 1:47–71 (1968).

284. Clayton, R. M., Thomson, I. & de Pomerai, D. I. Relationship between crystallin mRNA expression in retina cells and their capacity to re-differentiate into lens cells. *Nature* **282**, 628–9 (1979).

285. Ingolia, T. D. & Craig, E. A. Four small *Drosophila* heat shock proteins are related to each other and to mammalian α-crystallin. *Proc Natl Acad Sci U S A* **79**, 2360–4 (1982).

286. Klemenz, R., Frohli, E., Steiger, R. H., Schafer, R. & Aoyama, A. αB-crystallin is a small heat shock protein. *Proc Natl Acad Sci U S A* **88**, 3652–6 (1991).

287. Horwitz, J. α-crystallin can function as a molecular chaperone. *Proc Natl Acad Sci U S A* **89**, 10449–53 (1992).

288. Bhat, S. P. & Nagineni, C. N. αB subunit of lens-specific protein α-crystallin is present in other ocular and non-ocular tissues. *Biochem Biophys Res Commun* **158**, 319–25 (1989).

289. Dubin, R. A., Wawrousek, E. F. & Piatigorsky, J. Expression of the murine αB-crystallin gene is not restricted to the lens. *Mol Cell Biol* **9**, 1083–91 (1989).

290. Sax, C. M. & Piatigorsky, J. Expression of the α-crystallin/small heat-shock protein/molecular chaperone genes in the lens and other tissues. *Adv Enzymol Relat Areas Mol Biol* **69**, 155–201 (1994).

291. de Jong, W. W., Leunissen, J. A. & Voorter, C. E. Evolution of the α-crystallin/small heat-shock protein family. *Mol Biol Evol* **10**, 103–26 (1993).

292. Caspers, G. J., Leunissen, J. A. & de Jong, W. W. The expanding small heat-shock protein family, and structure predictions of the conserved "α-crystallin domain." *J Mol Evol* **40**, 238–48 (1995).

293. Horwitz, J., Emmons, T. & Takemoto, L. The ability of lens α crystallin to protect against heat-induced aggregation is age-dependent. *Curr Eye Res* **11**, 817–22 (1992).

294. Brady, J. P. et al. Targeted disruption of the mouse αA-crystallin gene induces cataract and cytoplasmic inclusion bodies containing the small heat shock protein αB-crystallin. *Proc Natl Acad Sci U S A* **94**, 884–9 (1997).

295. Brady, J. P. et al. αB-crystallin in lens development and muscle integrity: a gene knockout approach. *Invest Ophthalmol Vis Sci* **42**, 2924–34 (2001).

296. Nicholl, I. D. & Quinlan, R. A. Chaperone activity of α-crystallins modulates intermediate filament assembly. *Embo J* **13**, 945–53 (1994).

297. Scotting, P., McDermott, H. & Mayer, R. J. Ubiquitin-protein conjugates and αB crystallin are selectively present in cells undergoing major cytomorphological reorganisation in early chicken embryos. *FEBS Lett* **285**, 75–9 (1991).

297a. Iwaki, T., Iwaki, A., Liem, R. K. & Goldman, J. E. Expression of αB-crystallin in the developing rat kidney. *Kidney Int* **40**, 52–6 (1991).

298. Longoni, S., Lattonen, S., Bullock, G. & Chiesi, M. Cardiac α-crystallin. II. Intracellular localization. *Mol Cell Biochem* **97**, 121–8 (1990).

299. Atomi, Y., Yamada, S., Strohman, R. & Nonomura, Y. αB-crystallin in skeletal muscle: purification and localization. *J Biochem (Tokyo)* **110**, 812–22 (1991).

299a. Bennardini, F., Wrzosek, A. & Chiesi, M. αB-crystallin in cardiac tissue. Association with actin and desmin filaments. *Circ Res* **71**, 288–94 (1992).

299b. Golenhofen, N., Arbeiter, A., Koob, R. & Drenckhahn, D. Ischemia-induced association of the stress protein αB-crystallin with I-band portion of cardiac titin. *J Mol Cell Cardiol* **34**, 309–19 (2002).

299c. Ray, P. S. et al. Transgene overexpression of αB crystallin confers simultaneous protection against cardiomyocyte apoptosis and necrosis during myocardial ischemia and reperfusion. *Faseb J* **15**, 393–402 (2001).

299d. Morrison, L. E., Whittaker, R. J., Klepper, R. E., Wawrousek, E. F. & Glembotski, C. C. Roles for αB-crystallin and HSPB2 in protecting the myocardium from ischemia-reperfusion-induced damage in a KO mouse model. *Am J Physiol Heart Circ Physiol* **286**, H847–55 (2004).

299e. Golenhofen, N., Redel, A., Wawrousek, E. F. & Drenckhahn, D. Ischemia-induced increase of stiffness of αB-crystallin/HSPB2-deficient myocardium. *Pflugers Arch* **451**, 518–25 (2006).

300. Srinivasan, A. N., Nagineni, C. N. & Bhat, S. P. αA-crystallin is expressed in non-ocular tissues. *J Biol Chem* **267**, 23337–41 (1992).

301. Thyagarajan, T. & Kulkarni, A. B. Transforming growth factor-β1 negatively regulates crystallin expression in teeth. *J Bone Miner Res* **17**, 1710–7 (2002).

302. Clark, J. I. & Muchowski, P. J. Small heat-shock proteins and their potential role in human disease. *Curr Opin Struct Biol* **10**, 52–9 (2000).

302a. Moyano, J. V., Evans, J. R., Chen, F. et al. αB-crystallin is a novel oncoprotein that predicts poor clinical outcome in breast cancer. *J Clin Invest* **116**, 261–70 (2006).

303. Andley, U. P., Song, Z., Wawrousek, E. F. & Bassnett, S. The molecular chaperone αA-crystallin enhances lens epithelial cell growth and resistance to UVA stress. *J Biol Chem* **273**, 31252–61 (1998).

304. Andley, U. P., Song, Z., Wawrousek, E. F., Fleming, T. P. & Bassnett, S. Differential protective activity of αA- and αB-crystallin in lens epithelial cells. *J Biol Chem* **275**, 36823–31 (2000).

305. Andley, U. P. et al. Lens epithelial cells derived from αB-crystallin knockout mice demonstrate hyperproliferation and genomic instability. *Faseb J* **15**, 221–229 (2001).

306. Bai, F., Xi, J. H., Wawrousek, E. F., Fleming, T. P. & Andley, U. P. Hyperproliferation and p53 status of lens epithelial cells derived from αB-crystallin knockout mice. *J Biol Chem* (2003).

307. Mao, Y. W. et al. Human bcl-2 gene attenuates the ability of rabbit lens epithelial cells against H2O2-induced apoptosis through down-regulation of the αB-crystallin gene. *J Biol Chem* **276**, 43435–45 (2001).

308. Ray, P. S. et al. Transgene overexpression of αB crystallin confers simultaneous protection against cardiomyocyte apoptosis and necrosis during myocardial ischemia and reperfusion. *Faseb J* **15**, 393–402 (2001).

309. Alge, C. S. et al. Retinal pigment epithelium is protected against apoptosis by αB-crystallin. *Invest Ophthalmol Vis Sci* **43**, 3575–82 (2002).

310. Mao, Y. W., Liu, J. P., Xiang, H. & Li, D. W. Human αA- and αB-crystallins bind to Bax and Bcl-X(S) to sequester their translocation during staurosporine-induced apoptosis. *Cell Death Differ* **11**, 512–26 (2004).

311. Kamradt, M. C., Chen, F. & Cryns, V. L. The small heat shock protein αB-crystallin negatively regulates cytochrome c- and caspase-8-dependent activation of caspase-3 by inhibiting its autoproteolytic maturation. *J Biol Chem* **276**, 16059–63 (2001).

312. Kamradt, M. C., Chen, F., Sam, S. & Cryns, V. L. The small heat shock protein αB-crystallin negatively regulates apoptosis during myogenic differentiation by inhibiting caspase-3 activation. *J Biol Chem* **277**, 38731–6 (2002).

312a. Morozov, V. & Wawrousek, E. F. Caspase-dependent secondary lens fiber cell disintegration in αA-/αB-crystallin double knockout mice. *Development* **132**, 813–21 (2006).

313. den Engelsman, J., Keijsers, V., de Jong, W. W. & Boelens, W. C. The small heat-shock protein αB-crystallin promotes FBX4-dependent ubiquitination. *J Biol Chem* **278**, 4699–704 (2003).

314. Gangalum, R. K., Schibler, M. J. & Bhat, S. P. Small heat shock protein αB-crystallin is part of cell cycle-dependent Golgi reorganization. *J Biol Chem* **279**, 43374–7 (2004).

315. Inaguma, Y., Ito, H., Iwamoto, I., Saga, S. & Kato, K. αB-crystallin phosphorylated at Ser-59 is localized in centrosomes and midbodies during mitosis. *Eur J Cell Biol* **80**, 741–8 (2001).

316. van Rijk, A. E., Stege, G. J., Bennink, E. J., May, A. & Bloemendal, H. Nuclear stain-

ing for the small heat shock protein αB-crystallin colocalizes with splicing factor SC35. *Eur J Cell Biol* **82**, 361–8 (2003).

317. van den IJssel, P., Wheelock, R., Prescott, A., Russell, P. & Quinlan, R. A. Nuclear speckle localisation of the small heat shock protein αB-crystallin and its inhibition by the R120G cardiomyopathy-linked mutation. *Exp Cell Res* **287**, 249–61 (2003).

318. den Engelsman, J. et al. Mimicking phosphorylation of the small heat-shock protein αB-crystallin recruits the F-box protein FBX4 to nuclear SC35 speckles. *Eur J Biochem* **271**, 4195–203 (2004).

319. Tsvetkova, N. M. et al. Small heat-shock proteins regulate membrane lipid polymorphism. *Proc Natl Acad Sci U S A* **99**, 13504–9 (2002).

320. Spector, A., Chiesa, R., Sredy, J. & Garner, W. cAMP-dependent phosphorylation of bovine lens α-crystallin. *Proc Natl Acad Sci U S A* **82**, 4712–6 (1985).

321. Voorter, C. E., Mulders, J. W., Bloemendal, H. & de Jong, W. W. Some aspects of the phosphorylation of α-crystallin A. *Eur J Biochem* **160**, 203–10 (1986).

322. Chiesa, R., Gawinowicz-Kolks, M. A. & Spector, A. The phosphorylation of the primary gene products of α-crystallin. *J Biol Chem* **262**, 1438–41 (1987).

323. den Engelsman, J. et al. Nuclear import of αB-crystallin is phosphorylation-dependent and hampered by hyperphosphorylation of the myopathy-related mutant R120G. *J Biol Chem* **280**, 37139–48 (2005).

324. Maddala, R. & Rao, V. P. α-Crystallin localizes to the leading edges of migrating lens epithelial cells. *Exp Cell Res* **306**, 203–15 (2005).

325. Kantorow, M. & Piatigorsky, J. α-Crystallin/small heat shock protein has autokinase activity. *Proc Natl Acad Sci U S A* **91**, 3112–6 (1994).

326. Kantorow, M. & Piatigorsky, J. Phosphorylations of αA- and αB-crystallin. *Int J Biol Macromol* **22**, 307–14 (1998).

327. Driessen, H. P., Herbrink, P., Bloemendal, H. & de Jong, W. W. Primary structure of the bovine β-crystallin Bp chain. Internal duplication and homology with gamma-crystallin. *Eur J Biochem* **121**, 83–91 (1981).

328. Lubsen, N. H., Aarts, H. J. & Schoenmakers, J. G. The evolution of lenticular proteins: the β- and γ-crystallin super gene family. *Prog Biophys Mol Biol* **51**, 47–76 (1988).

329. D'Alessio, G. The evolution of monomeric and oligomeric $\beta\gamma$-type crystallins. Facts and hypotheses. *Eur J Biochem* **269**, 3122–30 (2002).

330. Wistow, G. et al. γN-crystallin and the evolution of the $\beta\gamma$-crystallin superfamily in vertebrates. *Febs J* **272**, 2276–91 (2005).

331. Blundell, T. et al. The molecular structure and stability of the eye lens: x-ray analysis of γ-crystallin II. *Nature* **289**, 771–7 (1981).

332. Wistow, G. et al. X-ray analysis of the eye lens protein γ-II crystallin at 1.9 Å resolution. *J Mol Biol* **170**, 175–202 (1983).

333. Bax, B. et al. X-ray analysis of β B2-crystallin and evolution of oligomeric lens proteins. *Nature* **347**, 776–80 (1990).

334. Inana, G., Piatigorsky, J., Norman, B., Slingsby, C. & Blundell, T. Gene and protein structure of a β-crystallin polypeptide in murine lens: relationship of exons and structural motifs. *Nature* **302**, 310–5 (1983).

335. Piatigorsky, J. Lens crystallins and their gene families. *Cell* **38**, 620–1 (1984).

336. van Rens, G. L. et al. Structure of the bovine eye lens γs-crystallin gene (formerly βs). *Gene* **78**, 225–33 (1989).

337. Moormann, R. J. et al. Characterization of the rat γ-crystallin gene family and its expression in the eye lens. *J Mol Biol* **182**, 419–30 (1985).

338. Head, M. W., Peter, A. & Clayton, R. M. Evidence for the extralenticular expression of members of the β-crystallin gene family in the chick and a comparison with δ-crystallin during differentiation and transdifferentiation. *Differentiation* **48**, 147–56 (1991).

339. Head, M. W., Sedowofia, K. & Clayton, R. M. βB2-crystallin in the mammalian retina. *Exp Eye Res* **61**, 423–8 (1995).

340. Smolich, B. D., Tarkington, S. K., Saha, M. S. & Grainger, R. M. *Xenopus* γ-crystallin gene expression: evidence that the γ-crystallin gene family is transcribed in lens and nonlens tissues. *Mol Cell Biol* **14**, 1355–63 (1994).

341. Sinha, D. et al. Cloning and mapping the mouse Crygs gene and non-lens expression of [γ]S-crystallin. *Mol Vis* **4**, 8 (1998).

342. Dirks, R. P. H., Van Genesen, S. T., Krüse, J. J., Jorissen, L. & Lubsen, N. H. Extralenticular expression of the rodent βB2-crystallin gene. *Exp Eye Res* **66**, 267–9 (1998).

343. Jones, S. E., Jomary, C., Grist, J., Makwana, J. & Neal, M. J. Retinal expression of γ-crystallins in the mouse. *Invest Ophthalmol Vis Sci* **40**, 3017–20 (1999).

344. Magabo, K. S., Horwitz, J., Piatigorsky, J. & Kantorow, M. Expression of βB(2)-crystallin mRNA and protein in retina, brain, and testis. *Invest Ophthalmol Vis Sci* **41**, 3056–60 (2000).

345. Xi, J. et al. A comprehensive analysis of the expression of crystallins in mouse retina. *Mol Vis* **9**, 410–9 (2003).

346. Shimeld, S. M. et al. Urochordate βγ-crystallin and the evolutionary origin of the vertebrate eye lens. *Curr Biol* **15**, 1684–9 (2005).

346a. Piatigorsky, J. Seeing the light: the role of inherited developmental cascades in the origins of vertebrate lenses and their crystallins. *Heredity* (March 1) (2006).

346b. Delsuc, F., Brinkmann, H., Chourrout, D. & Philippe, H. Tunicates and not cephalochordates are the closest living relatives of vertebrates. *Nature* **439**, 965–8 (2006).

347. Wistow, G., Summers, L. & Blundell, T. *Myxococcus xanthus* spore coat protein S may have a similar structure to vertebrate lens βγ-crystallins. *Nature* **315**, 771–3 (1985).

348. Bernier, F., Lemieux, G. & Pallotta, D. Gene families encode the major encystment-specific proteins of *Physarum polycephalum* plasmodia. *Gene* **59**, 265–77 (1987).

349. Wistow, G. Evolution of a protein superfamily: relationships between vertebrate lens crystallins and microorganism dormancy proteins. *J Mol Evol* **30**, 140–5 (1990).

350. Rosinke, B., Renner, C., Mayr, E. M., Jaenicke, R. & Holak, T. A. Ca2+-loaded spherulin 3a from *Physarum polycephalum* adopts the prototype γ-crystallin fold in aqueous solution. *J Mol Biol* **271**, 645–55 (1997).

351. Clout, N. J., Kretschmar, M., Jaenicke, R. & Slingsby, C. Crystal structure of the calcium-loaded spherulin 3a dimer sheds light on the evolution of the eye lens βγ-crystallin domain fold. *Structure (Camb)* **9**, 115–24 (2001).

352. Jobby, M. K. & Sharma, Y. Calcium-binding crystallins from *Yersinia pestis*:

Characterization of two single $\beta\gamma$-crystallin domains of a putative exported protein. *J Biol Chem* **280**, 1209–16 (2005).

353. Parkhill, J. et al. Genome sequence of *Yersinia pestis,* the causative agent of plague. *Nature* **413**, 523–7 (2001).

354. Wistow, G., Jaworski, C. & Rao, P. V. A non-lens member of the $\beta\gamma$-crystallin superfamily in a vertebrate, the amphibian Cynops. *Exp Eye Res* **61**, 637–9 (1995).

355. Takabatake, T., Takahashi, T. C., Takeshima, K. & Takata, K. Protein synthesis during neural and epidermal diferentiation in *Cynops* embryo. *Develop Growth & Differentiation* **33**, 277–82 (1991).

356. Takabatake, T., Takahashi, T. C. & Takeshima, K. Cloning of an epidermis-specific *Cynops* cDNA from neurula library. *Develop Growth & Differentiation* **34**, 277–83 (1992).

357. Chan, C. W., Saimi, Y. & Kung, C. A new multigene family encoding calcium-dependent calmodulin-binding membrane proteins of *Paramecium tetraurelia. Gene* **231**, 21–32 (1999).

358. Fan, J. et al. γE-crystallin recruitment to the plasma membrane by specific interaction between lens MIP/aquaporin-0 and γE-crystallin. *Invest Ophthalmol Vis Sci* **45**, 863–71 (2004).

359. Fan, J. et al. Specific interaction between lens MIP/Aquaporin-0 and two members of the γ-crystallin family. *Mol Vis* **11**, 76–87 (2005).

360. Krasko, A., Muller, I. M. & Muller, W. E. Evolutionary relationships of the metazoan $\beta\gamma$-crystallins, including that from the marine sponge Geodia cydonium. *Proc R Soc Lond B Biol Sci* **264**, 1077–84 (1997).

361. Di Maro, A., Pizzo, E., Cubellis, M. V. & D'Alessio, G. An intron-less $\beta\gamma$-crystallin-type gene from the sponge *Geodia cydonium. Gene* **299**, 79–82 (2002).

362. Antuch, W., Guntert, P. & Wuthrich, K. Ancestral $\beta\gamma$-crystallin precursor structure in a yeast killer toxin. *Nat Struct Biol* **3**, 662–5 (1996).

363. Ohno, A. et al. NMR structure of the *Streptomyces metalloproteinase* inhibitor, SMPI, isolated from *Streptomyces nigrescens* TK-23: another example of an ancestral $\beta\gamma$-crystallin precursor structure. *J Mol Biol* **282**, 421–33 (1998).

364. Karaolis, D. K. et al. A *Vibrio cholerae* pathogenicity island associated with epidemic and pandemic strains. *Proc Natl Acad Sci U S A* **95**, 3134–9 (1998).

365. Ray, M. E., Wistow, G., Su, Y. A., Meltzer, P. S. & Trent, J. M. AIM1, a novel non-lens member of the $\beta\gamma$-crystallin superfamily, is associated with the control of tumorigenicity in human malignant melanoma. *Proc Natl Acad Sci U S A* **94**, 3229–34 (1997).

366. Rajini, B., Graham, C., Wistow, G. & Sharma, Y. Stability, homodimerization, and calcium-binding properties of a single, variant $\beta\gamma$-crystallin domain of the protein absent in melanoma 1 (AIM1). *Biochemistry* **42**, 4552–9 (2003).

367. Teichmann, U. et al. Cloning and tissue expression of the mouse ortholog of AIM1, a $\beta\gamma$-crystallin superfamily member. *Mamm Genome* **9**, 715–20 (1998).

368. Wistow, G. Lens crystallins: gene recruitment and evolutionary dynamism. *Trends Biochem Sci* **18**, 301–6 (1993).

369. Wistow, G. J. et al. τ-Crystallin/α-enolase: one gene encodes both an enzyme and a lens structural protein. *J Cell Biol* **107**, 2729–36 (1988).

370. Wistow, G. J. & Piatigorsky, J. Gene conversion and splice-site slippage in the argininosuccinate lyases/δ-crystallins of the duck lens: members of an enzyme superfamily. *Gene* **96**, 263–70 (1990).

371. Stapel, S. O. & de Jong, W. W. Lamprey 48-kDa lens protein represents a novel class of crystallins. *FEBS Lett* **162**, 305–9 (1983).

372. Williams, L. A., Ding, L., Horwitz, J. & Piatigorsky, J. τ-Crystallin from the turtle lens: purification and partial characterization. *Exp Eye Res* **40**, 741–9 (1985).

373. Jaffe, N. S. & Horwitz, J. *Lens and Cataract* (eds. Podos, S. M. & Yanoff, M.) (Gower Medical Publishing, New York, 1992).

374. Ishikura, S., Usami, N., Araki, M. & Hara, A. Structural and functional characterization of rabbit and human L-gulonate 3-dehydrogenase. *J Biochem (Tokyo)* **137**, 303–14 (2005).

375. Mulders, J. W., Hendriks, W., Blankesteijn, W. M., Bloemendal, H. & de Jong, W. W. λ-Crystallin, a major rabbit lens protein, is related to hydroxyacyl-coenzyme A dehydrogenases. *J Biol Chem* **263**, 15462–6 (1988).

376. Suzuki, T. et al. λ-Crystallin related to dehydroascorbate reductase in the rabbit lens. *Jpn J Ophthalmol* **47**, 437–43 (2003).

377. Huang, Q. L., Russell, P., Stone, S. H. & Zigler, J. S., Jr. ζ-Crystallin, a novel lens protein from the guinea pig. *Curr Eye Res* **6**, 725–32 (1987).

378. Garland, D., Rao, P. V., Del Corso, A., Mura, U. & Zigler, J. S., Jr. ζ-Crystallin is a major protein in the lens of *Camelus dromedarius*. *Arch Biochem Biophys* **285**, 134–6 (1991).

379. Fujii, Y. et al. Taxon-specific ζ-crystallin in Japanese tree frog (*Hyla japonica*) lens. *J Biol Chem* **276**, 28134–9 (2001).

380. Wistow, G. & Kim, H. Lens protein expression in mammals: taxon-specificity and the recruitment of crystallins. *J Mol Evol* **32**, 262–9 (1991).

381. Graham, C., Hodin, J. & Wistow, G. A retinaldehyde dehydrogenase as a structural protein in a mammalian eye lens. Gene recruitment of η-crystallin. *J Biol Chem* **271**, 15623–8 (1996).

382. Kim, R. Y., Gasser, R. & Wistow, G. J. μ-Crystallin is a mammalian homologue of Agrobacterium ornithine cyclodeaminase and is expressed in human retina. *Proc Natl Acad Sci U S A* **89**, 9292–6 (1992).

383. Segovia, L., Horwitz, J., Gasser, R. & Wistow, G. Two roles for μ-crystallin: a lens structural protein in diurnal marsupials and a possible enzyme in mammalian retinas. *Mol Vis* **3**, 9 (1997).

384. Roll, B., van Boekel, M. A., Amons, R. & de Jong, W. W. ρB-crystallin, an aldose reductase-like lens protein in the gecko *Lepidodactylus lugubris*. *Biochem Biophys Res Commun* **217**, 452–8 (1995).

385. van Boekel, M. A., van Aalten, D. M., Caspers, G. J., Roll, B. & de Jong, W. W. Evolution of the aldose reductase-related gecko eye lens protein ρB-crystallin: a sheep in wolf's clothing. *J Mol Evol* **52**, 239–48 (2001).

386. Jimenez-Asensio, J., Gonzalez, P., Zigler, J. S., Jr. & Garland, D. L. Glyceraldehyde 3-phosphate dehydrogenase is an enzyme-crystallin in diurnal geckos of the genus *Phelsuma*. *Biochem Biophys Res Commun* **209**, 796–802 (1995).

387. Fujii, Y. et al. Purification and characterization of ρ-crystallin from Japanese common bullfrog lens. *J Biol Chem* **265**, 9914–23 (1990).

388. Lu, S. F., Pan, F. M. & Chiou, S. H. Sequence analysis of frog ρ-crystallin by cDNA cloning and sequencing: a member of the aldo-keto reductase family. *Biochem Biophys Res Commun* **214**, 1079–88 (1995).

389. Chiou, S. H., Lee, H. J., Chu, H., Lai, T. A. & Chang, G. G. Screening and kinetic analysis of δ-crystallins with endogenous argininosuccinate lyase activity in the lenses of vertebrates. *Biochem Int* **25**, 705–13 (1991).

390. Chiou, S. H. et al. Ostrich crystallins. Structural characterization of δ-crystallin with enzymic activity. *Biochem J* **273(Pt 2)**, 295–300 (1991).

391. Chiou, S. H., Hung, C. C. & Lin, C. W. Biochemical characterization of crystallins from pigeon lenses: structural and sequence analysis of pigeon δ-crystallin. *Biochim Biophys Acta* **1160**, 317–24 (1992).

392. Zigler, J. S., Jr. & Rao, P. V. Enzyme/crystallins and extremely high pyridine nucleotide levels in the eye lens. *Faseb J* **5**, 223–5 (1991).

393. Rao, P. V. & Zigler, J. S., Jr. Extremely high levels of NADPH in guinea pig lens: correlation with ζ-crystallin concentration. *Biochem Biophys Res Commun* **167**, 1221–8 (1990).

394. Rao, P. V. & Zigler, J. S., Jr. ζ-Crystallin from guinea pig lens is capable of functioning catalytically as an oxidoreductase. *Arch Biochem Biophys* **284**, 181–5 (1991).

395. Rao, C. M. & Zigler, J. S., Jr. Levels of reduced pyridine nucleotides and lens photodamage. *Photochem Photobiol* **56**, 523–8 (1992).

396. Werten, P. J., Roll, B., van Aalten, D. M. & de Jong, W. W. Gecko ι-crystallin: how cellular retinol-binding protein became an eye lens ultraviolet filter. *Proc Natl Acad Sci U S A* **97**, 3282–7 (2000).

397. Janssens, H. & Gehring, W. J. Isolation and characterization of drosocrystallin, a lens crystallin gene of *Drosophila melanogaster*. *Dev Biol* **207**, 204–14 (1999).

398. Edwards, J. S. & Meyer, M. R. Conservation of antigen 3G6: a crystalline cone constituent in the compound eye of arthropods. *J Neurobiol* **21**, 441–52 (1990).

399. Tomarev, S. I., Zinovieva, R. D. & Piatigorsky, J. Characterization of squid crystallin genes. Comparison with mammalian glutathione S-transferase genes. *J Biol Chem* **267**, 8604–12 (1992).

400. Tomarev, S. I., Zinovieva, R. D., Guo, K. & Piatigorsky, J. Squid glutathione S-transferase. Relationships with other glutathione S-transferases and S-crystallins of cephalopods. *J Biol Chem* **268**, 4534–42 (1993).

401. Tomarev, S. I., Chung, S. & Piatigorsky, J. Glutathione S-transferase and S-crystallins of cephalopods: evolution from active enzyme to lens-refractive proteins. *J Mol Evol* **41**, 1048–56 (1995).

402. Chiou, S. H. A novel crystallin from octopus lens. *FEBS Lett* **241**, 261–4 (1988).

403. Montgomery, M. K. & McFall-Ngai, M. J. The muscle-derived lens of a squid bioluminescent organ is biochemically convergent with the ocular lens. Evidence for recruitment of aldehyde dehydrogenase as a predominant structural protein. *J Biol Chem* **267**, 20999–1003 (1992).

404. Piatigorsky, J. & Kozmik, Z. Cubozozoan jellyfish: an evo/devo model for eyes and other sensory systems. *Int J Dev Biol* **48**, 719–29 (2004).

405. Treisman, J. E. How to make an eye. *Development* **131**, 3823–7 (2004).

405a. Castellano, S., Lobanov, A. V., Chapple, C., Novoselov, S. V., Albrecht, M., Hua, D., Lescure, A., Lengauer, T., Krol, A., Gladyshev, V. N. & Guigó, R. Diversity and functional plasticity of eukaryotic selenoproteins: identification and characterization of the SelJ family. *Proc Natl Acad Sci U S A* **102**, 16188–93 (2005).

406. Kishimoto, Y., Hiraiwa, M. & O'Brien, J. S. Saposins: structure, function, distribution, and molecular genetics. *J Lipid Res* **33**, 1255–67 (1992).

407. Furst, W. & Sandhoff, K. Activator proteins and topology of lysosomal sphingolipid catabolism. *Biochim Biophys Acta* **1126**, 1–16 (1992).

408. Vaccaro, A. M., Salvioli, R., Tatti, M. & Ciaffoni, F. Saposins and their interaction with lipids. *Neurochem Res* **24**, 307–14 (1999).

409. de Jong, W. W., Lubsen, N. H. & Kraft, H. J. Molecular evolution of the eye lens. *Progress in Retinal and Eye Research* **13**, 391–442 (1994).

410. Piatigorsky, J. & Zelenka, P. S. Transcriptional regulation of crystallin genes: cis elements, trans-factors, and signal transductin systems in the lens. *Advances in Developmental Biochemistry* **1**, 211–56 (1992).

411. Cvekl, A. & Piatigorsky, J. Lens development and crystallin gene expression: many roles for Pax-6. *Bioessays* **18**, 621–30 (1996).

412. Reza, H. M. & Yasuda, K. Lens differentiation and crystallin regulation: a chick model. *Int J Dev Biol* **48**, 805–17 (2004).

413. Kondoh, H., Uchikawa, M. & Kamachi, Y. Interplay of Pax6 and SOX2 in lens development as a paradigm of genetic switch mechanisms for cell differentiation. *Int J Dev Biol* **48**, 819–27 (2004).

414. Cvekl, A., Yang, Y., Chauhan, B. K. & Cveklova, K. Regulation of gene expression by Pax6 in ocular cells: a case of tissue-preferred expression of crystallins in lens. *Int J Dev Biol* **48**, 829–44 (2004).

415. Yang, Y. & Cvekl, A. Tissue-specific regulation of the mouse αA-crystallin gene in lens via recruitment of Pax6 and c-Maf to its promoter. *J Mol Biol* **351**, 453–69 (2005).

416. Kondoh, H., Yasuda, K. & Okada, T. S. Tissue-specific expression of a cloned chick δ-crystallin gene in mouse cells. *Nature* **301**, 440–2 (1983).

417. Kondoh, H. et al. Specific expression of the chicken δ-crystallin gene in the lens and the pyramidal neurons of the piriform cortex in transgenic mice. *Dev Biol* **120**, 177–85 (1987).

418. Hayashi, S., Goto, K., Okada, T. S. & Kondoh, H. Lens-specific enhancer in the third intron regulates expression of the chicken δ1-crystallin gene. *Genes Dev* **1**, 818–28 (1987).

419. Li, X., Cvekl, A., Bassnett, S. & Piatigorsky, J. Lens-preferred activity of chicken δ1- and δ2-crystallin enhancers in transgenic mice and evidence for retinoic acid-responsive regulation of the δ1-crystallin gene. *Dev Genet* **20**, 258–66 (1997).

420. Parker, D. S., Wawrousek, E. F. & Piatigorsky, J. Expression of the δ-crystallin genes in the embryonic chicken lens. *Dev Biol* **126**, 375–81 (1988).

421. Thomas, G., Zelenka, P. S., Cuthbertson, R. A., Norman, B. L. & Piatigorsky, J. Differential expression of the two δ-crystallin/argininosuccinate lyase genes in lens, heart, and brain of chicken embryos. *New Biol* **2**, 903–14 (1990).

422. Klement, J. F., Wawrousek, E. F. & Piatigorsky, J. Tissue-specific expression of the chicken α A-crystallin gene in cultured lens epithelia and transgenic mice. *J Biol Chem* **264**, 19837–44 (1989).

423. Duncan, M. K., Li, X., Ogino, H., Yasuda, K. & Piatigorsky, J. Developmental regulation of the chicken βB1-crystallin promoter in transgenic mice. *Mech Dev* **57**, 79–89 (1996).

424. Chen, Q., Dowhan, D. H., Liang, D., Moore, D. D. & Overbeek, P. A. CREB-binding protein/p300 co-activation of crystallin gene expression. *J Biol Chem* **277**, 24081–9 (2002).

425. Duncan, M. K., Haynes, J. I., 2nd, Cvekl, A. & Piatigorsky, J. Dual roles for Pax-6: a transcriptional repressor of lens fiber cell-specific β-crystallin genes. *Mol Cell Biol* **18**, 5579–86 (1998).

426. Cui, W., Tomarev, S. I., Piatigorsky, J., Chepelinsky, A. B. & Duncan, M. K. Mafs, Prox1, and Pax6 can regulate chicken βB1-crystallin gene expression. *J Biol Chem* **279**, 11088–95 (2004).

427. Yang, Y., Chauhan, B. K., Cveklova, K. & Cvekl, A. Transcriptional regulation of mouse αB- and γF-crystallin genes in lens: opposite promoter-specific interactions between Pax6 and large Maf transcription factors. *J Mol Biol* **344**, 351–68 (2004).

428. Cvekl, A., Sax, C. M., Bresnick, E. H. & Piatigorsky, J. A complex array of positive and negative elements regulates the chicken αA-crystallin gene: involvement of Pax-6, USF, CREB and/or CREM, and AP-1 proteins. *Mol Cell Biol* **14**, 7363–76 (1994).

429. Ilagan, J. G., Cvekl, A., Kantorow, M., Piatigorsky, J. & Sax, C. M. Regulation of αA-crystallin gene expression. Lens specificity achieved through the differential placement of similar transcriptional control elements in mouse and chicken. *J Biol Chem* **274**, 19973–8 (1999).

430. Overbeek, P. A., Chepelinsky, A. B., Khillan, J. S., Piatigorsky, J. & Westphal, H. Lens-specific expression and developmental regulation of the bacterial chloramphenicol acetyltransferase gene driven by the murine αA-crystallin promoter in transgenic mice. *Proc Natl Acad Sci U S A* **82**, 7815–9 (1985).

431. Gopal-Srivastava, R., Cvekl, A. & Piatigorsky, J. Pax-6 and αB-crystallin/small heat shock protein gene regulation in the murine lens. Interaction with the lens-specific regions, LSR1 and LSR2. *J Biol Chem* **271**, 23029–36 (1996).

432. Gopal-Srivastava, R., Cvekl, A. & Piatigorsky, J. Involvement of retinoic acid/ retinoid receptors in the regulation of murine αB-crystallin/small heat shock protein gene expression in the lens. *J Biol Chem* **273**, 17954–61 (1998).

433. Chauhan, B. K., Yang, Y., Cveklova, K. & Cvekl, A. Functional interactions between alternatively spliced forms of Pax6 in crystallin gene regulation and in haploinsufficiency. *Nucleic Acids Res* **32**, 1696–709 (2004).

434. Dubin, R. A., Gopal-Srivastava, R., Wawrousek, E. F. & Piatigorsky, J. Expression of the murine αB-crystallin gene in lens and skeletal muscle: identification of a muscle-preferred enhancer. *Mol Cell Biol* **11**, 4340–9 (1991).

435. Iwaki, A., Nagano, T., Nakagawa, M., Iwaki, T. & Fukumaki, Y. Identification and characterization of the gene encoding a new member of the α-crystallin/small hsp family, closely linked to the αB-crystallin gene in a head-to-head manner. *Genomics* **45**, 386–94 (1997).

436. Suzuki, A. et al. MKBP, a novel member of the small heat shock protein family, binds and activates the myotonic dystrophy protein kinase. *J Cell Biol* **140**, 1113–24 (1998).

437. Doerwald, L. et al. Sequence and functional conservation of the intergenic region between the head-to-head genes encoding the small heat shock proteins αB-crystallin and HspB2 in the mammalian lineage. *J Mol Evol* **59**, 674–86 (2004).

438. Adachi, N. & Lieber, M. R. Bidirectional gene organization: a common architectural feature of the human genome. *Cell* **109**, 807–9 (2002).

439. Trinklein, N. D. et al. An abundance of bidirectional promoters in the human genome. *Genome Res* **14**, 62–6 (2004).

440. Koyanagi, K. O., Hagiwara, M., Itoh, T., Gojobori, T. & Imanishi, T. Comparative genomics of bidirectional gene pairs and its implications for the evolution of a transcriptional regulation system. *Gene* **353**, 169–76 (2005).

441. Swamynathan, S. K. & Piatigorsky, J. Orientation-dependent influence of an intergenic enhancer on the promoter activity of the divergently transcribed mouse *Shsp/αB-crystallin* and *Mkbp/HspB2* genes. *J Biol Chem* **277**, 49700–6 (2002).

442. Sanyal, S., Jansen, H. G., de Grip, W. J., Nevo, E. & de Jong, W. W. The eye of the blind mole rat, *Spalax ehrenbergi*. Rudiment with hidden function? *Invest Ophthalmol Vis Sci* **31**, 1398–404 (1990).

443. Nevo, E. *Mosaic Evolution of Subterranean Mammals. Regression, Progression, and Global Convergence* (Oxford University Press, New York, 1999).

444. Hendriks, W., Leunissen, J., Nevo, E., Bloemendal, H. & de Jong, W. W. The lens protein αA-crystallin of the blind mole rat, *Spalax ehrenbergi:* evolutionary change and functional constraints. *Proc Natl Acad Sci U S A* **84**, 5320–4 (1987).

445. Avivi, A., Joel, A. & Nevo, E. The lens protein α-B-crystallin of the blind subterranean mole-rat: high homology with sighted mammals. *Gene* **264**, 45–9 (2001).

446. Hough, R. B. et al. Adaptive evolution of small heat shock protein/αB-crystallin promoter activity of the blind subterranean mole rat, *Spalax ehrenbergi. Proc Natl Acad Sci U S A* **99**, 8145–50 (2002).

447. Tomarev, S. I., Duncan, M. K., Roth, H. J., Cvekl, A. & Piatigorsky, J. Convergent evolution of crystallin gene regulation in squid and chicken: the AP-1/ARE connection. *J Mol Evol* **39**, 134–43 (1994).

448. Roth, H. J., Das, G. C. & Piatigorsky, J. Chicken βB1-crystallin gene expression:

presence of conserved functional polyomavirus enhancer-like and octamer binding-like promoter elements found in non-lens genes. *Mol Cell Biol* **11**, 1488–99 (1991).

449. Carosa, E., Kozmik, Z., Rall, J. E. & Piatigorsky, J. Structure and expression of the scallop Ω-crystallin gene. Evidence for convergent evolution of promoter sequences. *J Biol Chem* **277**, 656–64 (2002).

450. Kostrouch, Z. et al. Retinoic acid X receptor in the diploblast, *Tripedalia cystophora*. *Proc Natl Acad Sci U S A* **95**, 13442–7 (1998).

451. Kozmik, Z. et al. Role of Pax genes in eye evolution. A Cnidarian *PaxB* gene uniting Pax2 and Pax6 functions. *Dev Cell* **5**, 773–85 (2003).

452. Kamachi, Y., Uchikawa, M., Tanouchi, A., Sekido, R. & Kondoh, H. Pax6 and SOX2 form a co-DNA-binding partner complex that regulates initiation of lens development. *Genes Dev* **15**, 1272–86 (2001).

453. Blanco, J., Girard, F., Kamachi, Y., Kondoh, H. & Gehring, W. J. Functional analysis of the chicken δ1-crystallin enhancer activity in *Drosophila* reveals remarkable evolutionary conservation between chicken and fly. *Development* **132**, 1895–905 (2005).

454. Piatigorsky, J. Gene sharing in lens and cornea: facts and implications. *Prog Retin Eye Res* **17**, 145–74 (1998).

454a. Jester, J. V., Budge, A., Fisher, S. & Huang, J. Corneal keratocytes: phenotypic and species differences in abundant protein expression and *in vitro* light-scattering. *Invest Ophthalmol Vis Sci* **46**, 2369–78 (2005).

455. Piatigorsky, J. Enigma of the abundant water-soluble cytoplasmic proteins of the cornea: the "refracton" hypothesis. *Cornea* **20**, 853–8 (2001).

456. Maurice, D. M. The structure and transparency of the cornea. *J Physiol* **136**, 263–86 (1957).

457. Holt, W. S. & Kinoshita, J. H. The soluble proteins of the bovine cornea. *Invest Ophthalmol* **12**, 114–26 (1973).

458. Alexander, R. J., Silverman, B. & Henley, W. L. Isolation and characterization of BCP 54, the major soluble protein of bovine cornea. *Exp Eye Res* **32**, 205–16 (1981).

459. Silverman, B., Alexander, R. J. & Henley, W. L. Tissue and species specificity of BCP 54, the major soluble protein of bovine cornea. *Exp Eye Res* **33**, 19–29 (1981).

460. Cuthbertson, R. A., Tomarev, S. I. & Piatigorsky, J. Taxon-specific recruitment of enzymes as major soluble proteins in the corneal epithelium of three mammals, chicken, and squid. *Proc Natl Acad Sci U S A* **89**, 4004–8 (1992).

461. Rabaey, M. & Segers, J. Changes in polypeptide composition of the bovine corneal epithelium during development, in *Congress of the European Society of Ophthalmology* (ed. Trevor-Roper, P. D.) 41–44 (Academic Press, London, 1981).

462. Abedinia, M., Pain, T., Algar, E. M. & Holmes, R. S. Bovine corneal aldehyde dehydrogenase: the major soluble corneal protein with a possible dual protective role for the eye. *Exp Eye Res* **51**, 419–26 (1990).

463. Verhagen, C., Hoekzema, R., Verjans, G. M. & Kijlstra, A. Identification of bovine corneal protein 54 (BCP 54) as an aldehyde dehydrogenase. *Exp Eye Res* **53**, 283–4 (1991).

464. Cooper, D. L., Baptist, E. W., Enghild, J. J., Isola, N. R. & Klintworth, G. K. Bovine

corneal protein 54K (BCP54) is a homologue of the tumor-associated (class 3) rat aldehyde dehydrogenase (RATALD). *Gene* **98**, 201–7 (1991).

465. Downes, J. E., Swann, P. G. & Holmes, R. S. Differential corneal sensitivity to ultraviolet light among inbred strains of mice. Correlation of ultraviolet B sensitivity with aldehyde dehydrogenase deficiency. *Cornea* **13**, 67–72 (1994).

466. Pappa, A., Chen, C., Koutalos, Y., Townsend, A. J. & Vasiliou, V. Aldh3a1 protects human corneal epithelial cells from ultraviolet- and 4-hydroxy-2-nonenal-induced oxidative damage. *Free Radic Biol Med* **34**, 1178–89 (2003).

467. Jester, J. V. et al. The cellular basis of corneal transparency: evidence for "corneal crystallins." *J Cell Sci* **112 (Pt 5)**, 613–22 (1999).

468. Nees, D. W., Wawrousek, E. F., Robison, W. G., Jr. & Piatigorsky, J. Structurally normal corneas in aldehyde dehydrogenase 3a1-deficient mice. *Mol Cell Biol* **22**, 849–55 (2002).

469. Mitchell, J. & Cenedella, R. J. Quantitation of ultraviolet light-absorbing fractions of the cornea. *Cornea* **14**, 266–72 (1995).

470. Manzer, R. et al. Ultraviolet radiation decreases expression and induces aggregation of corneal ALDH3A1. *Chem Biol Interact* **143–144**, 45–53 (2003).

471. Pappa, A. et al. Human aldehyde dehydrogenase 3A1 inhibits proliferation and promotes survival of human corneal epithelial cells. *J Biol Chem* **280**, 27998–8006 (2005).

472. Sax, C. M. et al. Transketolase is a major protein in the mouse cornea. *J Biol Chem* **271**, 33568–74 (1996).

473. Xu, Z. P., Wawrousek, E. F. & Piatigorsky, J. Transketolase haploinsufficiency reduces adipose tissue and female fertility in mice. *Mol Cell Biol* **22**, 6142–7 (2002).

474. Nees, D. W., Fariss, R. N. & Piatigorsky, J. Serum albumin in mammalian cornea: implications for clinical application. *Invest Ophthalmol Vis Sci* **44**, 3339–45 (2003).

475. Guo, J., Sax, C. M., Piatigorsky, J. & Xu, F. X. Heterogeneous expression of transketolase in ocular tissues. *Curr Eye Res* **16**, 467–74 (1997).

476. Sax, C. M. et al. Transketolase gene expression in the cornea is influenced by environmental factors and developmentally controlled events. *Cornea* **19**, 833–41 (2000).

477. Sun, L., Sun, T. T. & Lavker, R. M. Identification of a cytosolic NADP$^+$-dependent isocitrate dehydrogenase that is preferentially expressed in bovine corneal epithelium. A corneal epithelial crystallin. *J Biol Chem* **274**, 17334–41 (1999).

477a. Jo, S-H., Lee, S-H., Chun, H. S., Lee, S. M., Koh, H-J., Lee, S-E., Chun, J-S., Park, J-W. & Huh, T-L. Cellular defense against UVB-induced phototoxicity by cytosolic NADP$^+$-dependent isocitrate dehydrogenase. *Biochem Biophys Res Comm* **292**, 542–9 (2002).

477b. Lee, S. M., Koh, H-J., Park, D-C., Song, B. J., Huh, T-L. & Park, J-W. Cytosolic NADP$^+$-dependent isocitrate dehydrogenase status modulates oxidative damage to cells. *Rad Biol Med* **32**, 1185–96 (2002).

477c. Koh, H-J., Lee, S-M., Son, B-G., Lee, S-H., Ryoo, Z. Y., Chang, K-T., Park, J-W., Park, D-C., Song, B. J., Veech, R. L., Song, H. & Huh, T-L. Cytosolic NADP$^+$-dependent isocitrate dehydrogenase plays a key role in lipid metabolism. *J Biol Chem* **279**, 39968–74 (2004).

477d. Lee, S.-H., Jo, S.-H., Lee, S.-M., Koh, H.-J., Song, H., Park, J.-W. & Huh, T.-L. Role of NADP$^+$-dependent isocitrate dehydrogenase (NADP+-ICDH) on cellular defence against oxidative injury by gamma-rays. *Int J Radiat Biol* **80**, 635–42 (2004).

478. Elzinga, S. D., Bednarz, A. L., van Oosterum, K., Dekker, P. J. & Grivell, L. A. Yeast mitochondrial NAD(+)-dependent isocitrate dehydrogenase is an RNA-binding protein. *Nucleic Acids Res* **21**, 5328–31 (1993).

479. Xu, Y. S., Kantorow, M., Davis, J. & Piatigorsky, J. Evidence for gelsolin as a corneal crystallin in zebrafish. *J Biol Chem* **275**, 24645–52 (2000).

480. Kwiatkowski, D. J. Functions of gelsolin: motility, signaling, apoptosis, cancer. *Curr Opin Cell Biol* **11**, 103–8 (1999).

481. McGough, A. M., Staiger, C. J., Min, J. K. & Simonetti, K. D. The gelsolin family of actin regulatory proteins: modular structures, versatile functions. *FEBS Lett* **552**, 75–81 (2003).

482. Silacci, P. et al. Gelsolin superfamily proteins: key regulators of cellular functions. *Cell Mol Life Sci* **61**, 2614–23 (2004).

483. Kanungo, J., Swamynathan, S. K. & Piatigorsky, J. Abundant corneal gelsolin in Zebrafish and the "four-eyed" fish, *Anableps anableps:* possible analogy with multifunctional lens crystallins. *Exp Eye Res* (2004).

484. Ubels, J. L., Edelhauser, H. F. & Austin, K. H. Healing of experimental corneal wounds treated with topically applied retinoids. *Am J Ophthalmol* **95**, 353–8 (1983).

485. Kanungo, J., Kozmik, Z., Swamynathan, S. K. & Piatigorsky, J. Gelsolin is a dorsalizing factor in zebrafish. *Proc Natl Acad Sci U S A* **100**, 3287–92 (2003).

486. Harland, R. & Gerhart, J. Formation and function of Spemann's organizer. *Annu Rev Cell Dev Biol* **13**, 611–67 (1997).

487. Dale, L. & Wardle, F. C. A gradient of BMP activity specifies dorsal-ventral fates in early *Xenopus* embryos. *Semin Cell Dev Biol* **10**, 319–26 (1999).

488. Hammerschmidt, M. & Mullins, M. C. Dorsoventral patterning in the zebrafish: bone morphogenetic proteins and beyond. *Results Probl Cell Differ* **40**, 72–95 (2002).

489. Manseau, L., Calley, J. & Phan, H. Profilin is required for posterior patterning of the *Drosophila* oocyte. *Development* **122**, 2109–16 (1996).

490. Manseau, L. J. & Schupbach, T. Cappuccino and spire: two unique maternal-effect loci required for both the anteroposterior and dorsoventral patterns of the *Drosophila* embryo. *Genes Dev* **3**, 1437–52 (1989).

491. Quinlan, M. E., Heuser, J. E., Kerkhoff, E. & Mullins, R. D. *Drosophila* Spire is an actin nucleation factor. *Nature* **433**, 382–8 (2005).

492. Kays, W. T. & Piatigorsky, J. Aldehyde dehydrogenase class 3 expression: identification of a cornea-preferred gene promoter in transgenic mice. *Proc Natl Acad Sci U S A* **94**, 13594–9 (1997).

493. Liu, C. Y. et al. The cloning of mouse keratocan cDNA and genomic DNA and the characterization of its expression during eye development. *J Biol Chem* **273**, 22584–8 (1998).

494. Liu, C., Arar, H., Kao, C. & Kao, W. W. Identification of a 3.2 kb 5′-flanking region

of the murine keratocan gene that directs β-galactosidase expression in the adult corneal stroma of transgenic mice. *Gene* **250**, 85–96 (2000).

495. Hough, R. B. & Piatigorsky, J. Preferential transcription of rabbit Aldh1a1 in the cornea: implication of hypoxia-related pathways. *Mol Cell Biol* **24**, 1324–40 (2004).

496. Augusteyn, R. C. The effect of light deprivation on the mouse lens. *Exp Eye Res* **66**, 669–74 (1998).

497. Downes, J. & Holmes, R. Development of aldehyde dehydrogenase and alcohol dehydrogenase in mouse eye: evidence for light-induced changes. *Biol Neonate* **61**, 118–23 (1992).

498. Boesch, J. S., Miskimins, R., Miskimins, W. K. & Lindahl, R. The same xenobiotic response element is required for constitutive and inducible expression of the mammalian aldehyde dehydrogenase-3 gene. *Arch Biochem Biophys* **361**, 223–30 (1999).

499. Gu, Y. Z., Hogenesch, J. B. & Bradfield, C. A. The PAS superfamily: sensors of environmental and developmental signals. *Annu Rev Pharmacol Toxicol* **40**, 519–61 (2000).

500. Reisdorph, R. & Lindahl, R. Hypoxia exerts cell-type-specific effects on expression of the class 3 aldehyde dehydrogenase gene. *Biochem Biophys Res Commun* **249**, 709–12 (1998).

501. Reisdorph, R. & Lindahl, R. Aldehyde dehydrogenase 3 gene regulation: studies on constitutive and hypoxia-modulated expression. *Chem Biol Interact* **130–132**, 227–33 (2001).

502. Antonsson, C., Arulampalam, V., Whitelaw, M. L., Pettersson, S. & Poellinger, L. Constitutive function of the basic helix-loop-helix/PAS factor Arnt. Regulation of target promoters via the E box motif. *J Biol Chem* **270**, 13968–72 (1995).

503. Huffman, J. L., Mokashi, A., Bachinger, H. P. & Brennan, R. G. The basic helix-loop-helix domain of the aryl hydrocarbon receptor nuclear transporter (ARNT) can oligomerize and bind E-box DNA specifically. *J Biol Chem* **276**, 40537–44 (2001).

504. Liu, J. J., Kao, W. W. & Wilson, S. E. Corneal epithelium-specific mouse keratin K12 promoter. *Exp Eye Res* **68**, 295–301 (1999).

505. Shiraishi, A. et al. Identification of the cornea-specific keratin 12 promoter by in vivo particle-mediated gene transfer. *Invest Ophthalmol Vis Sci* **39**, 2554–61 (1998).

506. Gopal-Srivastava, R., Kays, W. T. & Piatigorsky, J. Enhancer-independent promoter activity of the mouse αB-crystallin/small heat shock protein gene in the lens and cornea of transgenic mice. *Mech Dev* **92**, 125–34 (2000).

507. Masters, C. Cellular differentiation and the microcompartmentation of glycolysis. *Mech Ageing Dev* **61**, 11–22 (1991).

508. Masters, C. Microenvironmental factors and the binding of glycolytic enzymes to contractile filaments. *Int J Biochem* **24**, 405–10 (1992).

509. Uyeda, K. Interactions of glycolytic enzymes with cellular membranes. *Curr Top Cell Regul* **33**, 31–46 (1992).

510. Knull, H. R. & Walsh, J. L. Association of glycolytic enzymes with the cytoskeleton. *Curr Top Cell Regul* **33**, 15–30 (1992).

511. Wilson, J. E. Isozymes of mammalian hexokinase: structure, subcellular localization and metabolic function. *J Exp Biol* **206**, 2049–57 (2003).

512. Sebastian, S., Edassery, S. & Wilson, J. E. The human gene for the type III isozyme of hexokinase: structure, basal promoter, and evolution. *Arch Biochem Biophys* **395**, 113–20 (2001).

513. Pedersen, P. L., Mathupala, S., Rempel, A., Geschwind, J. F. & Ko, Y. H. Mitochondrial bound type II hexokinase: a key player in the growth and survival of many cancers and an ideal prospect for therapeutic intervention. *Biochim Biophys Acta* **1555**, 14–20 (2002).

514. Preller, A. & Wilson, J. E. Localization of the type III isozyme of hexokinase at the nuclear periphery. *Arch Biochem Biophys* **294**, 482–92 (1992).

515. Postic, C., Shiota, M. & Magnuson, M. A. Cell-specific roles of glucokinase in glucose homeostasis. *Recent Prog Horm Res* **56**, 195–217 (2001).

516. Hashimoto, M. & Wilson, J. E. Membrane potential-dependent conformational changes in mitochondrially bound hexokinase of brain. *Arch Biochem Biophys* **384**, 163–73 (2000).

517. Pastorino, J. G., Shulga, N. & Hoek, J. B. Mitochondrial binding of hexokinase II inhibits Bax-induced cytochrome c release and apoptosis. *J Biol Chem* **277**, 7610–8 (2002).

518. van Loo, G. et al. The role of mitochondrial factors in apoptosis: a Russian roulette with more than one bullet. *Cell Death Differ* **9**, 1031–42 (2002).

519. Azoulay-Zohar, H., Israelson, A., Abu-Hamad, S. & Shoshan-Barmatz, V. In self-defence: hexokinase promotes voltage-dependent anion channel closure and prevents mitochondria-mediated apoptotic cell death. *Biochem J* **377**, 347–55 (2004).

520. Yanagawa, T., Funasaka, T., Tsutsumi, S., Watanabe, H. & Raz, A. Novel roles of the autocrine motility factor/phosphoglucose isomerase in tumor malignancy. *Endocr Relat Cancer* **11**, 749–59 (2004).

521. Faik, P., Walker, J. I., Redmill, A. A. & Morgan, M. J. Mouse glucose-6-phosphate isomerase and neuroleukin have identical 3′ sequences. *Nature* **332**, 455–7 (1988).

522. Xu, W., Seiter, K., Feldman, E., Ahmed, T. & Chiao, J. W. The differentiation and maturation mediator for human myeloid leukemia cells shares homology with neuroleukin or phosphoglucose isomerase. *Blood* **87**, 4502–6 (1996).

523. Watanabe, H., Takehana, K., Date, M., Shinozaki, T. & Raz, A. Tumor cell autocrine motility factor is the neuroleukin/phosphohexose isomerase polypeptide. *Cancer Res* **56**, 2960–3 (1996).

524. Numata, O. Multifunctional proteins in *Tetrahymena*: 14-nm filament protein/citrate synthase and translation elongation factor-1 α. *Int Rev Cytol* **164**, 1–35 (1996).

525. Kojima, H., Chiba, J., Watanabe, Y. & Numata, O. Citrate synthase purified from *Tetrahymena* mitochondria is identical with *Tetrahymena* 14-nm filament protein. *J Biochem (Tokyo)* **118**, 189–95 (1995).

526. Takeda, T., Watanabe, Y. & Numata, O. Direct demonstration of the bifunctional property of *Tetrahymena* 14-nm filament protein/citrate synthase following expression of the gene in *Escherichia coli*. *Biochem Biophys Res Commun* **237**, 205–10 (1997).

527. Kojima, H. & Numata, O. Enzymatic form and cytoskeletal form of bifunctional *Tetrahymena* 49kDa protein is regulated by phosphorylation. *Zoolog Sci* **19**, 37–42 (2002).

528. Takeda, T., Kurasawa, Y., Watanabe, Y. & Numata, O. Polymerization of highly purified *Tetrahymena* 14-nm filament protein/citrate synthase into filaments and its possible role in regulation of enzymatic activity. *J Biochem (Tokyo)* **117**, 869–74 (1995).

529. Takeda, T., Yoshihama, I. & Numata, O. Identification of *Tetrahymena* hsp60 as a 14-nm filament protein/citrate synthase-binding protein and its possible involvement in the oral apparatus formation. *Genes Cells* **6**, 139–49 (2001).

530. Ueno, H., Gonda, K., Takeda, T. & Numata, O. Identification of elongation factor-1α as a Ca2+/calmodulin-binding protein in *Tetrahymena* cilia. *Cell Motil Cytoskeleton* **55**, 51–60 (2003).

531. Markert, C. L. Developmental genetics. *Harvey Lect* **59**, 187–218 (1965).

532. Li, S. S. Human and mouse lactate dehydrogenase genes A (muscle), B (heart), and C (testis): protein structure, genomic organization, regulation of expression, and molecular evolution. *Prog Clin Biol Res* **344**, 75–99 (1990).

533. Tsuji, S., Qureshi, M. A., Hou, E. W., Fitch, W. M. & Li, S. S. Evolutionary relationships of lactate dehydrogenases (LDHs) from mammals, birds, an amphibian, fish, barley, and bacteria: LDH cDNA sequences from *Xenopus,* pig, and rat. *Proc Natl Acad Sci U S A* **91**, 9392–6 (1994).

534. Markert, C. L., Amet, T. M. & Goldberg, E. Human testis-specific lactate dehydrogenase-C promoter drives overexpression of mouse lactate dehydrogenase-1 cDNA in testes of transgenic mice. *J Exp Zool* **282**, 171–8 (1998).

535. Pioli, P. A., Hamilton, B. J., Connolly, J. E., Brewer, G. & Rigby, W. F. Lactate dehydrogenase is an AU-rich element-binding protein that directly interacts with AUF1. *J Biol Chem* **277**, 35738–45 (2002).

536. Shaw, G. & Kamen, R. A conserved AU sequence from the 3′ untranslated region of GM-CSF mRNA mediates selective mRNA degradation. *Cell* **46**, 659–67 (1986).

537. Jarzembowski, J. A. & Malter, J. S. Cytoplasmic fate of eukaryotic mRNA: identification and characterization of AU-binding proteins. *Prog Mol Subcell Biol* **18**, 141–72 (1997).

538. Gallouzi, I. E. et al. HuR binding to cytoplasmic mRNA is perturbed by heat shock. *Proc Natl Acad Sci U S A* **97**, 3073–8 (2000).

539. Laroia, G., Cuesta, R., Brewer, G. & Schneider, R. J. Control of mRNA decay by heat shock-ubiquitin-proteasome pathway. *Science* **284**, 499–502 (1999).

540. Laroia, G., Sarkar, B. & Schneider, R. J. Ubiquitin-dependent mechanism regulates rapid turnover of AU-rich cytokine mRNAs. *Proc Natl Acad Sci U S A* **99**, 1842–6 (2002).

541. Kajita, Y., Nakayama, J., Aizawa, M. & Ishikawa, F. The UUAG-specific RNA binding protein, heterogeneous nuclear ribonucleoprotein D0. Common modular structure and binding properties of the 2xRBD-Gly family. *J Biol Chem* **270**, 22167–75 (1995).

542. Dempsey, L. A., Li, M. J., DePace, A., Bray-Ward, P. & Maizels, N. The human HNRPD locus maps to 4q21 and encodes a highly conserved protein. *Genomics* **49**, 378–84 (1998).

543. Wagner, B. J., DeMaria, C. T., Sun, Y., Wilson, G. M. & Brewer, G. Structure and genomic organization of the human AUF1 gene: alternative pre-mRNA splicing generates four protein isoforms. *Genomics* **48**, 195–202 (1998).

544. Loflin, P., Chen, C. Y. & Shyu, A. B. Unraveling a cytoplasmic role for hnRNP D in the in vivo mRNA destabilization directed by the AU-rich element. *Genes Dev* **13**, 1884–97 (1999).

545. Dempsey, L. A., Hanakahi, L. A. & Maizels, N. A specific isoform of hnRNP D interacts with DNA in the LR1 heterodimer: canonical RNA binding motifs in a sequence-specific duplex DNA binding protein. *J Biol Chem* **273**, 29224–9 (1998).

546. Dempsey, L. A., Sun, H., Hanakahi, L. A. & Maizels, N. G4 DNA binding by LR1 and its subunits, nucleolin and hnRNP D, A role for G-G pairing in immunoglobulin switch recombination. *J Biol Chem* **274**, 1066–71 (1999).

547. Brys, A. & Maizels, N. LR1 regulates c-myc transcription in B-cell lymphomas. *Proc Natl Acad Sci U S A* **91**, 4915–9 (1994).

548. Bulfone-Paus, S., Dempsey, L. A. & Maizels, N. Host factors LR1 and Sp1 regulate the Fp promoter of Epstein-Barr virus. *Proc Natl Acad Sci U S A* **92**, 8293–7 (1995).

549. Eversole, A. & Maizels, N. In vitro properties of the conserved mammalian protein hnRNP D suggest a role in telomere maintenance. *Mol Cell Biol* **20**, 5425–32 (2000).

550. Ronai, Z. Glycolytic enzymes as DNA binding proteins. *Int J Biochem* **25**, 1073–6 (1993).

551. Williams, K. R., Reddigari, S. & Patel, G. L. Identification of a nucleic acid helix-destabilizing protein from rat liver as lactate dehydrogenase-5. *Proc Natl Acad Sci U S A* **82**, 5260–4 (1985).

552. Sharief, F. S., Wilson, S. H. & Li, S. S. Identification of the mouse low-salt-eluting single-stranded DNA-binding protein as a mammalian lactate dehydrogenase-A isoenzyme. *Biochem J* **233**, 913–6 (1986).

553. Zheng, L., Roeder, R. G. & Luo, Y. S phase activation of the histone H2B promoter by OCA-S, a coactivator complex that contains GAPDH as a key component. *Cell* **114**, 255–66 (2003).

554. McKnight, S. Gene switching by metabolic enzymes—how did you get on the invitation list? *Cell* **114**, 150–2 (2003).

555. Mangia, F., Erickson, R. P. & Epstein, C. J. Synthesis of LDH-1 during mamalian oogenesis and early development. *Dev Biol* **54**, 146–50 (1976).

556. Whitt, G. S. Genetic, developmental and evolutionary aspects of the lactate dehydrogenase isozyme system. *Cell Biochem Funct* **2**, 134–9 (1984).

557. Hentze, M. W. Enzymes as RNA-binding proteins: a role for (di)nucleotide-binding domains? *Trends Biochem Sci* **19**, 101–3 (1994).

558. Kaptain, S. et al. A regulated RNA binding protein also possesses aconitase activity. *Proc Natl Acad Sci U S A* **88**, 10109–13 (1991).

559. Melefors, O. & Hentze, M. W. Translational regulation by mRNA/protein interactions in eukaryotic cells: ferritin and beyond. *Bioessays* **15**, 85–90 (1993).

560. Klausner, R. D. & Rouault, T. A. A double life: cytosolic aconitase as a regulatory RNA binding protein. *Mol Biol Cell* **4**, 1–5 (1993).

561. Sirover, M. A. Minireview. Emerging new functions of the glycolytic protein, glyceraldehyde-3-phosphate dehydrogenase, in mammalian cells. *Life Sci* **58**, 2271–7 (1996).

562. Sirover, M. A. New insights into an old protein: the functional diversity of mammalian glyceraldehyde-3-phosphate dehydrogenase. *Biochim Biophys Acta* **1432**, 159–84 (1999).

563. Pancholi, V. & Fischetti, V. A. A major surface protein on group A streptococci is a glyceraldehyde-3-phosphate-dehydrogenase with multiple binding activity. *J Exp Med* **176**, 415–26 (1992).

564. Modun, B., Morrissey, J. & Williams, P. The staphylococcal transferrin receptor: a glycolytic enzyme with novel functions. *Trends Microbiol* **8**, 231–7 (2000).

565. Pancholi, V. & Fischetti, V. A. Cell-to-cell signalling between group A streptococci and pharyngeal cells. Role of streptococcal surface dehydrogenase (SDH). *Adv Exp Med Biol* **418**, 499–504 (1997).

566. Pancholi, V. & Fischetti, V. A. Regulation of the phosphorylation of human pharyngeal cell proteins by group A streptococcal surface dehydrogenase: signal transduction between streptococci and pharyngeal cells. *J Exp Med* **186**, 1633–43 (1997).

567. Pancholi, V. & Fischetti, V. A. Glyceraldehyde-3-phosphate dehydrogenase on the surface of group A streptococci is also an ADP-ribosylating enzyme. *Proc Natl Acad Sci U S A* **90**, 8154–8 (1993).

568. Kawamoto, R. M. & Caswell, A. H. Autophosphorylation of glyceraldehydephosphate dehydrogenase and phosphorylation of protein from skeletal muscle microsomes. *Biochemistry* **25**, 657–61 (1986).

569. Goudot-Crozel, V., Caillol, D., Djabali, M. & Dessein, A. J. The major parasite surface antigen associated with human resistance to schistosomiasis is a 37-kD glyceraldehyde-3P-dehydrogenase. *J Exp Med* **170**, 2065–80 (1989).

570. Gozalbo, D. et al. The cell wall-associated glyceraldehyde-3-phosphate dehydrogenase of *Candida albicans* is also a fibronectin and laminin binding protein. *Infect Immun* **66**, 2052–9 (1998).

571. Delgado, M. L. et al. The glyceraldehyde-3-phosphate dehydrogenase polypeptides encoded by the *Saccharomyces cerevisiae* TDH1, TDH2 and TDH3 genes are also cell wall proteins. *Microbiology* **147**, 411–7 (2001).

572. Gil, M. L., Delgado, M. L. & Gozalbo, D. The *Candida albicans* cell wall-associated glyceraldehyde-3-phosphate dehydrogenase activity increases in response to starvation and temperature upshift. *Med Mycol* **39**, 387–94 (2001).

573. Mazzola, J. L. & Sirover, M. A. Subcellular localization of human glyceraldehyde-3-phosphate dehydrogenase is independent of its glycolytic function. *Biochim Biophys Acta* **1622**, 50–6 (2003).

574. Schmitz, H. D. & Bereiter-Hahn, J. Glyceraldehyde-3-phosphate dehydrogenase associates with actin filaments in serum deprived NIH 3T3 cells only. *Cell Biol Int* **26**, 155–64 (2002).

575. Meyer-Siegler, K. et al. A human nuclear uracil DNA glycosylase is the 37-kDa subunit of glyceraldehyde-3-phosphate dehydrogenase. *Proc Natl Acad Sci U S A* **88**, 8460–4 (1991).

576. Wang, X., Sirover, M. A. & Anderson, L. E. Pea chloroplast glyceraldehyde-3-phosphate dehydrogenase has uracil glycosylase activity. *Arch Biochem Biophys* **367**, 348–53 (1999).

577. Huitorel, P. & Pantaloni, D. Bundling of microtubules by glyceraldehyde-3-phosphate dehydrogenase and its modulation by ATP. *Eur J Biochem* **150**, 265–9 (1985).

578. Tisdale, E. J. Glyceraldehyde-3-phosphate dehydrogenase is required for vesicular transport in the early secretory pathway. *J Biol Chem* **276**, 2480–6 (2001).

579. Tisdale, E. J. Glyceraldehyde-3-phosphate dehydrogenase is phosphorylated by protein kinase Ciota /lambda and plays a role in microtubule dynamics in the early secretory pathway. *J Biol Chem* **277**, 3334–41 (2002).

580. Glaser, P. E., Han, X. & Gross, R. W. Tubulin is the endogenous inhibitor of the glyceraldehyde 3-phosphate dehydrogenase isoform that catalyzes membrane fusion: Implications for the coordinated regulation of glycolysis and membrane fusion. *Proc Natl Acad Sci U S A* **99**, 14104–9 (2002).

581. Nagy, E. & Rigby, W. F. Glyceraldehyde-3-phosphate dehydrogenase selectively binds AU-rich RNA in the NAD(+)-binding region (Rossmann fold). *J Biol Chem* **270**, 2755–63 (1995).

582. Perucho, M., Salas, J. & Salas, M. L. Identification of the mammalian DNA-binding protein P8 as glyceraldehyde-3-phosphate dehydrogenase. *Eur J Biochem* **81**, 557–62 (1977).

583. Perucho, M., Salas, J. & Salas, M. L. Study of the interaction of glyceraldehyde-3-phosphate dehydrogenase with DNA. *Biochim Biophys Acta* **606**, 181–95 (1980).

584. Karpel, R. L. & Burchard, A. C. A basic isozyme of yeast glyceraldehyde-3-phosphate dehydrogenase with nucleic acid helix-destabilizing activity. *Biochim Biophys Acta* **654**, 256–67 (1981).

585. Brune, B. & Lapetina, E. G. Glyceraldehyde-3-phosphate dehydrogenase: a target for nitric oxide signaling. *Adv Pharmacol* **34**, 351–60 (1995).

586. Mohr, S., Stamler, J. S. & Brune, B. Posttranslational modification of glyceraldehyde-3-phosphate dehydrogenase by S-nitrosylation and subsequent NADH attachment. *J Biol Chem* **271**, 4209–14 (1996).

587. Wu, K., Aoki, C., Elste, A., Rogalski-Wilk, A. A. & Siekevitz, P. The synthesis of ATP by glycolytic enzymes in the postsynaptic density and the effect of endogenously generated nitric oxide. *Proc Natl Acad Sci U S A* **94**, 13273–8 (1997).

588. Denu, J. M. Linking chromatin function with metabolic networks: Sir2 family of NAD(+)-dependent deacetylases. *Trends Biochem Sci* **28**, 41–8 (2003).

589. Rutter, J., Reick, M., Wu, L. C. & McKnight, S. L. Regulation of clock and NPAS2 DNA binding by the redox state of NAD cofactors. *Science* **293**, 510–4 (2001).

590. Rutter, J., Reick, M. & McKnight, S. L. Metabolism and the control of circadian rhythms. *Annu Rev Biochem* **71**, 307–31 (2002).

591. Turner, J. & Crossley, M. The CtBP family: enigmatic and enzymatic transcriptional co-repressors. *Bioessays* **23**, 683–90 (2001).

592. Chinnadurai, G. CtBP family proteins: more than transcriptional corepressors. *Bioessays* **25**, 9–12 (2003).

593. Schaeper, U. et al. Molecular cloning and characterization of a cellular phosphoprotein that interacts with a conserved C-terminal domain of adenovirus E1A involved in negative modulation of oncogenic transformation. *Proc Natl Acad Sci U S A* **92**, 10467–71 (1995).

594. Kumar, V. et al. Transcription corepressor CtBP is an NAD(+)-regulated dehydrogenase. *Mol Cell* **10**, 857–69 (2002).

595. Balasubramanian, P., Zhao, L. J. & Chinnadurai, G. Nicotinamide adenine dinucleotide stimulates oligomerization, interaction with adenovirus E1A and an intrinsic dehydrogenase activity of CtBP. *FEBS Lett* **537**, 157–60 (2003).

596. Shi, Y. et al. Coordinated histone modifications mediated by a CtBP co-repressor complex. *Nature* **422**, 735–8 (2003).

597. Fjeld, C. C., Birdsong, W. T. & Goodman, R. H. Differential binding of NAD+ and NADH allows the transcriptional corepressor carboxyl-terminal binding protein to serve as a metabolic sensor. *Proc Natl Acad Sci U S A* **100**, 9202–7 (2003).

598. Zhang, Q., Piston, D. W. & Goodman, R. H. Regulation of corepressor function by nuclear NADH. *Science* **295**, 1895–7 (2002).

599. Grooteclaes, M. et al. C-terminal-binding protein corepresses epithelial and proapoptotic gene expression programs. *Proc Natl Acad Sci U S A* **100**, 4568–73 (2003).

600. Pancholi, V. Multifunctional α-enolase: its role in diseases. *Cell Mol Life Sci* **58**, 902–20 (2001).

601. Johnstone, S. A., Waisman, D. M. & Rattner, J. B. Enolase is present at the centrosome of HeLa cells. *Exp Cell Res* **202**, 458–63 (1992).

601a. Iida, H. & Yahara, I. Yeast heat shock protein of Mr 48,000 is an isoprotein of enolase. *Nature* **315**, 688–90 (1985).

602. Takei, N. et al. Neuronal survival factor from bovine brain is identical to neuron-specific enolase. *J Neurochem* **57**, 1178–84 (1991).

603. Miles, L. A. et al. Role of cell-surface lysines in plasminogen binding to cells: identification of α-enolase as a candidate plasminogen receptor. *Biochemistry* **30**, 1682–91 (1991).

604. Nakajima, K. et al. Plasminogen binds specifically to α-enolase on rat neuronal plasma membrane. *J Neurochem* **63**, 2048–57 (1994).

605. Redlitz, A., Fowler, B. J., Plow, E. F. & Miles, L. A. The role of an enolase-related molecule in plasminogen binding to cells. *Eur J Biochem* **227**, 407–15 (1995).

606. Arza, B., Felez, J., Lopez-Alemany, R., Miles, L. A. & Munoz-Canoves, P. Identification of an epitope of α-enolase (a candidate plasminogen receptor) by phage display. *Thromb Haemost* **78**, 1097–103 (1997).

607. Pancholi, V. & Fischetti, V. A. α-Enolase, a novel strong plasmin(ogen) binding protein on the surface of pathogenic streptococci. *J Biol Chem* **273**, 14503–15 (1998).

608. Bergmann, S., Rohde, M., Chhatwal, G. S. & Hammerschmidt, S. α-Enolase of *Streptococcus pneumoniae* is a plasmin(ogen)-binding protein displayed on the bacterial cell surface. *Mol Microbiol* **40**, 1273–87 (2001).

609. Hughes, M. J. et al. Identification of major outer surface proteins of *Streptococcus agalactiae*. *Infect Immun* **70**, 1254–9 (2002).

610. Sha, J. et al. Differential expression of the enolase gene under in vivo versus in vitro growth conditions of *Aeromonas hydrophila. Microb Pathog* **34**, 195–204 (2003).

611. Lee, K. W. et al. Cloning of the gene for phosphoglycerate kinase from *Schistosoma mansoni* and characterization of its gene product. *Mol Biochem Parasitol* **71**, 221–31 (1995).

612. Hussain, M., Peters, G., Chhatwal, G. S. & Herrmann, M. A lithium chloride-extracted, broad-spectrum-adhesive 42-kilodalton protein of *Staphylococcus epidermidis* is ornithine carbamoyltransferase. *Infect Immun* **67**, 6688–90 (1999).

613. McManaman, J. L. & Bain, D. L. Structural and conformational analysis of the oxidase to dehydrogenase conversion of xanthine oxidoreductase. *J Biol Chem* **277**, 21261–8 (2002).

614. Mather, I. H. A review and proposed nomenclature for major proteins of the milk-fat globule membrane. *J Dairy Sci* **83**, 203–47 (2000).

615. McManaman, J. L., Neville, M. C. & Wright, R. M. Mouse mammary gland xanthine oxidoreductase: purification, characterization, and regulation. *Arch Biochem Biophys* **371**, 308–16 (1999).

616. Kurosaki, M., Zanotta, S., Li Calzi, M., Garattini, E. & Terao, M. Expression of xanthine oxidoreductase in mouse mammary epithelium during pregnancy and lactation: regulation of gene expression by glucocorticoids and prolactin. *Biochem J* **319** (**Pt 3**), 801–10 (1996).

617. McManaman, J. L., Palmer, C. A., Wright, R. M. & Neville, M. C. Functional regulation of xanthine oxidoreductase expression and localization in the mouse mammary gland: evidence of a role in lipid secretion. *J Physiol* **545**, 567–79 (2002).

618. Heid, H. W., Schnolzer, M. & Keenan, T. W. Adipocyte differentiation-related protein is secreted into milk as a constituent of milk lipid globule membrane. *Biochem J* **320** (**Pt 3**), 1025–30 (1996).

619. Heid, H. W., Moll, R., Schwetlick, I., Rackwitz, H. R. & Keenan, T. W. Adipophilin is a specific marker of lipid accumulation in diverse cell types and diseases. *Cell Tissue Res* **294**, 309–21 (1998).

620. McManaman, J. L., Zabaronick, W., Schaack, J. & Orlicky, D. J. Lipid droplet targeting domains of adipophilin. *J Lipid Res* **44**, 668–73 (2003).

621. Mather, I. H. & Keenan, T. W. Origin and secretion of milk lipids. *J Mammary Gland Biol Neoplasia* **3**, 259–73 (1998).

622. Vorbach, C., Scriven, A. & Capecchi, M. R. The housekeeping gene xanthine oxidoreductase is necessary for milk fat droplet enveloping and secretion: gene sharing in the lactating mammary gland. *Genes Dev* **16**, 3223–35 (2002).

623. Laurent, T. C., Moore, E. C. & Reichard, P. Enzymatic synthesis of deoxyribonucleotides. IV. Isolation and characterization of thioredoxin, the hydrogen donor from *Escherichia coli* B. *J Biol Chem* **239**, 3436–44 (1964).

624. Holmgren, A. Thioredoxin structure and mechanism: conformational changes on oxidation of the active-site sulfhydryls to a disulfide. *Structure* **3**, 239–43 (1995).

625. Arner, E. S. & Holmgren, A. Physiological functions of thioredoxin and thioredoxin reductase. *Eur J Biochem* **267**, 6102–9 (2000).

626. Hirota, K., Nakamura, H., Masutani, H. & Yodoi, J. Thioredoxin superfamily and thioredoxin-inducing agents. *Ann N Y Acad Sci* **957**, 189–99 (2002).

627. Nakamura, H. et al. Circulating thioredoxin suppresses lipopolysaccharide-induced neutrophil chemotaxis. *Proc Natl Acad Sci U S A* **98**, 15143–8 (2001).

628. Schenk, H., Vogt, M., Droge, W. & Schulze-Osthoff, K. Thioredoxin as a potent costimulus of cytokine expression. *J Immunol* **156**, 765–71 (1996).

629. Yoshida, S. et al. Involvement of thioredoxin in rheumatoid arthritis: its costimulatory roles in the TNF-α-induced production of IL-6 and IL-8 from cultured synovial fibroblasts. *J Immunol* **163**, 351–8 (1999).

630. Bertini, R. et al. Thioredoxin, a redox enzyme released in infection and inflammation, is a unique chemoattractant for neutrophils, monocytes, and T cells. *J Exp Med* **189**, 1783–9 (1999).

631. Nakamura, H., Nakamura, K. & Yodoi, J. Redox regulation of cellular activation. *Annu Rev Immunol* **15**, 351–69 (1997).

632. Saitoh, M. et al. Mammalian thioredoxin is a direct inhibitor of apoptosis signal-regulating kinase (ASK) 1. *Embo J* **17**, 2596–606 (1998).

633. Mark, D. F. & Richardson, C. C. *Escherichia coli* thioredoxin: a subunit of bacteriophage T7 DNA polymerase. *Proc Natl Acad Sci U S A* **73**, 780–4 (1976).

634. Huber, H. E., Russel, M., Model, P. & Richardson, C. C. Interaction of mutant thioredoxins of *Escherichia coli* with the gene 5 protein of phage T7. The redox capacity of thioredoxin is not required for stimulation of DNA polymerase activity. *J Biol Chem* **261**, 15006–12 (1986).

635. Kumar, J. K., Tabor, S. & Richardson, C. C. Role of the C-terminal residue of the DNA polymerase of bacteriophage T7. *J Biol Chem* **276**, 34905–12 (2001).

636. Tabor, S., Huber, H. E. & Richardson, C. C. *Escherichia coli* thioredoxin confers processivity on the DNA polymerase activity of the gene 5 protein of bacteriophage T7. *J Biol Chem* **262**, 16212–23 (1987).

637. Huber, H. E., Tabor, S. & Richardson, C. C. *Escherichia coli* thioredoxin stabilizes complexes of bacteriophage T7 DNA polymerase and primed templates. *J Biol Chem* **262**, 16224–32 (1987).

638. Adler, S. & Modrich, P. T7-induced DNA polymerase. Requirement for thioredoxin sulfhydryl groups. *J Biol Chem* **258**, 6956–62 (1983).

639. Himawan, J. S. & Richardson, C. C. Amino acid residues critical for the interaction between bacteriophage T7 DNA polymerase and *Escherichia coli* thioredoxin. *J Biol Chem* **271**, 19999–20008 (1996).

640. Davidson, J. F., Fox, R., Harris, D. D., Lyons-Abbott, S. & Loeb, L. A. Insertion of the T3 DNA polymerase thioredoxin binding domain enhances the processivity and fidelity of Taq DNA polymerase. *Nucleic Acids Res* **31**, 4702–9 (2003).

641. Russel, M. & Model, P. Thioredoxin is required for filamentous phage assembly. *Proc Natl Acad Sci U S A* **82**, 29–33 (1985).

642. Feng, J. N., Russel, M. & Model, P. A permeabilized cell system that assembles filamentous bacteriophage. *Proc Natl Acad Sci U S A* **94**, 4068–73 (1997).

643. Russel, M. Filamentous phage assembly. *Mol Microbiol* **5**, 1607–13 (1991).

644. Russel, M. & Model, P. The role of thioredoxin in filamentous phage assembly. Construction, isolation, and characterization of mutant thioredoxins. *J Biol Chem* **261**, 14997–5005 (1986).

645. Feng, J. N., Model, P. & Russel, M. A trans-envelope protein complex needed for filamentous phage assembly and export. *Mol Microbiol* **34**, 745–55 (1999).

646. Tonissen, K. et al. Site-directed mutagenesis of human thioredoxin. Identification of cysteine 74 as critical to its function in the "early pregnancy factor" system. *J Biol Chem* **268**, 22485–9 (1993).

647. Warren, G. & Wickner, W. Organelle inheritance. *Cell* **84**, 395–400 (1996).

648. Xu, Z. & Wickner, W. Thioredoxin is required for vacuole inheritance in *Saccharomyces cerevisiae. J Cell Biol* **132**, 787–94 (1996).

649. Xu, Z., Mayer, A., Muller, E. & Wickner, W. A heterodimer of thioredoxin and I(B)2 cooperates with Sec18p (NSF) to promote yeast vacuole inheritance. *J Cell Biol* **136**, 299–306 (1997).

650. Roberts, D. G., Lamb, M. R. & Dieckmann, C. L. Characterization of the EYE2 gene required for eyespot assembly in *Chlamydomonas reinhardtii. Genetics* **158**, 1037–49 (2001).

651. Kern, R., Malki, A., Holmgren, A. & Richarme, G. Chaperone properties of *Escherichia coli* thioredoxin and thioredoxin reductase. *Biochem J* **371**, 965–72 (2003).

652. Noiva, R. & Lennarz, W. J. Protein disulfide isomerase. A multifunctional protein resident in the lumen of the endoplasmic reticulum. *J Biol Chem* **267**, 3553–6 (1992).

653. Noiva, R. Protein disulfide isomerase: the multifunctional redox chaperone of the endoplasmic reticulum. *Semin Cell Dev Biol* **10**, 481–93 (1999).

654. Turano, C., Coppari, S., Altieri, F. & Ferraro, A. Proteins of the PDI family: unpredicted non-ER locations and functions. *J Cell Physiol* **193**, 154–63 (2002).

655. Cai, H., Wang, C. C. & Tsou, C. L. Chaperone-like activity of protein disulfide isomerase in the refolding of a protein with no disulfide bonds. *J Biol Chem* **269**, 24550–2 (1994).

656. Song, J. L. & Wang, C. C. Chaperone-like activity of protein disulfide-isomerase in the refolding of rhodanese. *Eur J Biochem* **231**, 312–6 (1995).

657. Quan, H., Fan, G. & Wang, C. C. Independence of the chaperone activity of protein disulfide isomerase from its thioredoxin-like active site. *J Biol Chem* **270**, 17078–80 (1995).

658. Ferrari, D. M. & Soling, H. D. The protein disulphide-isomerase family: unravelling a string of folds. *Biochem J* **339** (**Pt 1**), 1–10 (1999).

658a. Turano, C., Coppari, S., Altieri, F. & Ferraro, A. Proteins of the PDI family: unpredicted non-ER locations and functions. *J Cell Physiol* **193**, 154–63 (2002).

659. Koivu, J. et al. A single polypeptide acts both as the β subunit of prolyl 4-hydroxylase and as a protein disulfide-isomerase. *J Biol Chem* **262**, 6447–9 (1987).

660. Pihlajaniemi, T. et al. Molecular cloning of the β-subunit of human prolyl 4-

hydroxylase. This subunit and protein disulphide isomerase are products of the same gene. *Embo J* **6**, 643–9 (1987).

661. Wetterau, J. R., Combs, K. A., Spinner, S. N. & Joiner, B. J. Protein disulfide isomerase is a component of the microsomal triglyceride transfer protein complex. *J Biol Chem* **265**, 9801–7 (1990).

662. Berriot-Varoqueaux, N., Aggerbeck, L. P., Samson-Bouma, M. & Wetterau, J. R. The role of the microsomal triglyceride transfer protein in abetalipoproteinemia. *Annu Rev Nutr* **20**, 663–97 (2000).

663. Wells, W. W., Xu, D. P., Yang, Y. F. & Rocque, P. A. Mammalian thioltransferase (glutaredoxin) and protein disulfide isomerase have dehydroascorbate reductase activity. *J Biol Chem* **265**, 15361–4 (1990).

664. O'Brien, P. J. & Herschlag, D. Catalytic promiscuity and the evolution of new enzymatic activities. *Chem Biol* **6**, R91-R105 (1999).

665. Rafikova, O., Rafikov, R. & Nudler, E. Catalysis of S-nitrosothiols formation by serum albumin: the mechanism and implication in vascular control. *Proc Natl Acad Sci U S A* **99**, 5913–8 (2002).

666. Yang, J., Petersen, C. E., Ha, C. E. & Bhagavan, N. V. Structural insights into human serum albumin-mediated prostaglandin catalysis. *Protein Sci* **11**, 538–45 (2002).

667. Sogorb, M. A., Carrera, V., Benabent, M. & Vilanova, E. Rabbit serum albumin hydrolyzes the carbamate carbaryl. *Chem Res Toxicol* **15**, 520–6 (2002).

668. Hollfelder, F., Kirby, A. J., Tawfik, D. S., Kikuchi, K. & Hilvert, D. Characterization of proton-transfer catalysis by serum albumins. *J Am Chem Soc* **122**, 1022–9 (2000).

669. James, L. C. & Tawfik, D. S. Catalytic and binding poly-reactivities shared by two unrelated proteins: the potential role of promiscuity in enzyme evolution. *Protein Sci* **10**, 2600–7 (2001).

670. Hollfelder, F., Kirby, A. J. & Tawfik, D. S. On the magnitude and specificity of medium effects in enzyme-like catalysts for proton transfer. *J Org Chem* **66**, 5866–74 (2001).

671. Zhu, L. & Crouch, R. K. Albumin in the cornea is oxidized by hydrogen peroxide. *Cornea* **11**, 567–72 (1992).

672. Jarabak, R., Westley, J., Dungan, J. M. & Horowitz, P. A chaperone-mimetic effect of serum albumin on rhodanese. *J Biochem Toxicol* **8**, 41–8 (1993).

673. Sabah, J. R., Davidson, H., McConkey, E. N. & Takemoto, L. In vivo passage of albumin from the aqueous humor into the lens. *Mol Vis* **10**, 254–9 (2004).

674. Sabah, J., McConkey, E., Welti, R., Albin, K. & Takemoto, L. J. Role of albumin as a fatty acid carrier for biosynthesis of lens lipids. *Exp Eye Res* **80**, 31–6 (2005).

675. Yin, H. L. Gelsolin: calcium- and polyphosphoinositide-regulated actin-modulating protein. *Bioessays* **7**, 176–9 (1987).

676. Sun, H. Q., Yamamoto, M., Mejillano, M. & Yin, H. L. Gelsolin, a multifunctional actin regulatory protein. *J Biol Chem* **274**, 33179–82 (1999).

677. Archer, S. K., Claudianos, C. & Campbell, H. D. Evolution of the gelsolin family of actin-binding proteins as novel transcriptional coactivators. *Bioessays* **27**, 388–96 (2005).

678. Kothakota, S. et al. Caspase-3-generated fragment of gelsolin: effector of morphological change in apoptosis. *Science* **278**, 294–8 (1997).

679. Ohtsu, M. et al. Inhibition of apoptosis by the actin-regulatory protein gelsolin. *Embo J* **16**, 4650–6 (1997).

680. Kusano, H. et al. Human gelsolin prevents apoptosis by inhibiting apoptotic mitochondrial changes via closing VDAC. *Oncogene* **19**, 4807–14 (2000).

681. Nishimura, K. et al. Modulation of androgen receptor transactivation by gelsolin: a newly identified androgen receptor coregulator. *Cancer Res* **63**, 4888–94 (2003).

682. Ting, H. J., Yeh, S., Nishimura, K. & Chang, C. Supervillin associates with androgen receptor and modulates its transcriptional activity. *Proc Natl Acad Sci U S A* **99**, 661–6 (2002).

683. Ozanne, D. M. et al. Androgen receptor nuclear translocation is facilitated by the f-actin cross-linking protein filamin. *Mol Endocrinol* **14**, 1618–26 (2000).

684. Witke, W. et al. Hemostatic, inflammatory, and fibroblast responses are blunted in mice lacking gelsolin. *Cell* **81**, 41–51 (1995).

685. Azuma, T., Witke, W., Stossel, T. P., Hartwig, J. H. & Kwiatkowski, D. J. Gelsolin is a downstream effector of rac for fibroblast motility. *Embo J* **17**, 1362–70 (1998).

686. Arora, P. D., Glogauer, M., Kapus, A., Kwiatkowski, D. J. & McCulloch, C. A. Gelsolin mediates collagen phagocytosis through a rac-dependent step. *Mol Biol Cell* **15**, 588–99 (2004).

687. Chellaiah, M. et al. Gelsolin deficiency blocks podosome assembly and produces increased bone mass and strength. *J Cell Biol* **148**, 665–78 (2000).

688. Sagawa, N. et al. Gelsolin suppresses tumorigenicity through inhibiting PKC activation in a human lung cancer cell line, PC10. *Br J Cancer* **88**, 606–12 (2003).

689. Lind, S. E., Smith, D. B., Janmey, P. A. & Stossel, T. P. Depression of gelsolin levels and detection of gelsolin-actin complexes in plasma of patients with acute lung injury. *Am Rev Respir Dis* **138**, 429–34 (1988).

690. Herrmannsdoerfer, A. J. et al. Vascular clearance and organ uptake of G- and F-actin in the rat. *Am J Physiol* **265**, G1071–81 (1993).

691. Vouyiouklis, D. A. & Brophy, P. J. A novel gelsolin isoform expressed by oligodendrocytes in the central nervous system. *J Neurochem* **69**, 995–1005 (1997).

692. Lim, M. L., Lum, M. G., Hansen, T. M., Roucou, X. & Nagley, P. On the release of cytochrome c from mitochondria during cell death signaling. *J Biomed Sci* **9**, 488–506 (2002).

693. Jiang, X. & Wang, X. Cytochrome C-mediated apoptosis. *Annu Rev Biochem* **73**, 87–106 (2004).

694. Boehning, D., van Rossum, D. B., Patterson, R. L. & Snyder, S. H. A peptide inhibitor of cytochrome c/inositol 1,4,5-trisphosphate receptor binding blocks intrinsic and extrinsic cell death pathways. *Proc Natl Acad Sci U S A* **102**, 1466–71 (2005).

695. Strasser, A., O'Connor, L. & Dixit, V. M. Apoptosis signaling. *Annu Rev Biochem* **69**, 217–45 (2000).

696. Kluck, R. M. et al. Cytochrome c activation of CPP32-like proteolysis plays a critical role in a *Xenopus* cell-free apoptosis system. *Embo J* **16**, 4639–49 (1997).

697. Yu, T., Wang, X., Purring-Koch, C., Wei, Y. & McLendon, G. L. A mutational epitope for cytochrome C binding to the apoptosis protease activation factor-1. *J Biol Chem* **276**, 13034–8 (2001).

698. MacNichol, E. F., Jr., Kunz, Y. W., Levine, J. S., Harosi, F. I. & Collins, B. A. Ellipsosomes: organelles containing a cytochrome-like pigment in the retinal cones of certain fishes. *Science* **200**, 549–52 (1978).

699. Nag, T. C. & Bhattacharjee, J. Retinal ellipsosomes: morphology, development, identification, and comparison with oil droplets. *Cell Tissue Res* **279**, 633–7 (1995).

700. Kosak, S. T. & Groudine, M. Gene order and dynamic domains. *Science* **306**, 644–7 (2004).

701. Kosak, S. T. & Groudine, M. Form follows function: the genomic organization of cellular differentiation. *Genes Dev* **18**, 1371–84 (2004).

702. Misteli, T. Spatial positioning; a new dimension in genome function. *Cell* **119**, 153–6 (2004).

703. Carlson, M. Genetics of transcriptional regulation in yeast: connections to the RNA polymerase II CTD. *Annu Rev Cell Dev Biol* **13**, 1–23 (1997).

704. Naar, A. M., Lemon, B. D. & Tjian, R. Transcriptional coactivator complexes. *Annu Rev Biochem* **70**, 475–501 (2001).

705. Lemon, B. & Tjian, R. Orchestrated response: a symphony of transcription factors for gene control. *Genes Dev* **14**, 2551–69 (2000).

706. Glass, C. K. & Rosenfeld, M. G. The coregulator exchange in transcriptional functions of nuclear receptors. *Genes Dev* **14**, 121–41 (2000).

707. Koleske, A. J. & Young, R. A. An RNA polymerase II holoenzyme responsive to activators. *Nature* **368**, 466–9 (1994).

708. Kim, Y. J., Bjorklund, S., Li, Y., Sayre, M. H. & Kornberg, R. D. A multiprotein mediator of transcriptional activation and its interaction with the C-terminal repeat domain of RNA polymerase II. *Cell* **77**, 599–608 (1994).

709. Malik, S. & Roeder, R. G. Transcriptional regulation through Mediator-like coactivators in yeast and metazoan cells. *Trends Biochem Sci* **25**, 277–83 (2000).

710. Dotson, M. R. et al. Structural organization of yeast and mammalian mediator complexes. *Proc Natl Acad Sci U S A* **97**, 14307–10 (2000).

711. Bourbon, H. M. et al. A unified nomenclature for protein subunits of mediator complexes linking transcriptional regulators to RNA polymerase II. *Mol Cell* **14**, 553–7 (2004).

712. Sluder, A. E. & Maina, C. V. Nuclear receptors in nematodes: themes and variations. *Trends Genet* **17**, 206–13 (2001).

713. Li, H. et al. Cytochrome c release and apoptosis induced by mitochondrial targeting of nuclear orphan receptor TR3. *Science* **289**, 1159–64 (2000).

714. Bayaa, M., Booth, R. A., Sheng, Y. & Liu, X. J. The classical progesterone receptor mediates *Xenopus* oocyte maturation through a nongenomic mechanism. *Proc Natl Acad Sci U S A* **97**, 12607–12 (2000).

715. Simoncini, T. et al. Interaction of oestrogen receptor with the regulatory subunit of phosphatidylinositol-3-OH kinase. *Nature* **407**, 538–41 (2000).

716. Wyrick, J. J. & Young, R. A. Deciphering gene expression regulatory networks. *Curr Opin Genet Dev* **12**, 130–6 (2002).

717. Shi, Y. Metabolic enzymes and coenzymes in transcription—a direct link between metabolism and transcription? *Trends Genet* **20**, 445–52 (2004).

718. Turner, B. M. Memorable transcription. *Nat Cell Biol* **5**, 390–3 (2003).

719. Frommer, W. B., Schulze, W. X. & Lalonde, S. Plant science. Hexokinase, Jack-of-all-trades. *Science* **300**, 261–3 (2003).

720. Herrero, P., Martinez-Campa, C. & Moreno, F. The hexokinase 2 protein participates in regulatory DNA-protein complexes necessary for glucose repression of the SUC2 gene in *Saccharomyces cerevisiae*. *FEBS Lett* **434**, 71–6 (1998).

721. Moore, B. et al. Role of the Arabidopsis glucose sensor HXK1 in nutrient, light, and hormonal signaling. *Science* **300**, 332–6 (2003).

722. Hall, D. A. et al. Regulation of gene expression by a metabolic enzyme. *Science* **306**, 482–4 (2004).

723. Ottosen, S., Herrera, F. J. & Triezenberg, S. J. Transcription. Proteasome parts at gene promoters. *Science* **296**, 479–81 (2002).

724. Ferdous, A., Gonzalez, F., Sun, L., Kodadek, T. & Johnston, S. A. The 19S regulatory particle of the proteasome is required for efficient transcription elongation by RNA polymerase II. *Mol Cell* **7**, 981–91 (2001).

725. Ferdous, A., Kodadek, T. & Johnston, S. A. A nonproteolytic function of the 19S regulatory subunit of the 26S proteasome is required for efficient activated transcription by human RNA polymerase II. *Biochemistry* **41**, 12798–805 (2002).

726. Gonzalez, F., Delahodde, A., Kodadek, T. & Johnston, S. A. Recruitment of a 19S proteasome subcomplex to an activated promoter. *Science* **296**, 548–50 (2002).

727. Salghetti, S. E., Caudy, A. A., Chenoweth, J. G. & Tansey, W. P. Regulation of transcriptional activation domain function by ubiquitin. *Science* **293**, 1651–3 (2001).

728. Gillette, T. G., Gonzalez, F., Delahodde, A., Johnston, S. A. & Kodadek, T. Physical and functional association of RNA polymerase II and the proteasome. *Proc Natl Acad Sci U S A* **101**, 5904–9 (2004).

729. Wolffe, A. P., Tafuri, S., Ranjan, M. & Familari, M. The Y-box factors: a family of nucleic acid binding proteins conserved from *Escherichia coli* to man. *New Biol* **4**, 290–8 (1992).

730. Matsumoto, K. & Wolffe, A. P. Gene regulation by Y-box proteins: coupling control of transcription and translation. *Trends Cell Biol* **8**, 318–23 (1998).

731. Kohno, K., Izumi, H., Uchiumi, T., Ashizuka, M. & Kuwano, M. The pleiotropic functions of the Y-box-binding protein, YB-1. *Bioessays* **25**, 691–8 (2003).

732. Didier, D. K., Schiffenbauer, J., Woulfe, S. L., Zacheis, M. & Schwartz, B. D. Characterization of the cDNA encoding a protein binding to the major histocompatibility complex class II Y box. *Proc Natl Acad Sci U S A* **85**, 7322–6 (1988).

733. Ranjan, M., Tafuri, S. R. & Wolffe, A. P. Masking mRNA from translation in somatic cells. *Genes Dev* **7**, 1725–36 (1993).

734. Tafuri, S. R. & Wolffe, A. P. Selective recruitment of masked maternal mRNA from

messenger ribonucleoprotein particles containing FRGY2 (mRNP4). *J Biol Chem* **268**, 24255–61 (1993).

735. Meric, F., Matsumoto, K. & Wolffe, A. P. Regulated unmasking of in vivo synthesized maternal mRNA at oocyte maturation. A role for the chaperone nucleoplasmin. *J Biol Chem* **272**, 12840–6 (1997).

736. Fukuda, T. et al. Characterization of the 5′-untranslated region of YB-1 mRNA and autoregulation of translation by YB-1 protein. *Nucleic Acids Res* **32**, 611–22 (2004).

737. Ashizuka, M. et al. Novel translational control through an iron-responsive element by interaction of multifunctional protein YB-1 and IRP2. *Mol Cell Biol* **22**, 6375–83 (2002).

738. Wool, I. G. Extraribosomal functions of ribosomal proteins. *Trends Biochem Sci* **21**, 164–5 (1996).

739. Stickeler, E. et al. The RNA binding protein YB-1 binds A/C-rich exon enhancers and stimulates splicing of the CD44 alternative exon v4. *Embo J* **20**, 3821–30 (2001).

740. Wilkinson, M. F. & Shyu, A. B. Multifunctional regulatory proteins that control gene expression in both the nucleus and the cytoplasm. *Bioessays* **23**, 775–87 (2001).

741. Niessing, D., Blanke, S. & Jackle, H. Bicoid associates with the 5′-cap-bound complex of caudal mRNA and represses translation. *Genes Dev* **16**, 2576–82 (2002).

742. Nedelec, S. et al. Emx2 homeodomain transcription factor interacts with eukaryotic translation initiation factor 4E (eIF4E) in the axons of olfactory sensory neurons. *Proc Natl Acad Sci U S A* **101**, 10815–20 (2004).

743. Topisirovic, I. et al. The proline-rich homeodomain protein, PRH, is a tissue-specific inhibitor of eIF4E-dependent cyclin D1 mRNA transport and growth. *Embo J* **22**, 689–703 (2003).

744. Wilkinson, M. F. & Shyu, A. B. RNA surveillance by nuclear scanning? *Nat Cell Biol* **4**, E144–7 (2002).

745. Iborra, F. J., Escargueil, A. E., Kwek, K. Y., Akoulitchev, A. & Cook, P. R. Molecular cross-talk between the transcription, translation, and nonsense-mediated decay machineries. *J Cell Sci* **117**, 899–906 (2004).

746. Prochiantz, A. & Joliot, A. Can transcription factors function as cell-cell signalling molecules? *Nat Rev Mol Cell Biol* **4**, 814–9 (2003).

747. Desvoyes, B., Faure-Rabasse, S., Chen, M. H., Park, J. W. & Scholthof, H. B. A novel plant homeodomain protein interacts in a functionally relevant manner with a virus movement protein. *Plant Physiol* **129**, 1521–32 (2002).

747a. Andersson, U., Erlandsson-Harris, H., Yang, H. & Tracey, K. J. HMGB1 as a DNA-binding cytokine. *J Leukoc Biol* **72**, 1084–91 (2002).

748. Hofmann, W. A. et al. Actin is part of pre-initiation complexes and is necessary for transcription by RNA polymerase II. *Nat Cell Biol* **6**, 1094–101 (2004).

749. Misteli, T. Protein dynamics: implications for nuclear architecture and gene expression. *Science* **291**, 843–7 (2001).

750. Hughes, A. L. *Adaptive Evolution of Genes and Genomes* (Oxford University Press, New York, 1999).

751. DeVries, A. L. & Wohlschlag, D. E. Freezing resistance in some Antarctic fishes. *Science* **163**, 1073–5 (1969).

752. DeVries, A. L. Glycoproteins as biological antifreeze agents in antarctic fishes. *Science* **172**, 1152–5 (1971).

753. Cheng, C. H. Evolution of the diverse antifreeze proteins. *Curr Opin Genet Dev* **8**, 715–20 (1998).

754. Fletcher, G. L., Hew, C. L. & Davies, P. L. Antifreeze proteins of teleost fishes. *Annu Rev Physiol* **63**, 359–90 (2001).

755. Ewart, K. V. & Hew, C. L. (eds.) *Fish Antifreeze Proteins* (World Scientific, New Jersey, 2002).

756. Cheng, C. H. & Chen, L. Evolution of an antifreeze glycoprotein. *Nature* **401**, 443–4 (1999).

757. Jia, Z. & Davies, P. L. Antifreeze proteins: an unusual receptor-ligand interaction. *Trends Biochem Sci* **27**, 101–6 (2002).

758. Ko, T. P. et al. The refined crystal structure of an eel pout type III antifreeze protein RD1 at 0.62-Å resolution reveals structural microheterogeneity of protein and solvation. *Biophys J* **84**, 1228–37 (2003).

759. Atici, O. & Nalbantoglu, B. Antifreeze proteins in higher plants. *Phytochemistry* **64**, 1187–96 (2003).

760. Zbikowska, H. M. Fish can be first—advances in fish transgenesis for commercial applications. *Transgenic Res* **12**, 379–89 (2003).

761. Tyshenko, M. G. & Walker, V. K. Hyperactive spruce budworm antifreeze protein expression in transgenic *Drosophila* does not confer cold shock tolerance. *Cryobiology* **49**, 28–36 (2004).

762. Deng, G., Andrews, D. W. & Laursen, R. A. Amino acid sequence of a new type of antifreeze protein, from the longhorn sculpin *Myoxocephalus octodecimspinosis*. *FEBS Lett* **402**, 17–20 (1997).

763. Liou, Y. C., Tocilj, A., Davies, P. L. & Jia, Z. Mimicry of ice structure by surface hydroxyls and water of a β-helix antifreeze protein. *Nature* **406**, 322–4 (2000).

764. Graether, S. P. et al. β-Helix structure and ice-binding properties of a hyperactive antifreeze protein from an insect. *Nature* **406**, 325–8 (2000).

765. Duman, J. G. Antifreeze and ice nucleator proteins in terrestrial arthropods. *Annu Rev Physiol* **63**, 327–57 (2001).

766. Graether, S. P. & Sykes, B. D. Cold survival in freeze-intolerant insects: the structure and function of β-helical antifreeze proteins. *Eur J Biochem* **271**, 3285–96 (2004).

767. Raymond, J. A. & DeVries, A. L. Adsorption inhibition as a mechanism of freezing resistance in polar fishes. *Proc Natl Acad Sci U S A* **74**, 2589–93 (1977).

768. Yang, D. S. et al. Identification of the ice-binding surface on a type III antifreeze protein with a "flatness function" algorithm. *Biophys J* **74**, 2142–51 (1998).

769. Leinala, E. K., Davies, P. L. & Jia, Z. Crystal structure of β-helical antifreeze protein points to a general ice binding model. *Structure (Camb)* **10**, 619–27 (2002).

770. Fairley, K. et al. Type I shorthorn sculpin antifreeze protein: recombinant synthesis, solution conformation, and ice growth inhibition studies. *J Biol Chem* **277**, 24073–80 (2002).

771. Graether, S. P. et al. Spruce budworm antifreeze protein: changes in structure and dynamics at low temperature. *J Mol Biol* **327**, 1155–68 (2003).

772. Daley, M. E., Graether, S. P. & Sykes, B. D. Hydrogen bonding on the ice-binding face of a β-helical antifreeze protein indicated by amide proton NMR chemical shifts. *Biochemistry* **43**, 13012–7 (2004).

773. Marshall, C. B., Daley, M. E., Sykes, B. D. & Davies, P. L. Enhancing the activity of a β-helical antifreeze protein by the engineered addition of coils. *Biochemistry* **43**, 11637–46 (2004).

774. Baardsnes, J., Kuiper, M. J. & Davies, P. L. Antifreeze protein dimer: when two ice-binding faces are better than one. *J Biol Chem* **278**, 38942–7 (2003).

775. Marshall, C. B., Fletcher, G. L. & Davies, P. L. Hyperactive antifreeze protein in a fish. *Nature* **429**, 153 (2004).

776. Tyshenko, M. G., Doucet, D., Davies, P. L. & Walker, V. K. The antifreeze potential of the spruce budworm thermal hysteresis protein. *Nat Biotechnol* **15**, 887–90 (1997).

777. Leinala, E. K. et al. A β-helical antifreeze protein isoform with increased activity. Structural and functional insights. *J Biol Chem* **277**, 33349–52 (2002).

778. Scott, G. K., Fletcher, G. L. & Davies, P. L. Fish antifreeze proteins: recent evolution. *Can J Fish Aquat Sci* **43**, 1028–1034 (1986).

779. Logsdon, J. M., Jr. & Doolittle, W. F. Origin of antifreeze protein genes: a cool tale in molecular evolution. *Proc Natl Acad Sci U S A* **94**, 3485–7 (1997).

780. Li, Z., Lin, Q., Yang, D. S., Ewart, K. V. & Hew, C. L. The role of Ca2+-coordinating residues of herring antifreeze protein in antifreeze activity. *Biochemistry* **43**, 14547–54 (2004).

781. Baardsnes, J. & Davies, P. L. Sialic acid synthase: the origin of fish type III antifreeze protein? *Trends Biochem Sci* **26**, 468–9 (2001).

782. Chen, L., DeVries, A. L. & Cheng, C. H. Evolution of antifreeze glycoprotein gene from a trypsinogen gene in Antarctic notothenioid fish. *Proc Natl Acad Sci U S A* **94**, 3811–6 (1997).

783. Cheng, C. H., Chen, L., Near, T. J. & Jin, Y. Functional antifreeze glycoprotein genes in temperate-water New Zealand nototheniid fish infer an Antarctic evolutionary origin. *Mol Biol Evol* **20**, 1897–908 (2003).

784. Chen, L., DeVries, A. L. & Cheng, C. H. Convergent evolution of antifreeze glycoproteins in Antarctic notothenioid fish and Arctic cod. *Proc Natl Acad Sci U S A* **94**, 3817–22 (1997).

785. Gong, Z., Ewart, K. V., Hu, Z., Fletcher, G. L. & Hew, C. L. Skin antifreeze protein genes of the winter flounder, *Pleuronectes americanus,* encode distinct and active polypeptides without the secretory signal and prosequences. *J Biol Chem* **271**, 4106–12 (1996).

786. Worrall, D. et al. A carrot leucine-rich-repeat protein that inhibits ice recrystallization. *Science* **282**, 115–7 (1998).

787. Sidebottom, C. et al. Heat-stable antifreeze protein from grass. *Nature* **406**, 256 (2000).

788. Pudney, P. D. et al. The physico-chemical characterization of a boiling stable

antifreeze protein from a perennial grass (*Lolium perenne*). *Arch Biochem Biophys* **410**, 238–45 (2003).

789. Knight, C. A. Structural biology. Adding to the antifreeze agenda. *Nature* **406**, 249, 251 (2000).

790. Hon, W. C., Griffith, M., Mlynarz, A., Kwok, Y. C. & Yang, D. S. Antifreeze proteins in winter rye are similar to pathogenesis-related proteins. *Plant Physiol* **109**, 879–89 (1995).

791. Yeh, S. et al. Chitinase genes responsive to cold encode antifreeze proteins in winter cereals. *Plant Physiol* **124**, 1251–64 (2000).

792. Huang, T. & Duman, J. G. Cloning and characterization of a thermal hysteresis (antifreeze) protein with DNA-binding activity from winter bittersweet nightshade, *Solanum dulcamara*. *Plant Mol Biol* **48**, 339–50 (2002).

793. Hardison, R. Hemoglobins from bacteria to man: evolution of different patterns of gene expression. *J Exp Biol* **201** (**Pt 8**), 1099–117 (1998).

794. Goodman, M. et al. An evolutionary tree for invertebrate globin sequences. *J Mol Evol* **27**, 236–49 (1988).

795. Ermler, U., Siddiqui, R. A., Cramm, R. & Friedrich, B. Crystal structure of the flavohemoglobin from *Alcaligenes eutrophus* at 1.75 Å resolution. *Embo J* **14**, 6067–77 (1995).

796. Moens, L. et al. Globins in nonvertebrate species: dispersal by horizontal gene transfer and evolution of the structure-function relationships. *Mol Biol Evol* **13**, 324–33 (1996).

797. Hausladen, A., Gow, A. J. & Stamler, J. S. Nitrosative stress: metabolic pathway involving the flavohemoglobin. *Proc Natl Acad Sci U S A* **95**, 14100–5 (1998).

798. Gardner, P. R., Gardner, A. M., Martin, L. A. & Salzman, A. L. Nitric oxide dioxygenase: an enzymic function for flavohemoglobin. *Proc Natl Acad Sci U S A* **95**, 10378–83 (1998).

799. Stevanin, T. M. et al. Flavohemoglobin Hmp affords inducible protection for *Escherichia coli* respiration, catalyzed by cytochromes bo' or bd, from nitric oxide. *J Biol Chem* **275**, 35868–75 (2000).

800. Kim, S. O., Orii, Y., Lloyd, D., Hughes, M. N. & Poole, R. K. Anoxic function for the *Escherichia coli* flavohaemoglobin (Hmp): reversible binding of nitric oxide and reduction to nitrous oxide. *FEBS Lett* **445**, 389–94 (1999).

801. Hausladen, A., Gow, A. & Stamler, J. S. Flavohemoglobin denitrosylase catalyzes the reaction of a nitroxyl equivalent with molecular oxygen. *Proc Natl Acad Sci U S A* **98**, 10108–12 (2001).

802. Mukai, M., Mills, C. E., Poole, R. K. & Yeh, S. R. Flavohemoglobin, a globin with a peroxidase-like catalytic site. *J Biol Chem* **276**, 7272–7 (2001).

803. Lebioda, L. et al. An enzymatic globin from a marine worm. *Nature* **401**, 445 (1999).

804. Liu, L., Zeng, M., Hausladen, A., Heitman, J. & Stamler, J. S. Protection from nitrosative stress by yeast flavohemoglobin. *Proc Natl Acad Sci U S A* **97**, 4672–6 (2000).

805. Minning, D. M. et al. *Ascaris* haemoglobin is a nitric oxide-activated "deoxygenase." *Nature* **401**, 497–502 (1999).

806. Imai, K. The haemoglobin enzyme. *Nature* **401**, 437, 439 (1999).

807. Powell, R. & Gannon, F. The leghaemoglobins. *Bioessays* **9**, 117–21 (1988).

808. Gow, A. J., Luchsinger, B. P., Pawloski, J. R., Singel, D. J. & Stamler, J. S. The oxyhemoglobin reaction of nitric oxide. *Proc Natl Acad Sci U S A* **96**, 9027–32 (1999).

809. Luchsinger, B. P. et al. Routes to S-nitroso-hemoglobin formation with heme redox and preferential reactivity in the β subunits. *Proc Natl Acad Sci U S A* **100**, 461–6 (2003).

810. Gladwin, M. T. & Schechter, A. N. NO contest: nitrite versus S-nitroso-hemoglobin. *Circ Res* **94**, 851–5 (2004).

811. Gladwin, M. T., Lancaster, J. R., Jr., Freeman, B. A. & Schechter, A. N. Nitric oxide's reactions with hemoglobin: a view through the SNO-storm. *Nat Med* **9**, 496–500 (2003).

812. Gladwin, M. T. et al. Relative role of heme nitrosylation and β-cysteine 93 nitrosation in the transport and metabolism of nitric oxide by hemoglobin in the human circulation. *Proc Natl Acad Sci U S A* **97**, 9943–8 (2000).

813. Gladwin, M. T. et al. Role of circulating nitrite and S-nitrosohemoglobin in the regulation of regional blood flow in humans. *Proc Natl Acad Sci U S A* **97**, 11482–7 (2000).

814. Gladwin, M. T. et al. S-Nitrosohemoglobin is unstable in the reductive erythrocyte environment and lacks O$_2$/NO-linked allosteric function. *J Biol Chem* **277**, 27818–28 (2002).

815. Cosby, K. et al. Nitrite reduction to nitric oxide by deoxyhemoglobin vasodilates the human circulation. *Nat Med* **9**, 1498–505 (2003).

816. Schechter, A. N. & Gladwin, M. T. Hemoglobin and the paracrine and endocrine functions of nitric oxide. *N Engl J Med* **348**, 1483–5 (2003).

817. Schechter, A. N., Gladwin, M. T. & Cannon, R. O., 3rd. NO solutions? *J Clin Invest* **109**, 1149–51 (2002).

818. Xu, X. et al. Measurements of nitric oxide on the heme iron and β-93 thiol of human hemoglobin during cycles of oxygenation and deoxygenation. *Proc Natl Acad Sci U S A* **100**, 11303–8 (2003).

819. Joshi, M. S. et al. Nitric oxide is consumed, rather than conserved, by reaction with oxyhemoglobin under physiological conditions. *Proc Natl Acad Sci U S A* **99**, 10341–6 (2002).

820. Dejam, A., Hunter, C. J., Schechter, A. N. & Gladwin, M. T. Emerging role of nitrite in human biology. *Blood Cells Mol Dis* **32**, 423–9 (2004).

821. Burr, A. H. et al. A hemoglobin with an optical function. *J Biol Chem* **275**, 4810–5 (2000).

822. Ellenby, C. Haemoglobin in the "chromotrope" of an insect parasitic nematode. *Nature* **202**, 615–6 (1964).

823. Burr, A. H., Schiefke, R. & Bollerup, G. Properties of a hemoglobin from the chromatrope of the nematode *Mermis nigrescens. Biochim Biophys Acta* **405**, 404–11 (1975).

824. Burr, A. H. & Harosi, F. I. Naturally crystalline hemoglobin of the nematode *Mermis nigrescens.* An in situ microspectrophotometric study of chemical properties and dichroism. *Biophys J* **47**, 527–36 (1985).

825. Burr, A. H. J., Eggleton, D. K., Patterson, R. & Leutscher-Hazelhoff, J. T. The role of

hemoglobin in the phototaxis of the nematode *Mermis nigrescens. Photochem Photobiol* **49**, 89–95 (1989).

826. Burr, A. H. J., Babinszki, C. P. F. & Ward, A. J. Components of phototaxis of the nematode *Mermis nigrescens. J. Comp. Physiol. A* **167**, 245–255 (1990).

827. Burr, A. H. J. & Babinszki, C. P. F. Scanning motion, ocellar morphology and orientation mechanisms in the phototaxis of the nematode *Mermis nigrescens. J. Comp. Physiol. A* **167**, 257–268 (1990).

828. Burr, A. H., Wagar, D. & Sidhu, P. Ocellar pigmentation and phototaxis in the nematode *Mermis nigrescens:* changes during development. *J Exp Biol* **203 (Pt 8)**, 1341–50 (2000).

829. Wride, M. A. et al. Expression profiling and gene discovery in the mouse lens. *Mol Vis* **9**, 360–96 (2003).

830. Kihm, A. J. et al. An abundant erythroid protein that stabilizes free α-haemoglobin. *Nature* **417**, 758–63 (2002).

831. Brachat, A. et al. A microarray-based, integrated approach to identify novel regulators of cancer drug response and apoptosis. *Oncogene* **21**, 8361–71 (2002).

832. Bassnett, S. & McNulty, R. The effect of elevated intraocular oxygen on organelle degradation in the embryonic chicken lens. *J Exp Biol* **206**, 4353–61 (2003).

833. Mansergh, F. C. et al. Gene expression changes during cataract progression in Sparc null mice: Differential regulation of mouse globins in the lens. *Mol Vis* **10**, 490–511 (2004).

834. Liu, L., Zeng, M. & Stamler, J. S. Hemoglobin induction in mouse macrophages. *Proc Natl Acad Sci U S A* **96**, 6643–7 (1999).

835. Bridges, C. B. Salivary chromosome maps. *J. Hered.* **26**, 60–64 (1935).

836. Lewis, E. B. Pseudoallelism and gene evolution. *Cold Spring Harb Symp Quant Biol* **16**, 159–74 (1951).

837. Stephens, S. G. Possible significances of duplication in evolution. *Adv Genet* **4**, 247–65 (1951).

838. Ingram, V. M. Gene evolution and the haemoglobins. *Nature* **189**, 704–8 (1961).

839. Ohno, S. Ancient linkage groups and frozen accidents. *Nature* **244**, 259–62 (1973).

840. Ohno, S. Gene duplication and the uniqueness of vertebrate genomes circa 1970–1999. *Semin Cell Dev Biol* **10**, 517–22 (1999).

841. Ohno, S., Wolf, U. & Atkin, N. B. Evolution from fish to mammals by gene duplication. *Hereditas* **59**, 169–87 (1968).

842. Kimura, M. *The Neutral Theory of Molecular Evolution* (Cambrige University Press, Cambridge, 1983).

843. Li, W.-H. *Molecular Evolution* (Sinauer Associates, Sunderland, MA, 1997).

844. Hughes, A. L. Phylogenies of developmentally important proteins do not support the hypothesis of two rounds of genome duplication early in vertebrate history. *J Mol Evol* **48**, 565–76 (1999).

845. Furlong, R. F. & Holland, P. W. Were vertebrates octoploid? *Philos Trans R Soc Lond B Biol Sci* **357**, 531–44 (2002).

846. Wolfe, K. H. & Shields, D. C. Molecular evidence for an ancient duplication of the entire yeast genome. *Nature* **387**, 708–13 (1997).

847. Kellis, M., Birren, B. W. & Lander, E. S. Proof and evolutionary analysis of ancient genome duplication in the yeast *Saccharomyces cerevisiae*. *Nature* **428**, 617–24 (2004).

848. Vandepoele, K., De Vos, W., Taylor, J. S., Meyer, A. & Van De Peer, Y. Major events in the genome evolution of vertebrates: Paranome age and size differ considerably between ray-finned fishes and land vertebrates. *Proc Natl Acad Sci U S A* **101**, 1638–43 (2004).

849. Holland, P. W. More genes in vertebrates? *J Struct Funct Genomics* **3**, 75–84 (2003).

850. Friedman, R. & Hughes, A. L. Gene duplication and the structure of eukaryotic genomes. *Genome Res* **11**, 373–81 (2001).

851. Friedman, R. & Hughes, A. L. The temporal distribution of gene duplication events in a set of highly conserved human gene families. *Mol Biol Evol* **20**, 154–61 (2003).

852. Gu, Z., Cavalcanti, A., Chen, F. C., Bouman, P. & Li, W. H. Extent of gene duplication in the genomes of *Drosophila*, nematode, and yeast. *Mol Biol Evol* **19**, 256–62 (2002).

853. Bailey, J. A. et al. Recent segmental duplications in the human genome. *Science* **297**, 1003–7 (2002).

854. Lynch, M. Genomics. Gene duplication and evolution. *Science* **297**, 945–7 (2002).

855. Lynch, M. & Conery, J. S. The evolutionary demography of duplicate genes. *J Struct Funct Genomics* **3**, 35–44 (2003).

856. Lynch, M. & Conery, J. S. The evolutionary fate and consequences of duplicate genes. *Science* **290**, 1151–5 (2000).

857. Katju, V. & Lynch, M. The structure and early evolution of recently arisen gene duplicates in the *Caenorhabditis elegans* genome. *Genetics* **165**, 1793–803 (2003).

858. Lynch, M. & Force, A. The probability of duplicate gene preservation by subfunctionalization. *Genetics* **154**, 459–73 (2000).

859. Wagner, A. Birth and death of duplicated genes in completely sequenced eukaryotes. *Trends Genet* **17**, 237–9 (2001).

860. Walsh, J. B. How often do duplicated genes evolve new functions? *Genetics* **139**, 421–8 (1995).

861. Nadeau, J. H. & Sankoff, D. Comparable rates of gene loss and functional divergence after genome duplications early in vertebrate evolution. *Genetics* **147**, 1259–66 (1997).

862. Wagner, A. The fate of duplicated genes: loss or new function? *Bioessays* **20**, 785–8 (1998).

863. Ohta, T. Further examples of evolution by gene duplication revealed through DNA sequence comparisons. *Genetics* **138**, 1331–7 (1994).

864. Hughes, A. L. & Hughes, M. K. Adaptive evolution in the rat olfactory receptor gene family. *J Mol Evol* **36**, 249–54 (1993).

865. Clark, A. G. Invasion and maintenance of a gene duplication. *Proc Natl Acad Sci U S A* **91**, 2950–4 (1994).

866. Thomas, J. H. Thinking about genetic redundancy. *Trends Genet* **9**, 395–9 (1993).

867. Golding, G. B. & Dean, A. M. The structural basis of molecular adaptation. *Mol Biol Evol* **15**, 355–69 (1998).

868. Gu, Z. et al. Role of duplicate genes in genetic robustness against null mutations. *Nature* **421**, 63–6 (2003).

869. Trabesinger-Ruef, N. et al. Pseudogenes in ribonuclease evolution: a source of new biomacromolecular function? *FEBS Lett* **382**, 319–22 (1996).

870. Soucek, J., Chudomel, V., Potmesilova, I. & Novak, J. T. Effect of ribonucleases on cell-mediated lympholysis reaction and on GM-CFC colonies in bone marrow culture. *Nat Immun Cell Growth Regul* **5**, 250–8 (1986).

871. Benner, S. A. & Allemann, R. K. The return of pancreatic ribonucleases. *Trends Biochem Sci* **14**, 396–7 (1989).

872. Nei, M., Gu, X. & Sitnikova, T. Evolution by the birth-and-death process in multigene families of the vertebrate immune system. *Proc Natl Acad Sci U S A* **94**, 7799–806 (1997).

873. Ota, T. & Nei, M. Divergent evolution and evolution by the birth-and-death process in the immunoglobulin VH gene family. *Mol Biol Evol* **11**, 469–82 (1994).

874. Nei, M. & Rooney, A. P. Concerted and birth-and-death evolution of multigene families. *Annu Rev Genet* (2005).

875. Nei, M., Rogozin, I. B. & Piontkivska, H. Purifying selection and birth-and-death evolution in the ubiquitin gene family. *Proc Natl Acad Sci U S A* **97**, 10866–71 (2000).

876. Rooney, A. P., Piontkivska, H. & Nei, M. Molecular evolution of the nontandemly repeated genes of the histone 3 multigene family. *Mol Biol Evol* **19**, 68–75 (2002).

877. Piontkivska, H., Rooney, A. P. & Nei, M. Purifying selection and birth-and-death evolution in the histone H4 gene family. *Mol Biol Evol* **19**, 689–97 (2002).

878. Nam, J., dePamphilis, C. W., Ma, H. & Nei, M. Antiquity and evolution of the MADS-box gene family controlling flower development in plants. *Mol Biol Evol* **20**, 1435–47 (2003).

879. Nam, J. et al. Type I MADS-box genes have experienced faster birth-and-death evolution than type II MADS-box genes in angiosperms. *Proc Natl Acad Sci U S A* **101**, 1910–5 (2004).

880. Nikolaidis, N. & Nei, M. Concerted and nonconcerted evolution of the hsp70 gene superfamily in two sibling species of nematodes. *Mol Biol Evol* **21**, 498–505 (2004).

881. Dykhuizen, D. & Hartl, D. L. Selective neutrality of 6PGD allozymes in *E. coli* and the effects of genetic background. *Genetics* **96**, 801–17 (1980).

882. Yokoyama, S. Evaluating adaptive evolution. *Nat Genet* **30**, 350–1 (2002).

883. Benner, S. A. et al. Developing new synthetic catalysts. How nature does it. *Acta Chem Scand* **50**, 243–8 (1996).

884. Wilks, H. M. et al. A specific, highly active malate dehydrogenase by redesign of a lactate dehydrogenase framework. *Science* **242**, 1541–4 (1988).

885. Aarts, H. J., Lubsen, N. H. & Schoenmakers, J. G. Crystallin gene expression during rat lens development. *Eur J Biochem* **183**, 31–6 (1989).

886. Zhang, J., Rosenberg, H. F. & Nei, M. Positive Darwinian selection after gene duplication in primate ribonuclease genes. *Proc Natl Acad Sci U S A* **95**, 3708–13 (1998).

887. Zhang, J., Zhang, Y. P. & Rosenberg, H. F. Adaptive evolution of a duplicated pancreatic ribonuclease gene in a leaf-eating monkey. *Nat Genet* **30**, 411–5 (2002).

888. Zhang, J. Rates of conservative and radical nonsynonymous nucleotide substitutions in mammalian nuclear genes. *J Mol Evol* **50**, 56–68 (2000).

889. Force, A. et al. Preservation of duplicate genes by complementary, degenerative mutations. *Genetics* **151**, 1531–45 (1999).

890. Lynch, M. & Conery, J. S. The origins of genome complexity. *Science* **302**, 1401–4 (2003).

891. Nieto, M. A. The snail superfamily of zinc-finger transcription factors. *Nat Rev Mol Cell Biol* **3**, 155–66 (2002).

892. Manzanares, M., Locascio, A. & Nieto, M. A. The increasing complexity of the Snail gene superfamily in metazoan evolution. *Trends Genet* **17**, 178–81 (2001).

893. Holland, P. W., Garcia-Fernandez, J., Williams, N. A. & Sidow, A. Gene duplications and the origins of vertebrate development. *Dev Suppl*, 125–33 (1994).

894. Smith, S. F. et al. Analyses of the extent of shared synteny and conserved gene orders between the genome of *Fugu rubripes* and human 20q. *Genome Res* **12**, 776–84 (2002).

895. Kataoka, H. et al. A novel snail-related transcription factor Smuc regulates basic helix-loop-helix transcription factor activities via specific E-box motifs. *Nucleic Acids Res* **28**, 626–33 (2000).

896. Katoh, M. Identification and characterization of human SNAIL3 (SNAI3) gene in silico. *Int J Mol Med* **11**, 383–8 (2003).

897. Manzanares, M., Blanco, M. J. & Nieto, M. A. *Snail3* orthologues in vertebrates: divergent members of the *Snail* zinc-finger gene family. *Dev Genes Evol* **214**, 47–53 (2004).

898. Langeland, J. A., Tomsa, J. M., Jackman, W. R., Jr. & Kimmel, C. B. An amphioxus *snail* gene: expression in paraxial mesoderm and neural plate suggests a conserved role in patterning the chordate embryo. *Dev Genes Evol* **208**, 569–77 (1998).

899. Corbo, J. C., Erives, A., Di Gregorio, A., Chang, A. & Levine, M. Dorsoventral patterning of the vertebrate neural tube is conserved in a protochordate. *Development* **124**, 2335–44 (1997).

900. Locascio, A., Manzanares, M., Blanco, M. J. & Nieto, M. A. Modularity and reshuffling of *Snail* and *Slug* expression during vertebrate evolution. *Proc Natl Acad Sci U S A* **99**, 16841–6 (2002).

901. McClintock, J. M., Carlson, R., Mann, D. M. & Prince, V. E. Consequences of Hox gene duplication in the vertebrates: an investigation of the zebrafish Hox paralogue group 1 genes. *Development* **128**, 2471–84 (2001).

902. McClintock, J. M., Kheirbek, M. A. & Prince, V. E. Knockdown of duplicated zebrafish *hoxb1* genes reveals distinct roles in hindbrain patterning and a novel mechanism of duplicate gene retention. *Development* **129**, 2339–54 (2002).

903. He, X. & Zhang, J. Rapid subfunctionalization accompanied by prolonged and substantial neofunctionalization in duplicate gene evolution. *Genetics* **169**, 1157–64 (2005).

904. Piatigorsky, J. Crystallin genes: specialization by changes in gene regulation may precede gene duplication. *J Struct Funct Genomics* **3**, 131–7 (2003).

905. Barabasi, A. L. & Oltvai, Z. N. Network biology: understanding the cell's functional organization. *Nat Rev Genet* **5**, 101–13 (2004).

906. Wagner, A. How the global structure of protein interaction networks evolves. *Proc R Soc Lond B Biol Sci* **270**, 457–66 (2003).

907. Van Rheede, T., Amons, R., Stewart, N. & De Jong, W. W. Lactate dehydrogenase A as a highly abundant eye lens protein in platypus (*Ornithorhynchus anatinus*): Upsilon (*υ*)-crystallin. *Mol Biol Evol* **20**, 994–8 (2003).

908. Hawkins, J. W. et al. Confirmation of assignment of the human α1-crystallin gene (*CRYA1*) to chromosome 21 with regional localization to q22.3. *Hum Genet* **76**, 375–80 (1987).

909. Ngo, J. T. et al. Assignment of the αB-crystallin gene to human chromosome 11. *Genomics* **5**, 665–9 (1989).

910. Brakenhoff, R. H. et al. Human αB-crystallin (CRYA2) gene mapped to chromosome 11q12-q23. *Hum Genet* **85**, 237–40 (1990).

911. Li, X., Wistow, G. J. & Piatigorsky, J. Linkage and expression of the argininosuccinate lyase/δ-crystallin genes of the duck: insertion of a CR1 element in the intergenic spacer. *Biochim Biophys Acta* **1261**, 25–34 (1995).

912. Nickerson, J. M. & Piatigorsky, J. Sequence of a complete chicken δ-crystallin cDNA. *Proc Natl Acad Sci U S A* **81**, 2611–5 (1984).

913. Nickerson, J. M. et al. Sequence of the chicken δ2 crystallin gene and its intergenic spacer. Extreme homology with the δ1 crystallin gene. *J Biol Chem* **261**, 552–7 (1986).

914. Piatigorsky, J., Norman, B. & Jones, R. E. Conservation of δ-crystallin gene structure between ducks and chickens. *J Mol Evol* **25**, 308–17 (1987).

915. Barbosa, P., Wistow, G. J., Cialkowski, M., Piatigorsky, J. & O'Brien, W. E. Expression of duck lens δ-crystallin cDNAs in yeast and bacterial hosts. δ2-Crystallin is an active argininosuccinate lyase. *J Biol Chem* **266**, 22319–22 (1991).

916. Kondoh, H., Araki, I., Yasuda, K., Matsubasa, T. & Mori, M. Expression of the chicken "δ2-crystalline" gene in mouse cells: evidence for encoding of argininosuccinate lyase. *Gene* **99**, 267–71 (1991).

917. Li, X., Zelenka, P. S. & Piatigorsky, J. Differential expression of the two δ-crystallin genes in lens and non-lens tissues: shift favoring δ2 expression from embryonic to adult chickens. *Dev Dyn* **196**, 114–23 (1993).

918. Benassayag, C. et al. Evidence for a direct functional antagonism of the selector genes *proboscipedia* and *eyeless* in *Drosophila* head development. *Development* **130**, 575–86 (2003).

919. Piatigorsky, J. & Horwitz, J. Characterization and enzyme activity of argininosuccinate lyase/ δ-crystallin of the embryonic duck lens. *Biochim Biophys Acta* **1295**, 158–64 (1996).

920. Lee, H. J., Chiou, S. H. & Chang, G. G. Biochemical characterization and kinetic analysis of duck δ-crystallin with endogenous argininosuccinate lyase activity. *Biochem J* **283** (**Pt 2**), 597–603 (1992).

921. Lee, H. J., Lin, C. C., Chiou, S. H. & Chang, G. G. Characterization of the multiple forms of duck lens δ-crystallin with endogenous argininosuccinate lyase activity. *Arch Biochem Biophys* **314**, 31–8 (1994).

922. Lin, C. W. & Chiou, S. H. Sequence analysis of pigeon δ-crystallin gene and its deduced primary structure. Comparison of avian δ-crystallins with and without endogenous argininosuccinate lyase activity. *FEBS Lett* **311**, 276–80 (1992).

923. Yu, C. W. & Chiou, S. H. Facile cloning and sequence analysis of goose δ-crystallin gene based on polymerase chain reaction. *Biochem Biophys Res Commun* **192**, 948–53 (1993).

924. Abu-Abed, M. et al. Structural comparison of the enzymatically active and inactive forms of δ-crystallin and the role of histidine 91. *Biochemistry* **36**, 14012–22 (1997).

925. Williams, L. A. & Piatigorsky, J. Comparative and evolutionary aspects of δ-crystallin in the vertebrate lens. *Eur J Biochem* **100**, 349–57 (1979).

925a. Ji, X., von Rosenvinge, E. C., Johnson, W. W., Tomarev, S. I., Piatigorsky, J., Armstrong, R. N. & Gilliland, G. L. Three-dimensional structure, catalytic properties, and evolution of a sigma class glutathione transferase from squid, a progenitor of the lens S-crystallins of cephalopods. *Biochemistry* **34**, 5317–28 (1995).

926. Watts, D. J. & Strogatz, S. H. Collective dynamics of "small-world" networks. *Nature* **393**, 440–2 (1998).

927. Hartwell, L. H., Hopfield, J. J., Leibler, S. & Murray, A. W. From molecular to modular cell biology. *Nature* **402**, C47–52 (1999).

928. Oltvai, Z. N. & Barabasi, A. L. Systems biology. Life's complexity pyramid. *Science* **298**, 763–4 (2002).

929. Papin, J. A., Price, N. D., Wiback, S. J., Fell, D. A. & Palsson, B. O. Metabolic pathways in the post-genome era. *Trends Biochem Sci* **28**, 250–8 (2003).

930. Ge, H., Walhout, A. J. & Vidal, M. Integrating "omic" information: a bridge between genomics and systems biology. *Trends Genet* **19**, 551–60 (2003).

931. Stuart, J. M., Segal, E., Koller, D. & Kim, S. K. A gene-coexpression network for global discovery of conserved genetic modules. *Science* **302**, 249–55 (2003).

932. Quackenbush, J. Genomics. Microarrays—guilt by association. *Science* **302**, 240–1 (2003).

933. Bader, G. D. & Hogue, C. W. Analyzing yeast protein-protein interaction data obtained from different sources. *Nat Biotechnol* **20**, 991–7 (2002).

934. Bader, G. D., Betel, D. & Hogue, C. W. BIND: the Biomolecular Interaction Network Database. *Nucleic Acids Res* **31**, 248–50 (2003).

935. Yook, S. H., Oltvai, Z. N. & Barabasi, A. L. Functional and topological characterization of protein interaction networks. *Proteomics* **4**, 928–42 (2004).

936. Butland, G. et al. Interaction network containing conserved and essential protein complexes in *Escherichia coli. Nature* **433**, 531–7 (2005).

937. Milo, R. et al. Network motifs: simple building blocks of complex networks. *Science* **298**, 824–7 (2002).

938. Milo, R. et al. Superfamilies of evolved and designed networks. *Science* **303**, 1538–42 (2004).

939. Dobrin, R., Beg, Q. K., Barabasi, A. L. & Oltvai, Z. N. Aggregation of topological motifs in the *Escherichia coli* transcriptional regulatory network. *BMC Bioinformatics* **5**, 10 (2004).

940. Tong, A. H. et al. Global mapping of the yeast genetic interaction network. *Science* **303,** 808–13 (2004).

941. Li, S. et al. A map of the interactome network of the metazoan *C. elegans*. *Science* **303,** 540–3 (2004).

942. Han, J. D. et al. Evidence for dynamically organized modularity in the yeast protein-protein interaction network. *Nature* **430,** 88–93 (2004).

943. Jeong, H., Mason, S. P., Barabasi, A. L. & Oltvai, Z. N. Lethality and centrality in protein networks. *Nature* **411,** 41–2 (2001).

944. Hartwell, L. Genetics. Robust interactions. *Science* **303,** 774–5 (2004).

945. Wagner, A. *Robustness and evolvability in living systems.* (Princeton University Press, Princeton, 2005).

946. Fraser, H. B., Hirsh, A. E., Steinmetz, L. M., Scharfe, C. & Feldman, M. W. Evolutionary rate in the protein interaction network. *Science* **296,** 750–2 (2002).

947. Fraser, H. B., Wall, D. P. & Hirsh, A. E. A simple dependence between protein evolution rate and the number of protein-protein interactions. *BMC Evol Biol* **3,** 11 (2003).

948. Jordan, I. K., Wolf, Y. I. & Koonin, E. V. No simple dependence between protein evolution rate and the number of protein-protein interactions: only the most prolific interactors tend to evolve slowly. *BMC Evol Biol* **3,** 1 (2003).

949. Bloom, J. D. & Adami, C. Evolutionary rate depends on number of protein-protein interactions independently of gene expression level: response. *BMC Evol Biol* **4,** 14 (2004).

950. Barabasi, A. L. & Albert, R. Emergence of scaling in random networks. *Science* **286,** 509–12 (1999).

951. Eisenberg, E. & Levanon, E. Y. Preferential attachment in the protein network evolution. *Phys Rev Lett* **91,** 138701 (2003).

952. Qin, H., Lu, H. H., Wu, W. B. & Li, W. H. Evolution of the yeast protein interaction network. *Proc Natl Acad Sci U S A* **100,** 12820–4 (2003).

953. Kemmeren, P. & Holstege, F. C. Integrating functional genomics data. *Biochem Soc Trans* **31,** 1484–7 (2003).

954. Batada, N. N., Shepp, L. A. & Siegmund, D. O. Stochastic model of protein-protein interaction: why signaling proteins need to be colocalized. *Proc Natl Acad Sci U S A* **101,** 6445–9 (2004).

955. Ghaemmaghami, S. et al. Global analysis of protein expression in yeast. *Nature* **425,** 737–41 (2003).

956. Wohlschlegel, J. A. & Yates, J. R. Proteomics: where's Waldo in yeast? *Nature* **425,** 671–2 (2003).

957. Kirschner, M. & Gerhart, J. Evolvability. *Proc Natl Acad Sci U S A* **95,** 8420–7 (1998).

958. Earl, D. J. & Deem, M. W. Evolvability is a selectable trait. *Proc Natl Acad Sci U S A* **101,** 11531–6 (2004).

959. Radman, M., Matic, I. & Taddei, F. Evolution of evolvability. *Ann N Y Acad Sci* **870,** 146–55 (1999).

960. Copley, S. D. Enzymes with extra talents: moonlighting functions and catalytic promiscuity. *Curr Opin Chem Biol* **7,** 265–72 (2003).

961. Miller, B. G. & Raines, R. T. Identifying latent enzyme activities: substrate ambiguity within modern bacterial sugar kinases. *Biochemistry* **43,** 6387–92 (2004).

962. Bornscheuer, U. T. & Kazlauskas, R. J. Catalytic promiscuity in biocatalysis: using old enzymes to form new bonds and follow new pathways. *Angew Chem Int Ed Engl* **43,** 6032–40 (2004).

963. Matsumura, I. & Ellington, A. D. In vitro evolution of β-glucuronidase into a β-galactosidase proceeds through non-specific intermediates. *J Mol Biol* **305,** 331–9 (2001).

964. Wouters, M. A., Liu, K., Riek, P. & Husain, A. A despecialization step underlying evolution of a family of serine proteases. *Mol Cell* **12,** 343–54 (2003).

965. Taverna, D. M. & Goldstein, R. A. Why are proteins marginally stable? *Proteins* **46,** 105–9 (2002).

966. Taverna, D. M. & Goldstein, R. A. Why are proteins so robust to site mutations? *J Mol Biol* **315,** 479–84 (2002).

967. Dyson, H. J. & Wright, P. E. Intrinsically unstructured proteins and their functions. *Nat Rev Mol Cell Biol* **6,** 197–208 (2005).

968. Bloom, J. D., Wilke, C. O., Arnold, F. H. & Adami, C. Stability and the evolvability of function in a model protein. *Biophys J* **86,** 2758–64 (2004).

969. Lindorff-Larsen, K., Rogen, P., Paci, E., Vendruscolo, M. & Dobson, C. M. Protein folding and the organization of the protein topology universe. *Trends Biochem Sci* **30,** 13–9 (2005).

970. James, L. C., Roversi, P. & Tawfik, D. S. Antibody multispecificity mediated by conformational diversity. *Science* **299,** 1362–7 (2003).

971. Foote, J. Immunology. Isomeric antibodies. *Science* **299,** 1327–8 (2003).

972. James, L. C. & Tawfik, D. S. The specificity of cross-reactivity: promiscuous antibody binding involves specific hydrogen bonds rather than nonspecific hydrophobic stickiness. *Protein Sci* **12,** 2183–93 (2003).

973. Lindorff-Larsen, K., Best, R. B., Depristo, M. A., Dobson, C. M. & Vendruscolo, M. Simultaneous determination of protein structure and dynamics. *Nature* **433,** 128–32 (2005).

974. Socolich, M. et al. Evolutionary information for specifying a protein fold. *Nature* **437,** 512–8 (2005).

975. Kelly, J. W. Structural biology: form and function instructions. *Nature* **437,** 486–7 (2005).

976. Russ, W. P., Lowery, D. M., Mishra, P., Yaffe, M. B. & Ranganathan, R. Natural-like function in artificial WW domains. *Nature* **437,** 579–83 (2005).

977. Kondrashov, F. A. In search of the limits of evolution. *Nat Genet* **37,** 9–10 (2005).

978. Khaitovich, P. et al. Parallel patterns of evolution in the genomes and transcriptomes of humans and chimpanzees. *Science* **309,** 1850–4 (2005).

979. Khaitovich, P., Paabo, S. & Weiss, G. Toward a neutral evolutionary model of gene expression. *Genetics* **170,** 929–39 (2005).

980. Jordan, I. K., Marino-Ramirez, L. & Koonin, E. V. Evolutionary significance of gene expression divergence. *Gene* **345**, 119–26 (2005).

981. Nooren, I. M. & Thornton, J. M. Diversity of protein-protein interactions. *Embo J* **22**, 3486–92 (2003).

982. Beckett, D. Functional switches in transcription regulation; molecular mimicry and plasticity in protein-protein interactions. *Biochemistry* **43**, 7983–91 (2004).

983. Ye, L. & Haupt, K. Molecularly imprinted polymers as antibody and receptor mimics for assays, sensors and drug discovery. *Anal Bioanal Chem* **378**, 1887–97 (2004).

984. Shapiro, L. β-Catenin and its multiple partners: promiscuity explained. *Nat Struct Biol* **8**, 484–7 (2001).

985. Gooding, J. M., Yap, K. L. & Ikura, M. The cadherin-catenin complex as a focal point of cell adhesion and signalling: new insights from three-dimensional structures. *Bioessays* **26**, 497–511 (2004).

986. Fourneau, J. M., Bach, J. M., van Endert, P. M. & Bach, J. F. The elusive case for a role of mimicry in autoimmune diseases. *Mol Immunol* **40**, 1095–102 (2004).

986a. Volfson, D., Marciniak, J., Blake, W. J., Ostroff, N., Tsimring, L. S. & Hasty, J. Origins of extrinsic variability in eukaryotic gene expression. *Nature* **439**, 861–4 (2006).

987. Samoilov, M., Plyasunov, S. & Arkin, A. P. Stochastic amplification and signaling in enzymatic futile cycles through noise-induced bistability with oscillations. *Proc Natl Acad Sci U S A* **102**, 2310–5 (2005).

988. Raser, J. M. & O'Shea, E. K. Noise in gene expression: origins, consequences, and control. *Science* **309**, 2010–3 (2005).

989. Ozbudak, E. M., Thattai, M., Kurtser, I., Grossman, A. D. & van Oudenaarden, A. Regulation of noise in the expression of a single gene. *Nat Genet* **31**, 69–73 (2002).

990. Elowitz, M. B., Levine, A. J., Siggia, E. D. & Swain, P. S. Stochastic gene expression in a single cell. *Science* **297**, 1183–6 (2002).

991. Blake, W. J., M, K. A., Cantor, C. R. & Collins, J. J. Noise in eukaryotic gene expression. *Nature* **422**, 633–7 (2003).

992. Raser, J. M. & O'Shea, E. K. Control of stochasticity in eukaryotic gene expression. *Science* **304**, 1811–4 (2004).

993. Paulsson, J. Summing up the noise in gene networks. *Nature* **427**, 415–8 (2004).

994. Bird, A. P. Gene number, noise reduction and biological complexity. *Trends Genet* **11**, 94–100 (1995).

995. McAdams, H. H. & Arkin, A. It's a noisy business! Genetic regulation at the nanomolar scale. *Trends Genet* **15**, 65–9 (1999).

996. Holstege, F. C. et al. Dissecting the regulatory circuitry of a eukaryotic genome. *Cell* **95**, 717–28 (1998).

997. Holstege, F. C. & Young, R. A. Transcriptional regulation: contending with complexity. *Proc Natl Acad Sci U S A* **96**, 2–4 (1999).

998. Steger, D. J., Haswell, E. S., Miller, A. L., Wente, S. R. & O'Shea, E. K. Regulation of chromatin remodeling by inositol polyphosphates. *Science* **299**, 114–6 (2003).

999. Shen, X., Xiao, H., Ranallo, R., Wu, W. H. & Wu, C. Modulation of ATP-dependent chromatin-remodeling complexes by inositol polyphosphates. *Science* **299**, 112–4 (2003).

1000. Kemkemer, R., Schrank, S., Vogel, W., Gruler, H. & Kaufmann, D. Increased noise as an effect of haploinsufficiency of the tumor-suppressor gene neurofibromatosis type 1 in vitro. *Proc Natl Acad Sci U S A* **99**, 13783–8 (2002).

1001. Marino, M. Biography of Erin K. O'Shea. *Proc Natl Acad Sci U S A* (2004).

1002. Rifkin, S. A., Kim, J. & White, K. P. Evolution of gene expression in the *Drosophila melanogaster* subgroup. *Nat Genet* **33**, 138–44 (2003).

1003. Steinmetz, L. M. et al. Dissecting the architecture of a quantitative trait locus in yeast. *Nature* **416**, 326–30 (2002).

1004. Brem, R. B., Yvert, G., Clinton, R. & Kruglyak, L. Genetic dissection of transcriptional regulation in budding yeast. *Science* **296**, 752–5 (2002).

1005. Yvert, G. et al. Trans-acting regulatory variation in *Saccharomyces cerevisiae* and the role of transcription factors. *Nat Genet* **35**, 57–64 (2003).

1006. Oleksiak, M. F., Churchill, G. A. & Crawford, D. L. Variation in gene expression within and among natural populations. *Nat Genet* **32**, 261–6 (2002).

1007. Enard, W. et al. Intra- and interspecific variation in primate gene expression patterns. *Science* **296**, 340–3 (2002).

1008. Cheung, V. G. et al. Natural variation in human gene expression assessed in lymphoblastoid cells. *Nat Genet* **33**, 422–5 (2003).

1009. Yan, H., Yuan, W., Velculescu, V. E., Vogelstein, B. & Kinzler, K. W. Allelic variation in human gene expression. *Science* **297**, 1143 (2002).

1010. Pastinen, T. & Hudson, T. J. Cis-acting regulatory variation in the human genome. *Science* **306**, 647–50 (2004).

1011. Cox, N. J. Human genetics: an expression of interest. *Nature* **430**, 733–4 (2004).

1012. Morley, M. et al. Genetic analysis of genome-wide variation in human gene expression. *Nature* **430**, 743–7 (2004).

1013. Zuckerkandl, E. & Pauling, L. Evolutionary divergence and convergence in proteins, in *Evolving Genes and Proteins* (eds. Bryson, V. & Vogel, H. J.) 97–166 (Academic Press, New York, 1965).

1014. Margoliash, E. Primary structure and evolution of cytochrome C. *Proc Natl Acad Sci U S A* **50**, 672–9 (1963).

1015. Kimura, M. Evolutionary rate at the molecular level. *Nature* **217**, 624–6 (1968).

1016. Kimura, M. & Ota, T. On the rate of molecular evolution. *J Mol Evol* **1**, 1–17 (1971).

1017. Bromham, L. & Penny, D. The modern molecular clock. *Nat Rev Genet* **4**, 216–24 (2003).

1018. Easteal, S., Collet, C. & Betty, D. *The Mammalian Molecular Clock* (R.G. Landes Company, Austin, TX, 1995).

1019. Gillespie, J. H. *The Causes of Molecular Evolution* (Oxford University Press, New York, 1991).

1020. Simpson, G. G. *Tempo and Mode in Evolution* (Columbia University Press, New York, 1944).

1021. Douzery, E. J., Snell, E. A., Bapteste, E., Delsuc, F. & Philippe, H. The timing of eukaryotic evolution: Does a relaxed molecular clock reconcile proteins and fossils? *Proc Natl Acad Sci U S A* **101**, 15386–15391 (2004).

1022. Ayala, F. J. Molecular clock mirages. *Bioessays* **21**, 71–5 (1999).

1023. Nei, M. & Jin, L. Variances of the average numbers of nucleotide substitutions within and between populations. *Mol Biol Evol* **6**, 290–300 (1989).

1024. Arbogast, B. S., Edwards, S. V., Wakelely, J., Beerli, P. & Slowindski, J. B. Estimating divergence times from molecular data on phylogenetic and population genetic timescales. *Annu Rev Ecol Syst* **33**, 707–40 (2002).

1025. Jordan, I. K., Marino-Ramirez, L. & Koonin, E. V. Evolutionary significance of gene expression divergence. *Gene* (2004).

1026. Yanai, I., Graur, D. & Ophir, R. Incongruent expression profiles between human and mouse orthologous genes suggest widespread neutral evolution of transcription control. *Omics* **8**, 15–24 (2004).

1027. Khaitovich, P. et al. A neutral model of transcriptome evolution. *PLoS Biol* **2**, E132 (2004).

1027a. Cutler, D. J. Understanding the overdispersed molecular clock. *Genetics* **154**, 1403–17 (2000).

1028. Ayala, F. J., Barrio, E. & Kwiatowski, J. Molecular clock or erratic evolution? A tale of two genes. *Proc Natl Acad Sci U S A* **93**, 11729–34 (1996).

1029. Ayala, F. J. Vagaries of the molecular clock. *Proc Natl Acad Sci U S A* **94**, 7776–83 (1997).

1030. Fitch, W. M. & Markowitz, E. An improved method for determining codon variability in a gene and its application to the rate of fixation of mutations in evolution. *Biochem Genet* **4**, 579–93 (1970).

1031. Fitch, W. M. & Ayala, F. J. The superoxide dismutase molecular clock revisited. *Proc Natl Acad Sci U S A* **91**, 6802–7 (1994).

1032. Jolles, J. et al. Episodic evolution in the stomach lysozymes of ruminants. *J Mol Evol* **28**, 528–35 (1989).

1033. Messier, W. & Stewart, C. B. Episodic adaptive evolution of primate lysozymes. *Nature* **385**, 151–4 (1997).

1034. Rawls, A. et al. Overlapping functions of the myogenic bHLH genes *MRF4* and *MyoD* revealed in double mutant mice. *Development* **125**, 2349–58 (1998).

1035. Kassar-Duchossoy, L. et al. Mrf4 determines skeletal muscle identity in *Myf5:Myod* double-mutant mice. *Nature* **431**, 466–71 (2004).

1036. Ain, R., Dai, G., Dunmore, J. H., Godwin, A. R. & Soares, M. J. A prolactin family paralog regulates reproductive adaptations to a physiological stressor. *Proc Natl Acad Sci U S A* **101**, 16543–8 (2004).

1037. Green, J. A. Defining the function of a prolactin gene family member. *Proc Natl Acad Sci U S A* **101**, 16397–8 (2004).

1038. Conant, G. C. & Wagner, A. Duplicate genes and robustness to transient gene knock-downs in *Caenorhabditis elegans*. *Proc Biol Sci* **271**, 89–96 (2004).

1039. Cadigan, K. M. & Nusse, R. Wnt signaling: a common theme in animal development. *Genes Dev* **11**, 3286–305 (1997).

1040. Oyama, T. et al. A truncated β-catenin disrupts the interaction between E-cadherin and α-catenin: a cause of loss of intercellular adhesiveness in human cancer cell lines. *Cancer Res* **54**, 6282–7 (1994).

1041. Veeman, M. T., Axelrod, J. D. & Moon, R. T. A second canon. Functions and mechanisms of β-catenin-independent Wnt signaling. *Dev Cell* **5**, 367–77 (2003).

1042. Tian, Q. et al. Proteomic analysis identifies that 14–3–3ζ interacts with β-catenin and facilitates its activation by Akt. *Proc Natl Acad Sci U S A* **101**, 15370–5 (2004).

1043. Haegel, H. et al. Lack of β-catenin affects mouse development at gastrulation. *Development* **121**, 3529–37 (1995).

1044. Huelsken, J. et al. Requirement for β-catenin in anterior-posterior axis formation in mice. *J Cell Biol* **148**, 567–78 (2000).

1045. Funayama, N., Fagotto, F., McCrea, P. & Gumbiner, B. M. Embryonic axis induction by the armadillo repeat domain of β-catenin: evidence for intracellular signaling. *J Cell Biol* **128**, 959–68 (1995).

1046. Cattelino, A. et al. The conditional inactivation of the β-catenin gene in endothelial cells causes a defective vascular pattern and increased vascular fragility. *J Cell Biol* **162**, 1111–22 (2003).

1047. Soshnikova, N. et al. Genetic interaction between Wnt/β-catenin and BMP receptor signaling during formation of the AER and the dorsal-ventral axis in the limb. *Genes Dev* **17**, 1963–8 (2003).

1048. Barrow, J. R. et al. Ectodermal Wnt3/β-catenin signaling is required for the establishment and maintenance of the apical ectodermal ridge. *Genes Dev* **17**, 394–409 (2003).

1049. Mucenski, M. L. et al. β-Catenin is required for specification of proximal/distal cell fate during lung morphogenesis. *J Biol Chem* **278**, 40231–8 (2003).

1050. Xu, Y., Banerjee, D., Huelsken, J., Birchmeier, W. & Sen, J. M. Deletion of β-catenin impairs T cell development. *Nat Immunol* **4**, 1177–82 (2003).

1051. Cobas, M. et al. β-Catenin is dispensable for hematopoiesis and lymphopoiesis. *J Exp Med* **199**, 221–9 (2004).

1052. Bamji, S. X. et al. Role of β-catenin in synaptic vesicle localization and presynaptic assembly. *Neuron* **40**, 719–31 (2003).

1053. Austin, C. P. et al. The knockout mouse project. *Nat Genet* **36**, 921–4 (2004).

1054. Margulis, L. & Sagan, D. *Acquiring Genomes. A Theory of the Origins of Species.* (Basic Books, New York, 2002).

1055. Doolittle, W. F. Lateral genomics. *Trends Cell Biol* **9**, M5–8 (1999).

1056. Ochman, H., Lawrence, J. G. & Groisman, E. A. Lateral gene transfer and the nature of bacterial innovation. *Nature* **405**, 299–304 (2000).

1057. Lawrence, J. G. & Ochman, H. Reconciling the many faces of lateral gene transfer. *Trends Microbiol* **10**, 1–4 (2002).

1058. Daubin, V., Moran, N. A. & Ochman, H. Phylogenetics and the cohesion of bacterial genomes. *Science* **301**, 829–32 (2003).

1059. Boucher, Y. et al. Lateral gene transfer and the origins of prokaryotic groups. *Annu Rev Genet* **37**, 283–328 (2003).

1060. Beiko, R. G., Harlow, T. J. & Ragan, M. A. Highways of gene sharing in prokaryotes. *Proc Natl Acad Sci U S A* **102**, 14332–7 (2005).

1061. Woese, C. R. A new biology for a new century. *Microbiol Mol Biol Rev* **68**, 173–86 (2004).

1062. de Duve, C. The onset of selection. *Nature* **433**, 581–2 (2005).

1063. Stanhope, M. J. et al. Phylogenetic analyses do not support horizontal gene transfers from bacteria to vertebrates. *Nature* **411**, 940–4 (2001).

1064. Kurland, C. G., Canback, B. & Berg, O. G. Horizontal gene transfer: a critical view. *Proc Natl Acad Sci U S A* **100**, 9658–62 (2003).

1065. Doolittle, W. F. et al. How big is the iceberg of which organellar genes in nuclear genomes are but the tip? *Philos Trans R Soc Lond B Biol Sci* **358**, 39–57; discussion 57–8 (2003).

1066. Koonin, E. V. Horizontal gene transfer: the path to maturity. *Mol Microbiol* **50**, 725–7 (2003).

1067. Daubin, V. & Ochman, H. Start-up entities in the origin of new genes. *Curr Opin Genet Dev* **14**, 616–9 (2004).

1068. Britten, R. J. Coding sequences of functioning human genes derived entirely from mobile element sequences. *Proc Natl Acad Sci U S A* **101**, 16825–30 (2004).

1069. Hastings, P. J., Rosenberg, S. M. & Slack, A. Antibiotic-induced lateral transfer of antibiotic resistance. *Trends Microbiol* **12**, 401–4 (2004).

1070. Iyer, L. M., Aravind, L., Coon, S. L., Klein, D. C. & Koonin, E. V. Evolution of cell-cell signaling in animals: did late horizontal gene transfer from bacteria have a role? *Trends Genet* **20**, 292–9 (2004).

1071. Andersson, S. G., Karlberg, O., Canback, B. & Kurland, C. G. On the origin of mitochondria: a genomics perspective. *Philos Trans R Soc Lond B Biol Sci* **358**, 165–77; discussion 177–9 (2003).

1072. Tomitani, A. et al. Chlorophyll b and phycobilins in the common ancestor of cyanobacteria and chloroplasts. *Nature* **400**, 159–62 (1999).

1073. Rujan, T. & Martin, W. How many genes in *Arabidopsis* come from cyanobacteria? An estimate from 386 protein phylogenies. *Trends Genet* **17**, 113–20 (2001).

1074. Martin, W. et al. Evolutionary analysis of *Arabidopsis*, cyanobacterial, and chloroplast genomes reveals plastid phylogeny and thousands of cyanobacterial genes in the nucleus. *Proc Natl Acad Sci U S A* **99**, 12246–51 (2002).

1075. Bergthorsson, U., Adams, K. L., Thomason, B. & Palmer, J. D. Widespread horizontal transfer of mitochondrial genes in flowering plants. *Nature* **424**, 197–201 (2003).

1076. Bergthorsson, U., Richardson, A. O., Young, G. J., Goertzen, L. R. & Palmer, J. D. Massive horizontal transfer of mitochondrial genes from diverse land plant donors to the basal angiosperm *Amborella*. *Proc Natl Acad Sci U S A* **101**, 17747–52 (2004).

1077. Archibald, J. M., Rogers, M. B., Toop, M., Ishida, K. & Keeling, P. J. Lateral gene transfer and the evolution of plastid-targeted proteins in the secondary plastid-containing alga *Bigelowiella natans*. *Proc Natl Acad Sci U S A* **100**, 7678–83 (2003).

1078. Raymond, J. & Blankenship, R. E. Horizontal gene transfer in eukaryotic algal evolution. *Proc Natl Acad Sci U S A* **100**, 7419–20 (2003).

1079. Keeling, P. J., Archibald, J. M., Fast, N. M. & Palmer, J. D. Comment on "The evolution of modern eukaryotic phytoplankton". *Science* **306**, 2191; author reply 2191 (2004).

1080. Doolittle, W. F. You are what you eat: a gene transfer ratchet could account for bacterial genes in eukaryotic nuclear genomes. *Trends Genet* **14**, 307–11 (1998).

1081. Gray, M. W. Evolutionary biology: the hydrogenosome's murky past. *Nature* **434**, 29–31 (2005).

1082. Boxma, B. et al. An anaerobic mitochondrion that produces hydrogen. *Nature* **434**, 74–9 (2005).

1083. Andersson, J. O., Sjogren, A. M., Davis, L. A., Embley, T. M. & Roger, A. J. Phylogenetic analyses of diplomonad genes reveal frequent lateral gene transfers affecting eukaryotes. *Curr Biol* **13**, 94–104 (2003).

1084. Gelvin, S. B. Agricultural biotechnology: gene exchange by design. *Nature* **433**, 583–4 (2005).

1085. Broothaerts, W. et al. Gene transfer to plants by diverse species of bacteria. *Nature* **433**, 629–33 (2005).

1086. Zardoya, R., Ding, X., Kitagawa, Y. & Chrispeels, M. J. Origin of plant glycerol transporters by horizontal gene transfer and functional recruitment. *Proc Natl Acad Sci U S A* **99**, 14893–6 (2002).

1087. Matthysse, A. G. et al. A functional cellulose synthase from ascidian epidermis. *Proc Natl Acad Sci U S A* **101**, 986–91 (2004).

1088. Andersson, J. O. & Roger, A. J. Evolution of glutamate dehydrogenase genes: evidence for lateral gene transfer within and between prokaryotes and eukaryotes. *BMC Evol Biol* **3**, 14 (2003).

1089. Keeling, P. J. & Palmer, J. D. Lateral transfer at the gene and subgenic levels in the evolution of eukaryotic enolase. *Proc Natl Acad Sci U S A* **98**, 10745–50 (2001).

1090. Qian, Q. & Keeling, P. J. Diplonemid glyceraldehyde-3-phosphate dehydrogenase (GAPDH) and prokaryote-to-eukaryote lateral gene transfer. *Protist* **152**, 193–201 (2001).

1091. Wolf, Y. I. & Koonin, E. V. Origin of an animal mitochondrial DNA polymerase subunit via lineage-specific acquisition of a glycyl-tRNA synthetase from bacteria of the *Thermus-Deinococcus* group. *Trends Genet* **17**, 431–3 (2001).

1092. Wolf, Y. I., Aravind, L., Grishin, N. V. & Koonin, E. V. Evolution of aminoacyl-tRNA synthetases—analysis of unique domain architectures and phylogenetic trees reveals a complex history of horizontal gene transfer events. *Genome Res* **9**, 689–710 (1999).

1093. Woese, C. R., Olsen, G. J., Ibba, M. & Soll, D. Aminoacyl-tRNA synthetases, the genetic code, and the evolutionary process. *Microbiol Mol Biol Rev* **64**, 202–36 (2000).

1094. Koonin, E. V. & Aravind, L. Origin and evolution of eukaryotic apoptosis: the bacterial connection. *Cell Death Differ* **9**, 394–404 (2002).

1094a. Stach, R. *Kafka. The Decisive Years* (Harcourt, Orlando, Florida, 2005).

1095. Fisher, R. A. *The Genetical Theory of Natural Selection* (Clarendon Press, Oxford, 1930).

1096. Wright, S. Evolution in Mendelian populations. *Genetics* **16**, 97–159 (1931).

1097. Haldane, J. B. S. *The Causes of Evolution* (Longmans and Green, London, 1932).

1098. Dobzhansky, T. *Genetics and the Origin of Species* (Columbia University Press, New York, 1937).

1099. Huxley, J. S. *Evolution: The Modern Synthesis* (Allen and Unwin, London, 1942).

1100. Mayr, E. *Animal Species and Evolution* (Harvard University Press, Cambridge, Massachusetts, 1963).

1101. King, J. L. & Jukes, T. H. Non-Darwinian evolution. *Science* **164**, 788–98 (1969).

1102. Rutherford, S. L. & Lindquist, S. Hsp90 as a capacitor for morphological evolution. *Nature* **396**, 336–42 (1998).

1103. Queitsch, C., Sangster, T. A. & Lindquist, S. Hsp90 as a capacitor of phenotypic variation. *Nature* **417**, 618–24 (2002).

1104. Behe, M. *Darwin's Black Box: The Biochemical Challenge to Evolution.* (Free Press, New York, 1996).

1105. Orr, H. A. Annals of Science Master Planned: Why intelligent design isn't. *The New Yorker* (2005).

1106. Wolpert, L. *The Unnatural Nature of Science* (Harvard University Press, Cambridge, Massachusetts, 1992).

1107. Ruse, M. *The Evolution-Creation Struggle* (Harvard University Press, Cambridge, Massachusetts, 2005).

1108. Fishman, M. C. & Porter, J. A. Pharmaceuticals: a new grammar for drug discovery. *Nature* **437**, 491–3 (2005).

1109. Schlosser, G., Wagner, G. P. (eds). *Modularity in Development and Evolution* (The University of Chicago Press, Chicago, 2004).

1110. Piatigorsky, J. Multifunctional lens crystallins and corneal enzymes. More than meets the eye. *Ann N Y Acad Sci* **842**, 7–15 (1998).

1111. Hejtmancik, J. F., Kaiser, M. I. & Piatigorsky, J. Molecular biology and inherited disorders of the eye lens, in *The Metabolic & Molecular Bases of Inherited Disease* (eds. Scriver, C. R., Beaudet, A. L., Sly, S. S. & Valle, D.) 6033–61 (McGraw-Hill, New York, 2001).

1112. Fernald, R. D. & Wright, S. E. Maintenance of optical quality during crystalline lens growth. *Nature* **301**, 618–20 (1983).

1113. Haynes, J. I., 2nd, Duncan, M. K. & Piatigorsky, J. Spatial and temporal activity of the αB-crystallin/small heat shock protein gene promoter in transgenic mice. *Dev Dyn* **207**, 75–88 (1996).

1114. Cheng, C.-H. C. Freezing avoidance in polar fishes, in *Encyclopedia of Life Support Systems (EOLSS) - Theme 6.73 Extremophiles* (ed. Gerday, C.) (Developed under the Auspices of UNESCO, Eols Publishers, United Kingdom, 2003) (http://www.eolss.net).

1115. Zhang, D. Q. et al. Significance of conservative asparagine residues in the thermal hysteresis activity of carrot antifreeze protein. *Biochem J* **377**, 589–95 (2004).

1116. Kuiper, M. J., Davies, P. L. & Walker, V. K. A theoretical model of a plant antifreeze protein from *Lolium perenne*. *Biophys J* **81**, 3560–5 (2001).

1117. Taylor, J. S. & Raes, J. Duplication and divergence: The evolution of new genes and old ideas. *Annu Rev Genet* **38**, 615–43 (2004).

1118. Causier, B., Castillo, R., Zhou, J., Ingram, R., Xue, Y., Schwarz-Sommer, Z. & Davies, B. Evolution in action: following function in duplicated floral homeotic genes. *Curr Biol* **15**, 1508–12 (2005).

1119. Van de Peer, Y. Evolutionary genetics: when duplicated genes don't stick to the rules. *Heredity* **96**, 204–5 (2006).

1120. Qiu, J. Unfinished symphony. *Nature* **441**, 143–5 (2006)

1121. Pearson, H. What is a gene? *Nature* **441**, 399–401 (2006).

1122. David, L., Huber, W., Granovskaia, M., Toedling, J., Palm, C. J., Bofkin, L., Jones, T., Davis, R. W. & Steinmetz, L. M. A high-resolution map of transcription in the yeast genome. *Proc Natl Acad Sci U S A* **103**, 5320–5 (2006).

1123. Sanchez-Elsner, T., Gou, D., Kremmer, E. & Sauer, F. Noncoding RNAs of Trithorax response elements recruit *Drosophila* Ash1 to Ultrabithorax. *Science* **311**, 1118–23 (2006).

1124. Berezikov, E., Cuppen, E. & Plasterk, R. H. A. Approaches to microRNA discovery. *Nature Genetics Suppl* **38**, S2-S7 (2006).

1125. Giraldez, A. J., Mishima, Y., Rihel, J., Grocock, R. J., Van Dongen, S., Inoue, K., Enright, A. J. & Schier, A. F. Zebrafish miR-430 promotes deadenylation and clearance of maternal mRNAs. *Science* **312**, 75–9 (2006).

1126. Reichow, S. & Varani, G. RNA switches function. *Nature* **441**, 1054–5 (2006).

1127. Serganov, A., Polonskaia, A., Phan, A. T., Breaker, R. R. & Patel, D. J. Structural basis for gene regulation by a thiamine pyrophosphate-sensing riboswitch. *Nature* **441**, 1167–71 (2006).

1128. Batey, R. T. Structures for regulatory elements in mRNAs. *Curr Opin Struct Biol* **16**, 299–306 (2006).

1129. Chandler, V. L. & Stam, M. Chromatin conversations: Mechanisms and implications of paramutation. *Nature Rev Genet* **5**, 532–44 (2004).

1130. Rassoulzadegan, M., Grandjean, V., Gounon, P., Vincent, S., Gillot, I. & Cuzin, F. RNA-mediated non-mendelian inheritance of an epigenetic change in the mouse. *Nature* **441**, 469–74 (2006).

1131. Soloway, P. D. Paramutable possibilities. *Nature* **441**, 413–4 (2006).

1132. Bernstein, E. & Allis, C. D. RNA meets chromatin. *Genes Dev* **19**, 1635–55 (2006).

1133. Hawse, J. R., Cumming, J. R., Oppermann, B., Sheets, N. L., Reddy, V. N. & Kantorow, M. Activation of metallothioneins and a-crystallin/shsps in human lens epithelial cells by specific metals and the metal content of aging clear human lenses. *Invest Ophthalmol Vis Sci* **44**, 672–9 (2003).

1134. Bando, M., Oka, M., Kawai, K., Obazawa, H., Kobayashi, S. & Takehana, M. NADH binding properties of rabbit lens λ-crystallin. *Mol Vis* **12**, 692–97 (2006).

1135. Tümpel, S., Cambronero, F., Wiedemann, L. M. & Krumlauf, R. Evolution of cis elements in the differential expression of two *Hoxa2* coparalogous genes in pufferfish (*Takifugu rubripes*). *Proc Natl Acad Sci U S A* **103**, 5419–24.

1136. Bejerano, G. Lowe, C. B., Ahituv, N., King, B., Siepel, A., Salama, S. R., Rubin, E.

M. Kent, W. J. & Haussler, D. A distal enhancer and an ultraconserved exon are derived from a novel retroposon. *Nature* **441**, 87–90.

1137. Prud'homme, B., Gompel, N., Rokas, A., Kassner, V. A., Williams, T. M., Yeh, S.-D., True, J. R. & Carroll, S. B. Repeated morphological evolution through *cis*-regulatory changes in a pleiotrophic gene. *Nature* **440**, 1050–3 (2006).

1138. Nakamura, S., Sakurai, T. & Nonomura, Y. Differential expression of bovine adseverin in adrenal gland revealed by *in situ* hybridization. *J Biol Chem* **269**, 5890–6 (1994).

1139. Zhang, L., Marcu, M. G., Nau-Staudt, K. & Trifaró, J.-M. Recombinant scinderin enhances exocytosis, an effect blocked by two scinderin-derived actin-binding peptides and PIP_2. *Neuron* **17**, 287–96 (1996).

1140. Marcu, M. G., Zhang, L., Nau-Staudt, K. & Trifaró, J.-M. Recombinant scinderin, an F-actin severing protein, increases calcium-induced release of serotonin from permeabilized platelets, an effect blocked by two scinderin-derived actin-binding peptides and phosphatidylinositol 4,5-biphosphate. *Blood* **87**, 20–4 (1996).

1141. Dumitrescu Pene, T., Rosé, S. D., Lejen, T., Marcu, M. G. & Trifaró, J.-M. Expression of various scinderin domains in chromaffin cells indicates that this protein acts as a molecular switch in the control of actin filament dynamics and exocytosis. *J Neurochem* **92**, 780–9 (2005).

1142. Svensson, C. & Lundberg, K. Immune-specific up-regulation of adseverin gene expression by 2,3,7,8-tetrachlorodibenzo-p-dioxin. *Mol Pharmacol* **60**, 135–42 (2001).

1143. Svensson, C., Silverstone, A. E., Lai, Z.-W. & Lundberg, K. Dioxin-induced adseverin expression in the mouse thymus is strictly regulated and dependent on the aryl hydrocarbon receptor. *Biochem Biophys Res Comm* **291**, 1194–1200 (2002).

1144. Oberemm, A., Meckert, C., Brandenburger, L., Herzig, A., Lindner, Y., Kalenberg, K., Krause, E., Ittrich, C., Kopp-Schneider, A., Stahlmann, R., Richter-Reichhelm, H.-B. & Gundert-Remy, U. Differential signatures of protein expression in marmoset liver and thymus induced by single-dose TCDD treatment. *Toxicology* **206**, 33–48 (2005).

1145. Krishnamurthy, G., Vikram, R., Singh, S. B., Patel, N., Agarwal, S., Mukhopadhyay, G., Basu, S. K. & Mukhopadhyay, A. Hemoglobin receptor in *Leishmania* is a hexokinase located in the flagellar pocket. *J Biol Chem* **280**, 5884–91 (2005).

1146. Zhang, Q., Wang, S.-Y., Nottke, A. C., Rocheleau, J. V., Piston, D. W. & Goodman, R. H. Redox sensor CtBP mediates hypoxia-induced tumor cell migration. *Proc Natl Acad Sci U S A* **103**, 9029–33 (2006).

1147. Pennisi, E. Genes commute to factories before they start work. *Science* **312**, 1304 (2006).

1148. Angelo, M., Singel, D. J. & Stamler, J. S. An S-nitrosothiol (SNO) synthase function of hemoglobin that utilizes nitrite as a substrate. *Proc Natl Acad Sci U S A* **103**, 8366–71.

1149. Fraser, J., de Mello, L. V., Ward, D., Rees, H. H., Williams, D. R., Fang, Y., Brass, A., Gracey, A. Y. & Cossins, A. R. Hypoxia-inducible muyoglobin expression in nonmuscle tissues. *Proc Natl Acad Sci U S A* **103**, 2977–81 (2006).

1150. Riggs, A. F. & Gorr, T. A. A globin in every cell? *Proc Natl Acad Sci U S A* **103**, 2469–70 (2006).

1151. Christoffels, A., Koh, E. G. L., Chia, J.-M., Brenner, S., Aparicio, S. & Venkatesh, B. Fugu genome analysis provides evidence for a whole-genome duplication early during the evolution of ray-finned fishes. *Mol Biol Evol* **21**, 1146–51 (2004).

1152. Voit, E., Neves, A. R. & Santos, H. The intricate side of systems biology. *Proc Natl Acad Sci U S A* **103**, 9452–7 (2006).

1153. Spirin, V., Gelfand, M. S., Mironov, A. A. & Mirny, L. A. A metabolic network in the evolutionary context: multiscale structure and modularity. *Proc Natl Acad Sci U S A* **103**, 8774–9 (2006).

1154. Ohtsuki, H., Hauert, C., Lieberman, E. & Nowak, M. A. A simple rule for the evolution of cooperation on graphs and social networks. *Nature* **441**, 502–5 (2006).

1155. Gavin, A.-C. et al. Proteome survey reveals modularity of the yeast cell machinery. *Nature* **440**, 631–6 (2006).

1156. Krogan, N. J. et al. Global landscape of protein complexes in the yeast *Saccharomyces cerevisiae*. *Nature* **440**, 637–43 (2006).

1157. Borenstein, E. & Ruppin, E. Direct evolution of genetic robustness in microRNA. *Proc Natl Acad Sci U S A* **103**, 6593–8 (2006).

1158. Romero, P. R., Zaidi, S., Fang, Y. Y., Uversky, V. N., Radivojac, P., Oldfield, C. J., Cortese, M. S., Sickmeier, M., LeGall, T., Obradovic, Z. & Dunker, A. K. Alternative splicing in concert with protein intrinsic disorder enables increased functional diversity in multicellular organisms. *Proc Natl Acad Sci U S A* **103**, 8390–5 (2006).

1159. Williams, P. D., Pollock, D. D. & Goldstein, R. A. Functionality and the evolution of marginal stability in proteins: inferences from lattice simulations. *Evol Bioinform Online* **2**, 59–69 (2006).

1160. Bloom, J. D., Labthavikul, S. T., Otey, C. R. & Arnold, F. H. Protein stability promotes evolvability. *Proc Natl Acad Sci U S A* **103**, 5869–74 (2006).

1161. Cai, L., Friedman, N. & Xie, X. S. Stochastic protein expression in individual cells at the single molecule level. *Nature* **440**, 358–62 (2006).

1162. Newman, J. R. S., Ghaemmaghami, S., Ihmels, J., Breslow, D. K., Noble, M., DeRisi, J. L. & Weissman, J. S. Single-cell proteomic analysis of *S. cerevisiae* reveals the architecture of biological noise. *Nature* **441**, 840–6 (2006).

1163. Guido, N. J., Wang, X., Adalsteinsson, D., McMillen, D., Hasty, J., Cantor, C. R., Elston, T. C. & Collins, J. J. A bottom-up approach to gene regulation. *Nature* **439**, 856–60 (2006).

1164. Thomas, J. A., Welch, J. J., Woolfit, M. & Bromham, L. There is no universal molecular clock for invertebrates, but rate variation does not scale with body size. *Proc Natl Acad Sci U S A* **103**, 7366–71 (2006).

1165. Weinreich, D. M., DeLaney, N. F., DePristo, M. A. & Hartl, D. L. Darwinian evolution can follow only very few mutational paths to fitter proteins. *Science* **312**, 111–4 (2006).

Index

"Accidental specificity," 134

Accommodation, 60, 61, 77. *See also* Focusing/focused images

α-crystallins, 61, 63, 66, 67–71, 74, 75, 77, 189, 190, 192; αA-crystallin, 67–70, 89, 92, 190, 215; αB-crystallin, 66, 67–71, 86, 88, 89–91, 190; α-*crystallin* genes, 67, 94, 191; αA-*crystallin* gene, 67, 69, 70, 84, 85, 87, 89, 90, 190, 191, 215; αB-*crystallin* gene, 67, 69–71, 85, 88, 89–91, 164, 190, 191

Actin, 63, 103, 105, 107, 135–137, 151–152

Adaptability. *See* Evolvability/adaptability

Adaptationist view of evolution, 22–24

ADP-ribohydroglycolase, 64

Adseverin, 105, 137

α-enolase, 3, 62, 76, 123, 124, 189, 190

Aeromonas hydrophilia (bacterial pathogen), 124

AFGP (antifreeze glycoprotein), 154, 155, 157–161, 164; *AFGP* gene, 157, 160, 161

AFP (anti-freeze protein), 153–165; plant ice-active proteins, 161–163

African cichlid (*Haplochromis burtoni*), 65

Agriculture/crops, 154, 221; soil bacteria, 74, 221; carrot, 162, 163; winter rye (*Secale cereale*), 162, 163

Agrobacterium tumefaciens (soil bacterium), 221

AIM1 (absence in melanoma), 75

ALDH (aldehyde dehydrogenase), 17, 77, 80–81, 82, 91–92, 99–101, 103, 108; multiple functions of, 99–101; ALDH1, 81; ALDH1A1, 98, 100, 101, 108; ALDH1A9, 64, 80; ALDH1C1, 80; ALDH1C2, 80;

ALDH2, 81, 82; ALDH3A1, 98, 102, 107, 108, 134; retinaldehyde dehydrogenase, 62, 77, 81; *Aldh1a1* gene, 107, 108; *Aldh3a1* gene, 100, 107, 108

Aldose reductase, 62, 77

Algae, 56, 131, 165, 220; *Chlamydomonas,* 127, 131, 165

Alternative RNA splicing. *See* RNA splicing/ alternative RNA splicing

Amphibians, 37, 62, 192; frogs, 2, 24, 58, 62, 77, 79, 91, 93–94, 143, 148, 192, 218; newts, 74, 192

Amphioxus (cephalochordate), 58, 183, 184

Amyotrophic lateral sclerosis, 46

Anableps anableps (four-eyed fish), 60, 103, 104

Angiosperms, 178

Annelids, 56, 57, 60, 167

Antarctic eel pout (*Licodichthys dearborni*), 155

Antarctic Notothenioid toothfish (*Dissostichus mawsoni*), 153, 155, 157, 158, 159

Apoptosis, 17, 18, 68, 69–71, 105, 127, 128, 135, 136, 137, 138, 139–140, 143, 166, 171, 222, 237; antiapoptotic role, 69–71, 139, 215; pro-apoptotic role, 113, 139, 171. *See also* Apoptosis regulators; Cell death

Apoptosis regulators: Bak, 70; Bax, 68, 70, 113, 138; BCL2, 70; Bcl-2, 70, 138, 139, 171; Bcl-X$_s$, 68, 70; *BCL2* gene, 69, 70. *See also* α-crystallins

Aquaporin, 75, 221